BIOCHEMISTRY RESEARCH TRENDS

MELATONIN

MEDICAL USES AND ROLE IN HEALTH AND DISEASE

BIOCHEMISTRY RESEARCH TRENDS

Additional books in this series can be found on Nova's website under the Series tab.

Additional e-books in this series can be found on Nova's website under the eBooks tab.

BIOCHEMISTRY RESEARCH TRENDS

MELATONIN

MEDICAL USES AND ROLE IN HEALTH AND DISEASE

LORE CORREIA
AND
GERMAINE MAYERS
EDITORS

Copyright © 2018 by Nova Science Publishers, Inc.

All rights reserved. No part of this book may be reproduced, stored in a retrieval system or transmitted in any form or by any means: electronic, electrostatic, magnetic, tape, mechanical photocopying, recording or otherwise without the written permission of the Publisher.

We have partnered with Copyright Clearance Center to make it easy for you to obtain permissions to reuse content from this publication. Simply navigate to this publication's page on Nova's website and locate the "Get Permission" button below the title description. This button is linked directly to the title's permission page on copyright.com. Alternatively, you can visit copyright.com and search by title, ISBN, or ISSN.

For further questions about using the service on copyright.com, please contact:
Copyright Clearance Center
Phone: +1-(978) 750-8400   Fax: +1-(978) 750-4470   E-mail: info@copyright.com.

NOTICE TO THE READER

The Publisher has taken reasonable care in the preparation of this book, but makes no expressed or implied warranty of any kind and assumes no responsibility for any errors or omissions. No liability is assumed for incidental or consequential damages in connection with or arising out of information contained in this book. The Publisher shall not be liable for any special, consequential, or exemplary damages resulting, in whole or in part, from the readers' use of, or reliance upon, this material. Any parts of this book based on government reports are so indicated and copyright is claimed for those parts to the extent applicable to compilations of such works.

Independent verification should be sought for any data, advice or recommendations contained in this book. In addition, no responsibility is assumed by the publisher for any injury and/or damage to persons or property arising from any methods, products, instructions, ideas or otherwise contained in this publication.

This publication is designed to provide accurate and authoritative information with regard to the subject matter covered herein. It is sold with the clear understanding that the Publisher is not engaged in rendering legal or any other professional services. If legal or any other expert assistance is required, the services of a competent person should be sought. FROM A DECLARATION OF PARTICIPANTS JOINTLY ADOPTED BY A COMMITTEE OF THE AMERICAN BAR ASSOCIATION AND A COMMITTEE OF PUBLISHERS.

Additional color graphics may be available in the e-book version of this book.

Library of Congress Cataloging-in-Publication Data

ISBN: 978-1-53612-987-8

Published by Nova Science Publishers, Inc. † New York

Contents

Preface		vii
Chapter 1	Determination of Melatonin in Biological and Non-Biological Samples *T. Rosado, M. Barroso, A. P. Duarte and E. Gallardo*	1
Chapter 2	Melatonin as a Therapeutic Option for Optic Neuropathies *María F. González Fleitas, Marcos L. Aranda, Hernán H. Dieguez, Julián D. Devouassoux, Georgia A. Milne, Mónica Chianelli, María I. Keller Sarmiento, Pablo H. Sande, Damián Dorfman and Ruth E. Rosenstein*	69
Chapter 3	The Protective Role of Exogen Melatonin on the Prostate Gland under Experimental Models of Metabolic Diseases *Marina Guimarães Gobbo, Guilherme Henrique Tamarindo, Sebastião Roberto Taboga and Rejane Maira Góes*	103
Chapter 4	Melatonin Receptors, Behaviour and Brain Function *Olakunle J. Onaolapo and Adejoke Y. Onaolapo*	133
Chapter 5	The Immunomodulatory Properties of Melatonin *Sylwia Mańka and Ewa Majewska*	159
Chapter 6	How Melatonin Combats Ischemic Brain Injury *Eva Ramos, Paloma Patiño, Javier Egea and Alejandro Romero*	185
Chapter 7	Participation of Melatonin Receptors in Neurological Disorders *Anna Karynna A. A. Rocha, Edilson D. da Silva Júnior, Fernanda Amaral and Débora Amado*	211

Chapter 8	Epilepsy: Neuroprotective, Anti-Inflammatory, and Anticonvulsant Effects of Melatonin *Anna Karynna A. A. Rocha, José Cipolla-Neto and Débora Amado*	**233**
Chapter 9	Melatonin as a Treatment for Antipsychotic Induced Weight Gain: A Review of the Evidence *Trevor R. Norman*	**255**
Chapter 10	Melatonin: Beneficial Aspects and Underlying Mechanisms *Stephen C. Bondy*	**277**
Chapter 11	Melatonin and Oral Disorders: Potential Therapeutic Applications *Ana Capote-Moreno, Paloma Patiño and Alejandro Romero*	**295**
Index		**311**

PREFACE

In this compilation, the authors review pre-analytical procedures, as well as analytical conditions, to determine melatonin in both biological and non-biological samples. Several procedures for the extraction of melatonin have been described, including ultrasound-assisted extraction, liquid–liquid extraction, and solid phase extraction. Following this, a study is presented wherein melatonin administration in middle-age rats was able to recover prostatic morphology and strongly decrease amiloyd bodies deposition even after obesity induction. These results were probably related to total GST activity improvement. The authors maintain that melatonin should be used as benign prostate hyperplasia chemoprevention due to decreased cell proliferation, even in hyperlipidemic medium. The functions of melatonin receptors in health and disease are reviewed, with specific references to their involvement in mood, anxiety-related and neurodegenerative disorders; and the possibilities of melatonin receptors as mediators of melatonergic therapeutics. Next, the current knowledge regarding immunomodulatory properties of melatonin are reviewed in conjunction with its effect on non-specific, humoral and cellular immune responses with particular emphasis on the production of reactive oxygen species and cytokines as well as on apoptosis. The authors go on to underline the pleiotropic effects of melatonin to ameliorate the molecular and organ/tissue damage associated with brain ischemia based on preserving the functional integrity of the blood-brain barrier, inducing neurogenesis and cell proliferation through receptor-dependent mechanism, and improving synaptic transmission. Additionally, they examine the role of MT1 and MT2 receptors in neurological disorders and the therapeutic effects of melatonin receptors agonists on the central nervous system disorders. The book deliberates o the way clinical trials in antipsychotic medicated patients suggest some positive effects for melatonin, but the studies have substantial shortcomings. The mechanism by which melatonin exerts any clinical effects remains largely un-investigated. The strength of the clinical evidence for an effect of melatonin, like that of the pre-clinical studies, is equivocal. An overview of the range of disorders and conditions that they may be improved by melatonin administration is provided, followed by consideration of the processes that may underlie these changes. In the final chapter, the authors summarize the data documenting the use of melatonin and its beneficial effects in oral related-diseases, such as bisphosphonate-related osteonecrosis of the jaw, periodontitis, mucositis, cancer, and/or oral infection.

Chapter 1 - Melatonin has been found in many species of animals, and it was most extensively studied in vertebrates – mammals, birds, reptiles, amphibians, bony and cartilaginous fish, and cyclostomes. This endogenous neurohormone has also been identified in bacteria, algae, fungi, insects, and plants. Nowadays, melatonin is widely used in countries as a food supplement, however alternative or natural sources of melatonin have been a matter of interest.

Concentrations of melatonin in the different alternative sources may vary, ranging from picograms to micrograms, and have been subjected to evaluation by many authors who constantly update the levels found. Efficient analytical methods for the determination of melatonin, together with optimized extraction protocols, help confirming its presence and determine melatonin in the most diverse sources.

Melatonin can be detected by several methods, such as immunological techniques, radioimmunoassay and enzyme-immunoassay. However, in recent years, chromatographic methods have been the most widely used separation techniques in this field. Liquid chromatography techniques are more beneficial economically, and also more time-efficient when sample derivatization is not required prior to analysis; several detectors have been coupled to liquid chromatography, such as mass spectrometry, fluorescence and electrochemical detection. However, gas chromatographic applications have also been described, the most common being gas chromatography-mass spectrometry.

Sample pre-treatment represent the critical step before chromatographic measurement for the determination of melatonin. This becomes even more relevant when there are so many different sample sources which require a wide adaptation of the extraction procedures to solid samples (e.g., fruits, plants, mammal pineal gland, drug tablets) or liquid samples (e.g., urine, plasma, oral fluid, oils). Several procedures for the extraction of melatonin have been reported, and these include ultrasound-assisted extraction, liquid–liquid extraction, and solid phase extraction.

This chapter will review pre-analytical procedures, as well as analytical conditions, to determine melatonin in both biological and non-biological samples.

Chapter 2 - Glaucoma is a leading cause of irreversible blindness worldwide, and provokes progressive visual impairment attributable to the dysfunction and loss of retinal ganglion cells and optic nerve axons. Currently, clinical interventions for glaucoma are mainly restricted to the reduction of intraocular pressure, one of the major risk factors for the disease. However, lowering intraocular pressure is often insufficient to halt or reverse the progress of visual loss, underlining the need for the development of alternative treatment strategies.

Optic neuritis, the most common optic neuropathy affecting young adults, is a condition involving primary inflammation, demyelination, and axonal injury in the optic nerve which leads to retinal ganglion cell death and visual dysfunction. Although corticosteroids are the current mainstays of therapy for the treatment of optic neuritis, they fail to achieve long-term benefit on functional recovery, and provoke considerable side effects. Despite several differences between these diseases, glaucoma and optic neuritis are the two major degenerative causes of optic nerve damage, sharing a common feature which is retinal ganglion cell loss and axonal damage. Although the current management of glaucoma is mainly directed at the control of intraocular pressure, and the current therapy for optic neuritis mainly target the autoimmune response by using anti-inflammatory and immunosuppressive agents, a neuroprotective therapy able to avoid the death of retinal ganglion cells and optic nerve fiber loss should be the main goal of treatment for both diseases. In this chapter, the authors will discuss evidence supporting the neuroprotective effect of melatonin, a very safe compound (even at high doses) for human use, which prevents and slows the progression of retinal ganglion cell and optic nerve axon damage induced by experimental glaucoma or optic neuritis in rodents, and as such, it could be considered as a new therapeutic resource for the management of the most prevalent forms of optic neuropathy.

Chapter 3 - The prostate is an accessory gland of the male apparatus which is regulated by sexual steroids from its development phase. Several studies have reported that metabolic syndromes may also influence the prostate's physiology. Studies from the authors' research group as well as other studies from the literature revealed that diabetes and high lipid intake are unfavorable to the prostatic homeostasis and may raise histophysiological alterations that can lead to cancer. Prostate stromal remodeling, imbalance on the gland proliferation/apoptotic rate, testosterone/estrogen ratio and tissue antioxidant system impairment are important features observed under metabolic disturbances and are often associated with carcinogenesis. In the last decade, the therapeutic potential of melatonin has been investigated due to it exerting antioxidant properties and influencing cell dynamics. As well as prostate cells expressing MTR1 and MTR2 receptors, the gland responses to melatonin treatment may also be triggered by several pathways, including the androgen pathway, antioxidant enzymes modulation and, as recently reported by the authors' research group, changes in mitochondrial physiology. The authors have published articles describing conditions where tissue damage was induced, such as long-term diabetes, when animals treated with low melatonin doses showed an effective restoration of their prostate homeostasis and functions and managed to recover cell proliferation. It is worth mentioning that melatonin also raised testosterone serum levels, which were impaired after long-term diabetes induction. Hyperglycemia is often associated to an increase in oxidative stress, which leads to damage and cell death. Melatonin supplementation, even at low doses, promoted the normalization of activities of antioxidant enzymes on the prostate, which can be associated with reduced apoptosis indexes. In addition, the authors' experiments with human epithelial prostate cells with high proliferative potential proved that melatonin has antioxidant properties due to strongly reduced oxygen reactive species (ROS) generation associated to increased oxidative phosphorylation (OXPHOS). Although melatonin has been considered beneficial for the prostate gland under normal or diabetogenic situations, its effects on cell proliferation are still poorly understood. At pharmacological concentrations it may exert anti-clonogenic effects both under normal conditions and after a short period of pre-incubation of prostatic benign and cancerous cells in a hyperglycemic medium. On the other hand, with longer pre-incubation it may favor cell proliferation. Obesity is also considered a metabolic disorder and when associated with prostate aging had been reported to favor prostatic lesions. In this context, melatonin administration to middle-aged rats was able to recover prostatic morphology and strongly decrease amyloid bodies' deposition even after obesity exposure. These results were probably related to total glutathione *S-transferase* (GST) activity improvement. Interestingly, under pro-oxidant conditions, generated by docosahexaenoic acid incubation (omega-3 fatty acid), the indole not only reduced dramatically ROS generation and enhanced OXPHOS, but also restored cell capacity to survive stress situations independently of the sensitization of the membrane receptors. Furthermore, the authors' studies also suggest melatonin supplementation as a benign prostate hyperplasia chemoprevention method due to the decreased cell proliferation it promotes, even in a hyperlipidemic medium. The authors' studies revealed that melatonin exerts a protective role on the prostate gland under several metabolic conditions in rodents, human benign prostatic hyperplasia models and cancer cells.

Chapter 4 - Melatonin is a tryptophan-derived molecule that is critical to the transduction of circadian and seasonal information. It is also known to play crucial roles in several physiological processes, including the regulation of behavioural and cognitive processes in humans and rodents. There are evidences that a number of physiological and behavioural effects of melatonin in mammals are mediated by specific G-protein coupled receptors (GPCRs); melatonin (MT) $_1$ and $_2$, which are expressed in several locations in the mammalian central nervous system. In this chapter, the authors review the roles of melatonin receptors in health and disease, with specific references to their involvement in mood, anxiety-related and neurodegenerative disorders; and the possibilities of melatonin receptors as mediators of melatonergic therapeutics.

Chapter 5 - According to available literature, melatonin plays an important role in many aspects of human physiology, however this review will focus just on its effect on the immune system response. Despite a large number of studies proving a relationship between melatonin and the immune system, the definitive effect of its influence on immunity and a mechanism of action still remains arguable. Melatonin mediates between neurohormonal and immune systems by connecting the development of the immune reaction with melatonin secretion induced by inflammatory mediators. Melatonin is a kind of 'immunological buffer', which stimulates the immune processes, particularly in patients with immune disorders but also suppresses immunity in the case of its excessive activation. Furthermore, it modulates the apoptosis of leukocytes and the course of inflammatory responses, acting both as its activator and inhibitor. This immunomodulatory function, as well as its antioxidant properties make the hormone a potentially interesting therapeutic agent. A wide range of data suggests that melatonin might find widespread clinical use as a supportive therapy in the treatment of inflammatory conditions, bacterial and viral infections, cancer and autoimmune diseases.

Throughout the present chapter, the authors discuss the current knowledge about the immunomodulatory properties of melatonin and its effect on non-specific, humoral and cellular immune responses with particular emphasis on the production of reactive oxygen species and cytokines as well as on apoptosis. Finally, the authors briefly outline the clinical use of melatonin in several immune conditions, such as inflammation and cancer.

Chapter 6 - Stroke is the second leading cause of death and the main cause of disability worldwide. To date, there is no effective treatment to prevent the brain damage in ischemic stroke. Consequently, there is an obvious need to develop neuroprotective treatments for this pathology. The physiopathological events of the ischemic cascade; oxidative stress, Ca2+ dyshomeostasis, mitochondrial dysfunction, proinflammatory mediators, excitotoxicity and/or programmed neuronal cell death are regulated through multiple signaling pathways. In this chapter, the authors highlighted the pleiotropic effects of melatonin ameliorating the molecular, tissue and organ damage associated to brain ischemia. Its protective effects encompass from preserving the functional integrity of the blood-brain barrier, inducing neurogenesis and cell proliferation to improving synaptic transmission. This has been lately reinforced by the low cost of melatonin and its reduced toxicity. In fact, melatonin can be highly useful when combined with other therapies. However, additional investigations are necessary to determine clinical effectiveness, doses and the optimal timing of administration.

Chapter 7 - Melatonin regulates a variety of neurophysiological and neuroendocrine functions through the activation of two G protein-coupled melatonin receptors that have been identified in mammals as MT_1 and MT_2. Melatonin receptors are located in areas of the central nervous system, including the cerebral cortex, suprachiasmatic nucleus, cerebellum, thalamus, and hippocampus. On one hand, evidence has shown that MT_1 receptors can inhibit calcium influx, inducing inhibitory effects on neuronal electrical activity. On the other hand, MT_2 receptors are able to (a) decrease cAMP and cGMP accumulation, (b) mediate circadian rhythm, and (c) inhibit long-term potentiation and increase the activity of PKC. Some studies have demonstrated changes in MT_1 and MT_2 receptor expression in several neurological disorders such as epilepsy, Alzheimer's and Parkinson's diseases, and traumatic brain injury. The role of melatonin or melatonin receptor agonists as potent neuroprotective agents in these diseases has been considered both in humans and rodents. In addition, progress in understanding the role of melatonin receptors in the modulation of neurological disorders has led to the discovery of novel classes of drugs for treating several brain-related disorders including insomnia and mood disorders. In this review the authors will show the participation of MT_1 and MT_2 receptors in neurological disorders and the future therapeutic prospects of melatonin receptors agonists/antagonists on various central nervous system diseases.

Chapter 8 - The International League Against Epilepsy (ILAE) defined a seizure as "a transient occurrence of signs and/or symptoms due to abnormal excessive or synchronous neuronal activity in the brain." The neuropathology of epilepsy involves glutamate excitotoxicity, oxidative stress, and high inflammatory response. Anti-epileptic drugs (AEDs) are the main form of treatment. However, approximately 30% to 40% of individuals are unresponsive to them. Moreover, AEDs can produce undesirable symptoms, such as moderate disturbances of the central nervous system (CNS), liver failure, and bone health impairment. Melatonin has anti-inflammatory, antioxidant, anticonvulsant, and GABAergic activity on the CNS, and low toxicity at pharmacological doses. Melatonin may be a potential adjuvant for epilepsy treatment, as several studies have shown its administration can reduce oxidative stress and high inflammatory responses from occurring. In addition, melatonin has been safe and effective in decreasing seizure frequency, improving sleep quality, regulating the light-dark cycle, improving attention, memory, language, and anxiety in these patients. All of this demonstrates that melatonin has been effective in the treatment of epilepsy in experimental models as well as in patients. The authors present recent data from clinical and experimental studies, which show the neuroprotective, anti-inflammatory, and antioxidant effects of melatonin in many types of epilepsy; this should include positive future prospects for treatment.

Chapter 9 - A concerning aspect of the use of antipsychotic medication has been the issue of weight gain and the consequent increase of visceral adiposity leading to the development of the metabolic syndrome. The issue has reached prominence in recent years with the introduction of so-called second generation (SGAs) or 'atypical' antipsychotics. Because of their perceived benefits, with respect to neurological side effects, these medications have achieved first line status for the treatment of psychotic states in adults, adolescents and children. While behavioural strategies and life style modifications remain the corner stone for management of antipsychotic-induced weight gain, where such techniques fail adjunctive pharmacological treatments are used. Melatonin as a potential adjunctive treatment relies on the association between the development of diabetes and abnormal circadian rhythmicity. Limited pre-clinical studies suggest an effect of melatonin on visceral fat and other parameters associated with the metabolic syndrome but the effects on weight gain are equivocal. Clinical trials in antipsychotic medicated patients also suggest some positive effects for melatonin but the studies have significant shortcomings. The mechanism by which melatonin exerts any clinical effects remains largely un-investigated. The strength of the clinical evidence for an effect of melatonin, like that of the pre-clinical studies, is equivocal.

Chapter 10 - Melatonin has a reportedly beneficial effect in the treatment of numerous disorders. While it is not certain what the most important primary mechanism of melatonin's actions is, its application can lead to a variety of favorable outcomes in a range of pathological conditions. In addition to the prevention or amelioration of disease states, melatonin has attributes that can be of benefit to the normal healthy organism. The most notable of these, is the potential of melatonin to slow down the progression of several indices characterizing the aging process. The incidence of many secondary undesirable states associated with aging, may also be reduced simply by deceleration of normal aging. The utility of melatonin is further enhanced by its very low level of toxicity and absence of any capacity to induce or promote tumor development. As a natural product, melatonin cannot be patented and is thus inexpensive and readily available. Melatonin can be regarded either as a dietary supplement or as a remedy. A range of clinical reports supports the validation of the therapeutic use of melatonin for many disorders. These are buttressed by parallel descriptions of studies on experimental animal models. This review is intended firstly as a short summary of the range of disorders and conditions, where there is evidence that they may be improved by melatonin administration. This is followed by consideration of the processes that may underlie these advantageous changes. It is proposed that melatonin's ability to retard aging events may be the source of many of its positive attributes. Finally, the conclusion looks forward to future work focusing on the possibility of studies investing the possible advantages conferred by melatonin on the epigenetic profile.

Chapter 11 - In the last decade, melatonin has emerged as an important therapeutic alternative against numerous pathologies. Its capacity as free radical scavenger and antioxidant joined to its immunomodulatory action, oncostatic activity and anti-inflammatory properties, has been widely reported. Each of these actions are tightly controlled by the regulation of different signaling pathways. So, the purpose of this chapter is briefly summarize the data documenting the use of melatonin and its beneficial effects in oral related-diseases, such as bisphosphonate-related osteonecrosis of the jaw, periodontitis, mucositis, cancer, and/or oral infections. Furthermore, taking into account the low cost of melatonin and its reduced toxicity, the authors hypothesize on novel therapeutic strategies for the treatment of diseases of the oral cavity. Nevertheless, the molecular and cellular mechanisms underlying the actions of melatonin need further exploration and accordingly, new clinical studies should be conducted in human patients to assess its efficacy.

In: Melatonin: Medical Uses and Role in Health and Disease   ISBN: 978-1-53612-987-8
Editors: Lore Correia and Germaine Mayers   © 2018 Nova Science Publishers, Inc.

Chapter 1

DETERMINATION OF MELATONIN IN BIOLOGICAL AND NON-BIOLOGICAL SAMPLES

T. Rosado[1], M. Barroso[2], PhD, A. P. Duarte[1], PhD and E. Gallardo[1,], PhD*

[1]Centro de Investigação em Ciências da Saúde,
Universidade da Beira Interior (CICS-UBI), Covilhã, Portugal
[2]Instituto Nacional de Medicina Legal e Ciências Forenses,
Lisboa, Portugal

ABSTRACT

Melatonin has been found in many species of animals, and it was most extensively studied in vertebrates – mammals, birds, reptiles, amphibians, bony and cartilaginous fish, and cyclostomes. This endogenous neurohormone has also been identified in bacteria, algae, fungi, insects, and plants. Nowadays, melatonin is widely used in countries as a food supplement, however alternative or natural sources of melatonin have been a matter of interest.

Concentrations of melatonin in the different alternative sources may vary, ranging from picograms to micrograms, and have been subjected to evaluation by many authors who constantly update the levels found. Efficient analytical methods for the determination of melatonin, together with optimized extraction protocols, help confirming its presence and determine melatonin in the most diverse sources.

Melatonin can be detected by several methods, such as immunological techniques, radioimmunoassay and enzyme-immunoassay. However, in recent years, chromatographic methods have been the most widely used separation techniques in this field. Liquid chromatography techniques are more beneficial economically, and also more time-efficient when sample derivatization is not required prior to analysis; several

---

* Corresponding Author Email: egallardo@fcsaude.ubi.pt.

detectors have been coupled to liquid chromatography, such as mass spectrometry, fluorescence and electrochemical detection. However, gas chromatographic applications have also been described, the most common being gas chromatography-mass spectrometry.

Sample pre-treatment represent the critical step before chromatographic measurement for the determination of melatonin. This becomes even more relevant when there are so many different sample sources which require a wide adaptation of the extraction procedures to solid samples (e.g., fruits, plants, mammal pineal gland, drug tablets) or liquid samples (e.g., urine, plasma, oral fluid, oils). Several procedures for the extraction of melatonin have been reported, and these include ultrasound-assisted extraction, liquid–liquid extraction, and solid phase extraction.

This chapter will review pre-analytical procedures, as well as analytical conditions, to determine melatonin in both biological and non-biological samples.

Keywords: melatonin, extraction procedures, analytical techniques

1. INTRODUCTION

Melatonin (MLT) is a widely studied bio-molecule and its function has been investigated in bacteria, mammals, birds, amphibians, reptiles, fish, and plants. It is a low molecular weight molecule, ubiquitously present with pleiotropic biological activities (Hardeland et al. 2012; Nawaz et al. 2015). Since its discovery in bovine pineal gland in 1958 (Lerner et al. 1958; Nawaz et al. 2015), the number of reports about procedures and analytical techniques for levels determination has increased.

The extremely low levels of MLT during the daytime, particularly in plasma (where the hormone is diluted), have led researchers to develop accurate analytical methods (Muñoz et al. 2009). This accurate and sensitive determination of MLT allowed a better understanding on its physiological functions, biological synthesis and metabolic pathways, and the regulatory networks (Ye et al. 2017).

The presence of endogenous MLT in plants was proven by isolation and chemical identification, and it was shown that its levels during its cycle parallels those of the light–dark cycle, as it does in many other eukaryotic organisms. However, obtaining accurate measurements of this hormone's levels in plant constituents is recognized as a major challenge requiring special theoretical and analytical considerations (Van Tassel and O'Neill 2001). Whereas sampling animal blood or urine is simple, nondestructive, and results in relatively clean samples that can often be assayed directly for the presence of target compounds, plants must normally be destructively sampled by extraction followed by extensive purification before hormones can be assayed (Van Tassel and O'Neill 2001).

In the late years, the detection and measurement of this molecule in the most diverse samples has been performed using several different analytical strategies, such as enzyme immunoassay (EIA), radioimmunoassay (RIA), high-performance liquid chromatography with fluorescence (HPLC-FLD) or electrochemical detection (HPLC-ECD), gas chromatography–mass spectrometry (GC-MS) and liquid chromatography-mass spectrometry (LC-MS/MS) (Zhao et al. 2016; Ferreira et al. 2016).

Unfortunately, it is unpractical to directly analyze endogenous MLT in most sample tissues through analytical equipment due to its low concentrations and the complicated matrix. To overcome this difficulty, different sample purification techniques such as liquid–liquid extraction (LLE) or solid phase extraction (SPE) have been applied. These commonly used techniques suffer from disadvantages, including being time consuming, their high cost, and also the need of large volumes of samples and toxic organic solvents. On the other hand, SPE technique despite of the consumption of low amounts of organic solvents, is tedious, relatively expensive, and the analytes breakthrough when large sample volumes are analyzed (Talebianpoor et al. 2014). Miniaturized techniques have been developed and used for this purpose. Microextraction by packed sorbent (MEPS) (Mercolini, Mandrioli, and Raggi 2012) and dispersive liquid–liquid microextraction (DLLME) (Talebianpoor et al. 2014) have recently attracted much attention because they allow to overcome these limitations.

For MLT therapeutical properties, melatonin has been praised as a "panacea," leading to a great increase in pharmaceutical preparations containing it, and consequently the need to have analytical methods available for quality control is even greater (Pucci et al. 2003). There is a growth on reports about the quantification of this compound in pharmaceutical dosage forms, either alone or in association with other drugs. Although pre-concentrating melatonin does not appear to be a focus subject on these publications, different sample treatments have been reported.

The aim of this chapter is to describe pre-analytical procedures applied, as well as analytical conditions adopted, to determine MLT in both biological and non-biological samples.

2. MELATONIN IN BIOLOGICAL AND NON-BIOLOGICAL SAMPLES

Although *in vitro* studies can provide important information on the effects of melatonin in cell cultures or perfused tissues, studies of the fluctuations of melatonin concentrations in body fluids and tissues generally have the most relevance for understanding its function in organisms (de Almeida, Di Mascio, Harumi, Spence, et al. 2011).

MLT is a compound secreted mainly by the pineal gland, but it is synthesized also in many other tissues and cells, including the retina (Cardinali and Rosner 1971; Tosini and Menaker 1998; Liu et al. 2004), human and murine bone marrow cells (Conti, Conconi, Hertens, Skwarlo-Sonta, et al. 2000), platelets (Champier et al. 1997), gastrointestinal tract (Bubenik 2002), skin (Slominski et al. 2005), or lymphocytes (Carrillo-Vico et al. 2004; de Almeida, Di Mascio, Harumi, Spence, et al. 2011). All these samples have been carefully studied considering their application for MLT measurements, especially the pineal gland in mammals.

Regarding human monitoring of MLT, urine, oral fluid and plasma samples have been greatly chosen, though.

Urine sampling is an established method for field, clinical, and research studies, being considered the most feasible method of circadian phase assessment. In special populations such as infants and older adults with dementia, collection of urine by a caretaker may be the most practical approach. The primary urinary metabolite of MLT, 6-sulphatoxymelatonin (aMT6s), assumes great importance, and therefore its determination is recommended (Benloucif et al. 2008).

MLT has also been detected in oral fluid or saliva, and studies have proven to have a rhythm similar to that of blood MLT. One of the advantages of saliva testing is that the sample is collected under direct supervision without loss of privacy. In consequence, the risk of an invalid specimen being provided or sample adulteration and/or substitution (which are likely to occur in urine analysis, at least in drug testing scenarios) is reduced. In addition, monitoring of oral fluid may be especially advantageous and important when multiple serial samples are needed (Gallardo and Queiroz 2008). As a body fluid, saliva can be collected noninvasively in good quantity in a short time, and there are likely to be little social taboos or restrictions associated to its collection in a wide range of situations (Benloucif et al. 2008). Saliva sampling is a relatively practical and reliable method for field, clinical, and research trials, provided that samples are taken every 30 to 60 minutes under dim light (< 30 lux) for at least 1 hour prior to and throughout the expected rise in MLT. Saliva collection can be conducted at home (as part of a field study) if the subjects remain in dim light and follow instructions to avoid contamination of samples with food particles, food dyes, or blood (Benloucif et al. 2008).

However, the disruption or complete deprivation of sleep that occurs during frequent salivary sample collection limits its overnight use. Most assays require a volume of at least 0.4 mL per tube or a minimum of 1 mL for duplicates; young children and older adults may require monitoring or assistance to ensure compliance with the protocol and to obtain a sufficient quantity of sample for analysis (Benloucif et al. 2008). Also due to the low concentrations present in this biological matrix, the assays reported have been particularly technically challenging (Benloucif et al. 2008). Oral fluid contains several macromolecules (mucopolysaccharides and mucoproteins), which make it less easily pipetted than for instance urine, and may not be available from all individuals at all times, since there are drugs that can inhibit saliva secretion and cause dry mouth (Gallardo and Queiroz 2008).

Blood is also a very common biological specimen required in analysis. Regarding MLT determination, it's typically sampled at frequent intervals through the use of intravenous catheters. It is a common practice to insert the intravenous catheter at least 2 hours before sampling to ensure that any increase in adrenergic levels during catheter insertion does not affect melatonin levels. The addition of long-line tubing can allow frequent sampling during the evening and throughout the night without major disturbance of sleep. Sleep may be disrupted, however, if there is discomfort or pain associated with the intravenous line or if there is a need to replace the catheter during the night (Benloucif et al. 2008).

The plasma melatonin levels are about 3 times greater than those of saliva. Thus, when comparing results between studies, a 3 pg/mL salivary dim-light melatonin onset (DLMO) would be comparable to a plasma DLMO of approximately 10 pg/mL. The higher melatonin levels present in plasma allow greater resolution and sensitivity than those obtained when urine or saliva are sampled, particularly for individuals with low melatonin concentrations (Benloucif et al. 2008).

Afternoon-through-overnight or 24-hour plasma melatonin profiles can provide accurate measures of circadian phase, duration, and amplitude. Therefore, if blood is sampled at frequent intervals (e.g., every 20 to 30 minutes) throughout the night, plasma melatonin can be the most informative of these 3 sample types. However, the loss of a viable vein may become problematic in some subjects, especially in those who are overweight, old, and also in women. Because plasma sampling is made using an intravenous catheter, there are several potential problems: it is invasive and associated to a slight health risk, there may be data loss due to technical problems related to the intravenous catheter and long-line tubing, and it requires trained medical personnel. Therefore, plasma sampling is not recommended for field studies or routine clinical use (Benloucif et al. 2008).

MLT is also found in a large number of plant species. The roots, seeds, leaves, bulbs, and flowers were found to be rich sources of melatonin in most of the plant species examined (Nawaz et al. 2015). The concentration of melatonin in plants is much higher than levels in animals (Arnao and Hernandez-Ruiz 2014) However, the analysis of MLT in plants presents some difficulties. First, a wide range of concentrations among different types is observed (Feng et al. 2014). Thus, any analytical method must be sensitive to these variations. Second, the amphipatic characteristic of the molecule makes it difficult to choose a solvent yielding complete recovery and accurate results. Third, melatonin is a potent antioxidant (Tan et al. 2007; Garcia-Parrilla, Cantos, and Troncoso 2009) and reacts quickly with other food constituents; careful handling of the sample is thus a prerequisite. The analytical method must therefore take these constraints into account in a vegetal tissue by using an adequate matrix and by showing sufficient sensitivity and specificity (Garcia-Parrilla, Cantos, and Troncoso 2009).

One has to notice that plant extracts are chemically complex, which increases the potential for cross-reactivity with immunoassays and makes chromatographic resolution difficult. These extracts often contain large amounts of carbohydrates, lipids, and pigments – components that frequently interfere with assays and must be removed or reduced (Van Tassel et al. 2001; Van Tassel and O'Neill 2001). Elaborate and ingenious extraction protocols have been developed by plant biologists, but each step introduces additional opportunities for loss of the target molecule and increases the variability among samples. MLT's physicochemical properties make it exceedingly difficult to isolate from plant samples. For example, it coextracts and partitions with chlorophyll and phenolic compounds; it may sublime during vacuum drying and, like other indole alkaloids, it may be susceptible to destruction by impurities commonly found in organic solvents. It can be easily lost by binding to glass, polyvinylpolypyrrolidone, or nylon-membrane filters (Van Tassel and O'Neill 2001). One reason for these problems is that MLT is readily diffusible in both polar and nonpolar environments. This property is thought to allow melatonin to quickly diffuse throughout tissues and cells and is frequently cited as a distinguishing feature of melatonin (Poeggeler and Hardeland 1994; Van Tassel and O'Neill 2001).

MLT in plants has been reviewed in some literature (Huang and Mazza 2011; Kolář and Macháčková 2005; Paredes et al. 2008; Posmyk and Janas 2009; Reiter et al. 2001; Caniato et al. 2003; Tan 2015; Tan et al. 2011). In the recent years, the studies on melatonin in fruits have also significantly increased due to an increased interest in knowing the natural sources of MLT in human diets (Feng et al. 2014; González-Gómez et al. 2009; Johns et al. 2013). However, there is controversy in the literature over the stability of MLT in plants or fruits, with analytical platforms, extraction, and analysis methods varying greatly from one report to another; further, with time, temperature, light levels, extraction solvents and mechanical disruption among others all varying widely. This has in turn lead to conflicting results between labs, and has contributed to the difficulty in confirming and comparing the results across various labs (Erland et al. 2016).

Another potentially confounding factor in the field of phytomelatonin analysis, is the presence of MLT isomers in plant products. Recent studies have hypothesized that as many as forty isomers of melatonin may exist in plants, and the presence of these compounds may explain some of this inter-lab variability (Tan et al. 2011; Iriti and Vigentini 2015; Vigentini et al. 2015; Erland et al. 2016).

The scientific publications reporting analytical determination of MLT in pharmaceutical formulations were once limited to a few papers, but raising recently (Nunez-Vergara et al. 2001; Radi and Bekhiet 1998), The discovery of the therapeutical properties of MLT led to a great increase in the pharmaceutical preparation, especially of galenic nature, containing this drug, and consequently the need to have analytical methods at disposal for the quality control was even greater (Raggi, Bugamelli, and Pucci 2002). MLT is now widely available, where it is deemed to be a dietary supplement. Different formulations of melatonin are sold internationally, with both immediate-release and sustained-release options available. Appropriately timed administration of melatonin has chronobiotic properties and can assist with phase shifting the circadian system (either alone or, more typically, in combination with light exposure) (Hickie and Rogers 2011).

3. EXTRACTION AND PURIFICATION PROCEDURES

In recent years, the development of fast, precise, accurate and sensitive methodologies has become an important issue. Despite the great technological advances, most analytical instruments cannot handle sample matrices directly and, as a result, a sample-preparation step is commonly introduced. For organic trace analysis, this step mainly comprises extractions, which serve to isolate compounds of interest from a sample matrix. This leads to the concentration of target compounds being enhanced (enrichment) and the presence of matrix components being reduced (sample clean up) (Psillakis and Kalogerakis 2003).

Considering its low levels, an important issue of MLT measurements is the adequate extraction from samples (de Almeida, Di Mascio, Harumi, Warren Spence, et al. 2011). In any case, care with sample preparation can improve MLT signal. Pre-purification of MLT will decrease chromatogram noise and avoid the coalition with other compounds that can interfere with MLT peaks. Generally, fluorescence techniques are affected not only by co-elution of other fluorescent compounds also present in the sample, but also by quenchers. This should not be underrated since the majority of aromas absorbs light at wavelengths close to the excitation maximum of MLT. Therefore, samples should be tested in advance for quenching by adding known amounts of MLT (de Almeida, Di Mascio, Harumi, Warren Spence, et al. 2011).

One of the most common sample pre-treatments in an analytical process involve liquid–liquid extractions (LLE) that could be used, for instance, to increase selectivity, by isolating the analyte from matrix interfering species, or to enhance selectivity, by extracting and concentrating the analyte from a large volume of sample. Likewise what happens in other sample manipulations, the manual implementation of this mass transfer operation is usually labor-intensive and time-consuming demanding most of the time large amounts of chemicals that could be harmful to the operator, expensive, and environmentally hazardous (Silvestre et al. 2009).

MLT can be extracted by simple LLE, such as the addition of dichloromethane (1:1, v/v). Samples are then vigorously mixed and centrifuged to obtain aqueous and organic phases. With this procedure, melatonin is retained in the organic phases that are collected and dried under nitrogen atmosphere to concentrate melatonin. This yields a satisfactory recovery rate (generally more than 70%), and can be also applied to buffer-homogenized tissues. However, low precision and accuracy with single liquid–liquid extractions of MLT for high performance liquid chromatography (HPLC) coupled to fluorescence detector have been reported (Rizzo et al. 2002; de Almeida, Di Mascio, Harumi, Warren Spence, et al. 2011). For multiple analyses of MLT and its precursors or metabolites, more comprehensive liquid–liquid extractions have been described using a combination of different solvents (Harumi and Matsushima 2000; de Almeida, Di Mascio, Harumi, Warren Spence, et al. 2011).

In older investigations, chloroform (trichloromethane) was mostly used for MLT extraction and is still in use today. Although this method is effective, dichloromethane is preferred for reasons of lower toxicity. Generally, chlorinated methane should be of highest purity and protection from light and redox-active compounds is of utmost importance for avoiding formation of reactive intermediates which can destroy MLT (de Almeida, Di Mascio, Harumi, Warren Spence, et al. 2011). However, researches who apply LLE to pre-concentrate melatonin also report the use of ethyl acetate (Tomita et al. 2003; Tomita et al. 2001; Arnao and Hernández-Ruiz 2007), methylbutane (Lee et al. 2003) and acetonitrile or methanol (Helton et al. 1993) as extraction solvents.

Talebianpoor et al. (2014) (Talebianpoor et al. 2014) reports the use of miniaturized LLE that uses microliter volumes of the extraction solvent to quantify MLT, the dispersive liquid–liquid microextraction (DLLME). For DLLME, water-immiscible extraction solvent dissolved in a water-miscible dispersive solvent is rapidly injected into an aqueous solution by syringe. A cloudy solution containing fine droplets of extraction solvent dispersed entirely in the aqueous phase is formed. The analytes in the sample are extracted into the fine droplets, which were further separated by centrifugation and the enriched analytes in the sedimented phase are determined by either chromatographic or spectrometric methods. Regarding MLT determination, the authors showed that acetonitrile and carbon tetracholryde were the most suitable as disperser and as extractor solvents, respectively. This method has unique advantages such as simplicity of operation, speed, low extraction time, high extraction recovery and high enrichment factors (Talebianpoor et al. 2014).

The correct handling and maintenance of samples is also important. Samples of MLT should be kept constantly on ice and protected from light radiation, in order to avoid degradation. Despite its relative stability, MLT oxidation can occur over time, including reactions with singlet oxygen. The probability of this occurrence varies, and is dependent on oxygen availability and light incidence. For sample freezing, it is recommended that samples be dried and preferentially kept under vacuum or a nitrogen atmosphere (de Almeida, Di Mascio, Harumi, Warren Spence, et al. 2011).

Regarding most plants, fruits or foods, owing to expected interferences from the food matrix, a prior step of purification is generally deemed necessary. To achieve this purpose one must keep in mind that MLT is a tiny molecule with a molecular weight of 232.28, and its solubility is 0.1 mg/mL in water and 8 mg/mL in ethanol (Garcia-Parrilla, Cantos, and Troncoso 2009). Also the complex chemistries of the plants which often contain large amounts of carbohydrates, lipids and pigments, may induce false positive or false negative melatonin contents. And also the MLT may co-extract and partition with chlorophyll and phenolic compounds, as well as sublimating during vacuum drying and being susceptible to destruction by the impurities commonly found in organic solvents (Paredes et al. 2008; Van Tassel et al. 2001; Feng et al. 2014).

Published procedures involve an extraction with organic solvents from the food matrix which has normally been frozen and milled to powder. The extract can subsequently be submitted or not to a purification step. If MLT is going to be determined by RIA, then organic solvents, especially methanol, must be evaporated to avoid RIA antibody denaturation (Garcia-Parrilla, Cantos, and Troncoso 2009).

In fact, some additional recommendations are made for the recovery of MLT from plant tissues. Thus, to limit the potential enzymatic destruction of MLT it's suggested the use of 0.4 M perchloric acid (HClO$_4$) for protein precipitation (to prevent the potential enzymatic degradation of MLT) and the addition of a hydroxyl radical (•OH) scavenger. MLT rapidly undergoes structural modification in the presence of •OH because it is a highly efficient free radical scavenger. Additionally, the suggestion was made that the extraction procedures should be performed under dim light conditions to curtail photooxidation of MLT. A number of reports used acetone-, Tris- or tricine-based extraction mixtures which sometimes included antioxidants (e.g., butylated hydroxytoluene), HClO$_4$ or the chelator EDTA (Reiter et al. 2007).

It has been shown that homogenization ice-cold HClO$_4$ (Itoh et al. 1995) (Vieira et al. 1992), Tris-HCl buffer (Ramakrishna et al. 2012) (Iriti, Rossoni, and Faoro 2006), amongst other buffers can also represent an accurate means for MLT determination in tissues. In these cases, the homogenates are centrifuged and the resulting supernatant can be directly injected into the analytical system (Feng et al. 2014).

A subsequent solid phase extraction (SPE) can also avoid matrix interferences and achieve a lower detection limit. In SPE the analytes to be extracted are partitioned between a solid and a liquid (rather than between two immiscible liquids as in LLE) and these analytes must have a greater affinity for the solid phase than for the sample matrix (retention or adsorption step). Compounds retained on the solid phase can be removed at a later stage by eluting with a solvent with a greater affinity for the analytes (elution or desorption step) (Berrueta, Gallo, and Vicente 1995).
In the literature, SPE with C$_{18}$ cartridges has been widely applied to isolate MLT from several samples (Peniston-Bird et al. 1993; Kulczykowska and Iuvone 1998; Zhao et al. 2013; Ramakrishna et al. 2012; Pape and Lüning 2006). However, SPE may absorb MLT, and the high cost for MLT extraction is a serious disadvantage. A recent study showed that microextraction by packed sorbent (MEPS) sample pretreatment procedure is faster and lower cost than conventional SPE procedures and requires even lower volumes of sample and solvents for melatonin extraction (Mercolini, Mandrioli, and Raggi 2012; Feng et al. 2014).

One has to take into account that MLT is unstable in the extracts, the rapid generation of heat and pressure forces compounds from the matrix and produces MLT extracts with better recovery rates (Feng et al. 2014; Hemwimon, Pavasant, and Shotipruk 2007).

Table 1. Biological and non-biological sample homogenization and purification procedures to determine melatonin

Human samples

Sample	Homogenization	Separation	Supernatant	Extraction type	Extraction	Recoveries (%)	Analytical	Ref.
Prostate normal and cancer cell lines	Water bath at 37°C for 15 min.	n.s.	n.s.	LLE	1.5 volumes of dichloromethane	> 96.2	HPLC-UV	(Hevia et al. 2010)
Urine	500 μL of 0.2 M potassium phosphate buffer, pH 7.4 added to 0.5 ml urine.	n.s.	n.s.	SPE	Chem-Elute 1001	> 100.0	HPLC-FLD	(Minami et al. 2009)
Cadaver skin	1 mL of methanol, and homogenized for 5 min with an electric stirrer.	7000 rpm; 10 min	n.s.	n.s.	n.s.	n.s.	HPLC-UV	(Dubey, Mishra, and Jain 2007)
Oral fluid	200 μL of MeOH containing 687.5 μM ethyl analog.	4500 g; 5 min	200 μL	SPE	RP-select B	n.s.	HPLC-FLD	(Hamase et al. 2004)
Oral fluid	100 μL of a 1,4-dithio-dl-threitol [(±)-threo-1,4-dimercapto-2,3-butanediol] solution ($6.6 \cdot 10^{-2}$ M) added to the sample, and the mix incubated for 10 min at room temperature.	5000 rpm; 5 min	n.s.	SPE	Nexus SPE column	n.s.	LC-MS/MS	(Eriksso, Östin, and Levin 2003)
Skin fibroblasts *in vitro*	10 volumes of ice-chilled 0.4 M perchloric acid containing 2 mM EDTA.	15.000 g; 10 min	n.s.	LLE	2 x 2 mL of 2-methylbutane	n.s.	HPLC-UV	(Lee et al. 2003)
Serum	n.s.	n.s.	n.s.	SPE	Oasis® HLB	70.0	LC-MS/MS	(Yang et al. 2002)
Serum	1mL sample of serum diluted with 1 mL of phosphate-buffered saline (PBS; 50 mM NaH2PO4, 15 mM NaCl, pH 7.2).	0.22-μm Optex filter	n.s.	immunoaffinity chromatography	anti-melatonin immunoaffinity column	95.0	LC-MS	(Rolčík et al. 2002)

Human samples

Sample	Homogenization	Separation	Supernatant	Extraction type	Extraction	Recoveries (%)	Analytical	Ref.
Human bone marrow	Diluted in ice-cold phosphate-buffered saline (8.5g/L NaCl, sodium chloride; 1.36 g/L KH_2PO_4; dissolved in distilled water and adjusted to pH 7.4 with 10 N NaOH).	400 g; 10 min	2 mL	LLE	8 mL of dichloromethane	n.s.	HPLC-ECD	(Conti, Conconi, Hertens, Skwarlo-Sonta, et al. 2000)
Plasma and Oral fluid	pH 13 buffer solution (0.2 mL).	n.s.	n.s.	SPE	C8 Amprep cartridges	82.0	GC-MS	(Simonin, Bru, Lelièvre, et al. 1999)
Extracts of ovary	1.0–2.0 mL of ice-cold 10 mM phosphate-buffered saline (PBS) containing 50 mM disodium EDTA (pH 7.4).	10.000g; 10 min	n.s.	LLE	6–10 mL of chloroform	80.5	HPLC-FLD	(Itoh et al. 1999)
Plasma	100 µL of 1M sodium hydroxide.	n.s.	n.s.	LLE	5 mL methylene chloride	n.s.	GC-MS	(Covaci, Doneanu, Aboul-Enein, et al. 1999)
Plasma	1 mL of 0.1 N sodium hydroxide.	n.s.	n.s.	LLE	9 mL of methylene chloride	87.0	GC-MS	(Fourtillan et al. 1994)
Plasma	100 g/L activated charcoal added to a plasma pool, mixing at 4° C for 48 h.	30.000 g; 1 h	1-µm (pore size) filter paper	SPE	C18 octadecylsilyl (ODS)	> 70.0	HPLC-FLD	(Peniston-Bird et al. 1993)
Plasma	n.s.	n.s.	n.s.	LLE	(2 mL) were extracted with 8 mL dichloromethane	> 90.3	HPLC-ECD	(Vieira et al. 1992)

Table 1. (Continued)

Human samples

Sample	Homogenization	Separation	Supernatant	Extraction type	Extraction	Recoveries (%)	Analytical	Ref.
Oral fluid	n.s.	18 000g; 10 min	n.s.	LLE	chloroform (4 mL) vortexing lightly (1 min)	n.s.	HPLC-(n.s.)	(Vakkuri 1985)
Plasma	n.s.	n.s.	n.s.	LLE	(6:1) chloroform added; vortexed 10 s	71.0	GC-MS	(Skene et al. 1983)
Fish plasma	200 µL aliquot of plasma sample mixed (1:1, v:v) with 0.1M acetic acetate buffer (pH 4.6).	n.s.	n.s.	LLE	2mL chloroform	> 92.6	HPLC-FLD	(Muñoz et al. 2009)
Fish bile	100 µL aliquot sample was mixed (1:1, v:v) with 0..1M acetic acetate buffer (pH 4.6)	n.s.	n.s.	LLE	1mL chloroform	>87.1	HPLC-FLD	(Muñoz et al. 2009)
Fish intestine	(30–40 mg) were sonicated in 500 µL of 0.1M acetic acetate buffer (pH 4.6).	10 000g; 10 min.	n.s.	LLE	2.5 mL chloroform (1:5, v:v ratio)	> 86.1	HPLC-FLD	(Muñoz et al. 2009)
Hamster pineal gland	100 µL of ice-cold 0.1 M perchloric acid.	10 000g; 20 min.	90 µL mixed with 10 gl of 1 M sodium phosphate, pH 4.3, and filtered	n.s.	n.s.	n.s.	HPLC-ECD	(Harumi, Akutsu, and Matsushima 1996)
Rat pineal gland	400 µL of MeOH.	4500g; 5 min.	(50 µL), 25 nM of ethyl and isopropyl analogs	n.s.	n.s.	n.s.	HPLC-FLD	(Hamase et al. 2004)
Mice pineal gland	Homogenized on ice at 1000 rpm 20 times in 500 µL of CH$_3$OH solution.	4500g; 5 min.	400 µL	LLE	2 x 250 µL ethyl acetate	n.s.	HPLC-FLD	(Tomita et al. 2003)
Rat pineal gland	20 x in 500 µL ice cold MeOH.	4500g; 5 min.	50 µL	LLE	3 x 250 µL ethyl acetate	86.9	HPLC-FLD	(Tomita et al. 2001)

Animal, organisms samples

Sample	Homogenization	Separation	Supernatant	Extraction type	Extraction	Recoveries (%)	Analytical	Ref.
Mouse bone marrow	Cells re-suspended in 500 mL of mobile phase (10% methanol, 15% acetonitrile, 0.01 M citric acid, 0.001 M sodium acetate, pH 3.6).	14.000rpm; 15 min.	2 mL	LLE	8 mL of dichloromethane	n.s.	HPLC-ECD	(Conti, Conconi, Hertens, Skwarlo-Sonta, et al. 2000)
Rats and mice pineal gland	500 mL of methanol at 1000 rpm 20 times on ice.	4500g; 5 min.	100 µL	n.s.	n.s.	n.s.	HPLC-FLD	(Hamase et al. 2000)
Planarians	Sonication in 100 mL of ice-cold 0.15 M perchloric acid containing 0.1% ascorbic acid and 0.01% disodium EDTA.	20.000g; 20 min.	0.45-mm filter	n.s.	n.s.	n.s.	HPLC-FLD	(Itoh, Shinozawa, and Sumi 1999)
Bile of mammals	Mixed with one drop of a 1 M NaOH.	n.s.	n.s.	LLE	1 mL chloroform	40.0	HPLC-ECD	(Tan, Manchester, Reiter, Qi, Hanes, et al. 1999)
Slides of fresh bone marrow smears	Perfusion with normal saline, sonicated to disrupt the cells.	15.000g; 20 min.	50 µL	LLE	One drop 1 M NaOH, 1 mL chloroform	n.s.	HPLC-ECD	(Tan, Manchester, Reiter, Qi, Zhang, et al. 1999)
Rat pineal gland	Methanol (500 µL) added and homogenized at 1000 rpm 20 times.	4500g; 5 min.	100 µL	n.s.	n.s.	n.s.	HPLC-FLD	(Iinuma et al. 1999)
Mice pineal gland	500 ml of buffer (tricine, pH 6.0).	12.000rpm; 5 min	100 µL	LLE	5 mL dichloromethane	n.s.	GC-MS	(Vivien-Roels et al. 1998)
Flounder plasma	n.s.	n.s.	n.s.	SPE	C_{18} Bakerbond SPE cartridge	n.s.	HPLC-FLD	(Kulczy-kowska and Iuvone 1998)

Table 1. (Continued)

Animal, organisms samples

Sample	Homogenization	Separation	Supernatant	Extraction type	Extraction	Recoveries (%)	Analytical	Ref.
Rat, dog, and monkey plasma	n.s.	n.s.	n.s.	SPE	LC-CN cartridges	n.s.	HPLC-UV; LC-MS/MS	(Yeleswaram et al. 1997)
Rat ovary	1.0–1.5 mL of ice-cold 0.20 M sodium borate buffer (pH 10.0).	n.s.	n.s.	LLE	8–12 mL of chloroform	81.4	HPLC-FLD	(Itoh et al. 1997)
Hamster skin culture	0.1 M sodium phosphate buffer, pH 6.8, plus 1% Triton X-100.	10.000g; 15 min.	n.s.	LLE	Chloroform	n.s.	HPLC-UV; GC-MS	(Slominski et al. 1996)
Dinoflagellate cells	Mixed with 1 mL of 1 M Tris-HC1 buffer, pH 8.4, shock frozen with liquid nitrogen, and pulverized in a mortar.	n.s.	n.s.	LLE	300 µL of 0.4 N PCA containing 0.05% $Na_2S_2O_5$ as antioxidant and 0.1% EDTA	> 23.0	HPLC-FLD; HPLC-ECD	(Poeggeler and Hardeland 1994)
Different organs of adult female crickets (*Gryllus bimaculatus*)	Five times their volume of ice-cold 0.1 M perchloric acid containing 0.1% ascorbic acid and 0.01% disodium EDTA.	12.000g; 15 min	1 N potassium hydroxide	LLE	4 volumes of water-saturated chloroform	> 65.8	HPLC-FLD	(Itoh et al. 1995)
Rat plasma	n.s.	n.s.	50 - 200 µL	LLE	10 volumes of dichloromethane	n.s.	HPLC-ECD	(Raynaud et al. 1993)
Murine B16 melanoma cells	2 M acetic acid.	microfuge; 10 min.	n.s.	LLE	5% acetonitrile in 25% (v/v) methanol in water	n.s.	HPLC-UV	(Helton et al. 1993)
Chicks and Rats serum	n.s.	n.s.	n.s.	LLE	1 mL plus 2 mL chloroform	59.0	HPLC-ECD	(Huether et al. 1992)
Rat pineal gland	Sonicated in 100 µL of 0.1 M PCA.	12.000g; 10 min.	n.s.	n.s.	n.s.	n.s.	HPLC-ECD	(Vieira et al. 1992)
Chicks and Rats pineals and gut (upper duodenum)	0.4 N perchloric acid containing 0.1% EDTA and 0.05% sodium bisulphite.	n.s.	n.s.	n.s.	n.s.	91.0 (pineal), 39.0 (gut)	HPLC-ECD	(Huether et al. 1992)

Animal, organisms samples

Sample	Homogenization	Separation	Supernatant	Extraction type	Extraction	Recoveries (%)	Analytical	Ref.
Mice pineal gland	70 µl mobile phase.	90.000g; 30 min.	150 µL	n.s.	n.s.	n.s.	HPLC-ECD	(Conti and Maestroni 1996)

Plants, fruits and food samples

Sample	Homogenization	Separation	Supernatant	Extraction type	Extraction	Recoveries (%)	Analytical	Ref.
Cherry cultivars	2 g of each lyophilized cultivar. 10 mL of methanol added. These solutions were sonicated at 30°C during 30 min	9652 rpm; 30 min.	diluted with deionized water	SPE	Oasis® HLB	> 61.4	HPLC-ECD	(Rosado et al. 2017)
Fruits of sweet cherry	Frozen cherries were ground to a fine powder with liquid nitrogen in a mortar, suspended in 10 mL methanol and ultra-sonicated (80 Hz) for 35 min at 45°C.	10.000g; 15 min.	dissolved in 2 mL 5% methanol	SPE	C18 SPE cartridge	92.2	HPLC-FLD	(Zhao et al. 2013)
Grapes, must, wine and grappa	n.s.	1400g; 5 min.	syringe membrane filter with a pore size of 0.45 µm	MEPS	C8 sorbent	> 91.0	HPLC-FLD	(Mercolini, Mandrioli, and Raggi 2012)
Tissues of *Coffea canephora*	1 mL of Tris buffer (1 M Tris–HCL, pH 8.4) and homogenized in 500 µL of buffer [0.4 M perchloric acid, 0.05% sodium meta bisulfate, 0.1% ethylenediaminetetra-acetic acid (EDTA)].	12.000 rpm; 15 min.	resuspended in 500 µL methanol	n.s.	n.s.	n.s.	HPLC-FLD	(Ramakrishna et al. 2012)

Table 1. (Continued)

Plants, fruits and food samples

Sample	Homogenization	Separation	Supernatant	Extraction type	Extraction	Recoveries (%)	Analytical	Ref.
Rice grains	n.s.	n.s.	n.s.	MAE	195°C, 1000 W, 20 min, solvent 100% MeOH, and ratio of solvent to sample 10:1.	n.s.	HPLC-FLD	(Setyaningsih, Palma, and Barroso 2012)
Beans of Coffea species	Mixed with 5% trichloroacetic acid (7 mL) and submitted to agitation for 5 min in a vortex mixer.	10.000g; n.s.	n.s.	SPE	Sep-pak C18	n.s.	HPLC-FLD	(Ramakrishna et al. 2012)
Wine	Dried and re-diluted in methanol: water (1:1), up to a 3:1 concentration.	0.45-μm filters	n.s.	n.s.	n.s.	n.s.	LC-MS/MS	(Rodriguez-Naranjo et al. 2011)
Wine	n.s.	n.s.	n.s.	SPE	C18 SPE cartridges	98.0	HPLC-FLD	(Rodriguez-Naranjo et al. 2011)
Wine	n.s.	2000g; 5 min.	500 μL	SPE	BondElut C18 cartridges	> 91.0	HPLC-FLD	(Mercolini et al. 2008)
Purple wheat	Methanol acidified with HCL (1 N) (ratio 85:15, v/v) added to ground purple wheat (sample to solvent ratio of 1:8) and the pH adjusted to 1.0	1800rpm; 45 min.	n.s.	n.s.	n.s.	> 80.0	LC-MS	(Hosseinian, Li, and Beta 2008)
Roots in etiolated hypocotyls of Lupinus albus L.	Without homogenization.	n.s.	n.s.	LLE	15 mL ethyl acetate with butylated hydroxytoluene (5 μm), for 15 hr in darkness	n.s.	n.s.	(Arnao and Hernández-Ruiz 2007)

Plants, fruits and food samples

Sample	Homogenization	Separation	Supernatant	Extraction type	Extraction	Recoveries (%)	Analytical	Ref.
higher plants and other photoautotrophic organisms	1 g material and 2 g (1916 µL) Tris-HCL buffer (1 M, pH 8.4).	ground for 3 min.	2g of the resulting powder mixed with 2 g PCA (0.4 N)	SPE	C18 SPE cartridge	> 44.1	HPLC-FLD	(Pape and Lüning 2006)
Higher plants and other photoautotrophic organisms	To a mortar filled with liquid nitrogen a 1.5 g homogenous sample were added.	ground for 3 min.	6–10 mL extraction mixture (acetone:methanol:water 89:10:1) grinded for 2 min	SPE	C18 SPE cartridge	> 51.0	HPLC-FLD	(Pape and Lüning 2006)
Grape skin	10 mL of extraction buffer (1 mol L^{-1} Tris–HCL buffer, pH 8.4, 0.4 mol L^{-1} perchloric acid plus 0.1% EDTA, 0.05% Na$_2$S$_2$O$_5$ and 10 mmol L^{-1} ascorbic acid). Extraction was achieved Incubation on an orbital shaker for 1 h at room temperature, after 15 min of sonication.	10.000g; 10 min.	n.s.	SPE	two C18 Sep-Pak cartridges connected in series	n.s.	HPLC-FLD	(Iriti, Rossoni, and Faoro 2006)
Glycyrrhiza uralensis	1.0 mL of 80% (v/v) ethyl alcohol (EtOH) added immediately and pulverized at 30 Hz for 15 min.	700g; 10 min.	filtered through a 0.2 lm syringe filter	n.s.	n.s.	> 95	HPLC-UV	(Afreen, Zobayed, and Kozai 2006)

Table 1. (Continued)

Plants, fruits and food samples

Sample	Homogenization	Separation	Supernatant	Extraction type	Extraction	Recoveries (%)	Analytical	Ref.
Raw walnuts	Pulverized with a mortar and pestle by using 15 mL of methanol.	10.000g; 30 min.	Residues were redissolved in 1 mL (pH 7.4, 20 mM) of phosphate buffered saline	LLE	2 mL of chloroform	40	HPLC-ECD	(Reiter, Manchester, and Tan 2005)
Plant tissues	Etiolated tissue (2–4 g) cut into sections (3–5 mm) and, without homogenization.	n.s.	n.s.	LLE	15 mL ethyl acetate with butylated hydroxytoluene (5 µM), for 15 hr at 4 C in darkness	n.s.	HPLC-ECD; LC-MS/MS	(Hernández-Ruiz, Cano, and Arnao 2005)
Seeds of lupin	Ultra-Turrax T25 in 50 mM sodium phosphate buffer (pH 8.0) containing 5 µM butylated hydroxytoluene (BHT).	filtered under vacuum through a 0.45 µm nylon filter	n.s.	LLE	Ethyl acetate and 50 mM sodium phosphate buffer (first at pH 8.0 and second at pH 3.0)	n.s.	HPLC-ECD	(Hernandez-Ruiz, Cano, and Arnao 2004)
Herbal powder	20 mL methanol and ultrasonicated for 30 min at room temperature.	4000rpm; 10 min.	n.s.	SPE	C18 SPE cartridge	> 90.0	HPLC-FLD; LC-MS	(Chen et al. 2003)
Herbal powder	4g of frozen, powder with 10 mL of extraction solvent acetone–water–glycerol [80:15:5, v:v:v], plus 0.1 M tricine, 9 g/L NaCL, 1 g/L Na$_2$O$_5$S$_2$, 1 g/L dithiothreitol [DTT], 1 g/L butylated hydroxytoluene, pH to 8.0 with NaOH.	n.s.	n.s.	SPE	C18 SPE cartridge	> 34.0	HPLC-ECD /GC-MS	(Van Tassel et al. 2001)

Plants, fruits and food samples

Sample	Homogenization	Separation	Supernatant	Extraction type	Extraction	Recoveries (%)	Analytical	Ref.
Montmorency and Balaton tart cherries	1 mL of 0.05M potassium phosphate buffer (pH 8.0).	3000rpm; 5 min.	500 µL	LLE	25µL of a 1 M KOH, 700 µL of chloroform, shaken for 10 min	n.s.	HPLC-ECD	(Burkhardt et al. 2001)
Seeds of *Chenopodium rubrum L.*	15 mL of 1mol/L Tris-HCl buffer, pH 8.4 and 15 mL of 0.4 mol/L HClO$_4$ containing 0.1% Na$_2$EDTA, 0.05% Na$_2$S$_2$O$_5$, and 10 mmol/L ascorbic acid; orbital shaker 1h.	12.000g; 10 min.	n.s.	SPE	C18 cartridges in series (Sep-Pak Plus, Waters)	n.s.	LC-MS/MS	(Wolf et al. 2001)
In vitro cultured plants of St. John's wort	300 mL of buffer (0.4 mol L^{-1} perchloric acid, 0.05% sodium metabisulfate, 0.1% EDTA) dark, room temperature for 15 min.	12.000g; 15 min.	n.s.	n.s.	n.s.	n.s.	HPLC-ECD	(Murch, Campbell, and Saxena 2001)
Seeds	Pulverized with a mortar and pestle using 3mL of cold ethanol (per 100 mg tissue).	10.000rpm; 10 min.	n.s.	n.s.	n.s.	n.s.	HPLC-ECD	(Manchester et al. 2000)
St. John's wort	300 µL of buffer (0.4 M perchloric acid, 0.05% sodium metabissulfate, 0.1% EDTA) dark, room temperature for 15 min.	12.000 g; 15 min.	n.s.	n.s.	n.s.	n.s.	HPLC-ECD; HPLC-UV	(Murch, KrishnaRaj, and Saxena 2000)

Table 1. (Continued)

Plants, fruits and food samples

Sample	Homogenization	Separation	Supernatant	Extraction type	Extraction	Recoveries (%)	Analytical	Ref.
Chenopodium rubrum	1 M TRIS-HC1 buffer, pH 8.4, and 0.4 M perchloric acid (containing 0.1% EDTA, 0.05% sodium bisulphite and 10 mM ascorbic acid) shaker for 1 hr.	10.000g; 10 min.	n.s.	SPE	C 18 Sep-Pak cartridges	n.s.	LC-MS/MS	(Kolář et al. 1997)
Fruits or vegetables	Cut into small pieces, 10% sodium carbonate solution.	8.240g; n.s.	500 µL	LLE	3 mL diethylether	93.0	GC-MS	(Dubbels et al. 1995)

Comercial formulations

Sample	Homogenization	Separation	Supernatant	Extraction type	Extraction	Recoveries (%)	Analytical	Ref.
Lipidic microparticles (liposheres)	Dispersing the liposheres (30–40 mg) in ethanol under sonication (15 min).	Diluted to volume (10 mL), filtered (0.45-mmembrane filters)	n.s.	n.s.	n.s.	n.s.	HPLC-UV	(Tursilli et al. 2006)
Transdermal Patches	Adhesive layer of the patch (area 4 cm2) was dissolved in methanol.	n.s.	Diluted further with methanol	n.s.	n.s.	n.s.	HPLC-UV	(Kanikkannan et al. 2004)
Tablets	10 tablets finely ground, suspended in 3 mL methanol sonicated for 20 min.	Filtered through a Whatmann filter paper (N°4)	Diluted further with methanol	n.s.	n.s.	99.8	GC-MS	(Nunez-Vergara et al. 2001)
Tablets	Powdered sample, equivalent to 2 mg of melatonin, was treated with water-methanol 80:20 (v/v).	filtered through 0.45 µm Millipore filter	n.s.	n.s.	n.s.	> 98.7	HPLC-UV	(Andrisano et al. 2000)

Comercial formulations

Sample	Homogenization	Separation	Supernatant	Extraction type	Extraction	Recoveries (%)	Analytical	Ref.
Tablets	Powdered sample, equivalent to 2 mg of melatonin, was treated with water-methanol 80:20 (v/v).	filtered through 0.45 μm Millipore filter	n.s.	n.s.	n.s.	> 98.7	HPLC-UV	(Andrisano et al. 2000)
Tablets	One finely ground tablet was dissolved in 10 mL ethyl acetate and sonicated for 15 min.	filtered through a Whatmann (no.4)	n.s.	n.s.	n.s.	n.s.	GC-MS	(Aboul-Enein, Doneanu, and Covaci 1999)
Tablets	One finely ground tablet was dissolved in 10 mL ethyl acetate and sonicated for 15 min.	filtered through a Whatmann (no.4)	n.s.	n.s.	n.s.	n.s.	GC-MS	(Covaci, Doneanu, Aboul-Enein, et al. 1999)

*n.s. (not specified).

Sonication can induce cavitations which create microenvironment with high temperatures and high pressures, and in turn speed up the removal of analytes from the complex matrices (Feng et al. 2014; Stege et al. 2010). Hence, an ultrasound-assisted method has been applied before sample clean-up for MLT extraction (Feng et al. 2014). Setyaningsih, Palma, and Barroso (Setyaningsih, Palma, and Barroso 2012) developed a microwave-assisted extraction (MAE) method for MLT extraction from rice with high precision. They stated that the average amount of melatonin in short grain varieties (54.17 ± 13.48 ng/g) is much higher than the value (1.0 ± 0.06 ng/g) reported by Hattori et al.(1995) (Hattori et al. 1995) because of the new extraction method applied, justifying the great importance it should be given to this analytical step (Feng et al. 2014).

Table 1 reports an extensive review on biological and non-biological specimens pre-treatment to determine MLT by a wide range of chromatographic techniques. The extractions efficiencies of each sample clean-up, when possible, are also reported.

4. MELATONIN METHODS OF ANALYSIS

Sample preparation will also depend on the method used for analysis, since the presence of other compounds in the sample may interfere with the MLT signal (de Almeida, Di Mascio, Harumi, Warren Spence, et al. 2011).

Several quantitative approaches have been developed to analyze MLT content in samples. Immunoassays such as radioimmunoassay (RIA) and enzyme-linked immunosorbent assay (ELISA) have long been applied to MLT quantification (Muñoz et al. 2009). However, their results correlate poorly with those obtained by chromatographic techniques due to the cross-reactivity of MLT with similar compounds (Ishizaki, Uemura, and Kataoka 2017).

Various analytical methods are, then, described in literature for MLT determination in samples by high-performance liquid chromatography (HPLC) with fluorescence detection (FLD), electrochemical detection (ECD) and UV detection. Also, liquid chromatography coupled to mass spectrometry (LC-MS) and gas chromatography coupled to mass spectrometry (GC-MS) are reported, but less frequently (Martins, khalil, and Mainardes 2017).

4.1. Immunological Techniques

Formerly, studies of analysis of MLT were carried out to assess its presence in pineal gland, saliva or other biological fluids (Kennaway et al. 1977). Immunological techniques were considered appropriate, and were successfully applied to biological matrices. Therefore, a first approach consists in applying those available methods of common use in biological samples to food analysis. Main techniques involved both radioimmunoassay (RIA) and enzyme-immunoassay (EIA) (Garcia-Parrilla, Cantos, and Troncoso 2009).

The most crucial aspect of immunoassays is the preparation of the antiserum. Because MLT is too small to be capable of producing antisera on its own it must be coupled to an antigenic protein. In such a conjugate the small molecular weight substance is called a hapten. The resulting antiserum binds the protein and the hapten plus a portion of the adjacent protein. The hapten has few antigenic determinants relatively to the protein. Specificity studies of antisera produced by steroid–protein conjugates have shown that antisera are not able to discriminate structural differences in the hapten that are immediately at or close to the site of coupling (Grota et al. 1983; de Almeida, Di Mascio, Harumi, Warren Spence, et al. 2011). The choice of the hapten and conjugation reaction should therefore be determined by the type of discrimination that is required. Indolealkylamines have in common one ring nitrogen (position 1) and an adjacent carbon (position 2). Thus for melatonin, coupling via either position 1 or position 2 should allow the resulting antisera to discriminate different indoles that are commonly found in tissues. (de Almeida, Di Mascio, Harumi, Warren Spence, et al. 2011).

4.1.1. Enzyme-Immunoassay (EIA)

Enzyme-immunoassays (EIA), such as Enzyme-linked immunosorbent assays (ELISA) use antibodies and colorimetric enzymes. The enzymes catalyze color reactions with an enzyme-specific chromogen as an indicator. ELISA is specific and sensitive for melatonin measurement. However, commercial ELISA kits are usually very expensive and therefore not suitable for high throughput measurements. Customized MLT ELISAs usually involve development of MLT antibodies and melatonin MLT, which are generally non-trivial to produce and/or unavailable for most researchers (Li and Cassone 2015).

ELISA techniques offer a nonradioactive method for MLT measurement that, by allowing the direct use of serum, is simpler, faster, and more economical than most RIAs.

The first antibodies described in the literature were obtained after immunization of immunogens obtained from the coupling of a MLT hapten to a protein by Mannich condensation with formaldehyde (Simonin, Bru, Lelievre, et al. 1999).

Various ELISAs for MLT have been developed, one even allowing the direct use of human serum (Chegini et al. 1995). A commercial kit based on a competitive enzyme immunoassay (Melatonin ELISA, IBL-Hamburg) was applied to determine MLT in olive oil for the first time (de la Puerta et al. 2007). This same kit was used to determine MLT in grape skin extracts from different cultivars (Iriti, Rossoni, and Faoro 2006), wine (Guerrero, Martínez-Cruz, and Elorza 2008) and cow milk (Kollmann et al. 2008; Garcia-Parrilla, Cantos, and Troncoso 2009).

To determine the suitability of ELISA for routine application, one has to assess its major operational characteristics, i.e., stability, sensitivity, reproducibility, accuracy, and ease of handling (Chegini et al. 1995). Nevertheless, Johns et al. (2013) (Johns et al. 2013) reported that using ELISA technique should not affect the MLT levels in fruits due to the low cross-reactivity (a few percent). They also suggested the possibility of the coelution of another compound with the same retention time as MLT when using HPLC was ruled out by the ELISA test, as reported by Iriti et al.(2006) (Feng et al. 2014; Iriti, Rossoni, and Faoro 2006).

4.1.2. Radioimmunoassay (RIA)

With the raising of antisera for MLT in the mid-1970s, several laboratories have developed RIAs of MLT and refined this technique to achieve an acceptable degree of specificity, sensitivity, and precision (Chegini et al. 1995).

In RIAs, an anti-MLT antibody is typically mixed with samples or standards, and radioactively labeled ([125I] or [3H]) MLT or a MLT analog is used as tracer to compete for anti-MLT antibody binding. The antibody binding MLT and radioactive tracer are precipitated for measurement of radioactivity by scintillation or γ counts, depending on the tracer. The more MLT in samples binds to the antibody, the fewer binding sites are available to the tracer, and therefore the less radioactivity will be measured. RIA is widely used because it is sensitive, and the procedure is relatively simple; however, RIA is also expensive and the potential health risks for researchers and students preparing and handling the radioactive antigen are not inconsequential. (Li and Cassone 2015; Vakkuri et al. 1984).

Most of the RIAs were developed for MLT use [3H]-MLT as tracer (Fraser et al. 1983; Vakkuri et al. 1984; Webley, Mehl, and Willey 1985; Plebani et al. 1990). [3H]-MLThas only a low specific activity and the preparation is relatively complicated. Meanwhile, [125I]-iodination is preferred because of higher efficiency, the low radiolysis and the reduced hazards (Welp, Manz, and Peschke 2010).

A sensitive RIA for MLT has been developed by Rollag and Niswender (1976) (Rollag and Niswender 1976). This assay utilized rabbit antisera to a bovine serum albumin conjugate of N-succinyl-5-methoxytryptamine and N-3-(4-hydroxyphenyl)-propionyl-5 methoxytryptamine for radioiodination. Rollag and Niswender study was very important, hence have checked the RIA's ability to distinguish between MLT and compounds with similar structure (Gern, Owens, and Ralph 1978).

RIA is currently the method of choice for most laboratories concerned with pineal function, and it was also successfully applied to determine quantitatively melatonin in biological samples such as human serum, plasma and saliva (Leung 1991). However, the need for radioisotopes is an inherent disadvantage of RIA, precluding its utilization in laboratories that are not equipped for handling radioactive materials (Chegini et al. 1995).

A MLT RIA with no extraction step has been developed and this direct assay has been used to measure MLT in human, ovine, rat, and marsupial plasma and human saliva. Because the direct assay involves using radioiodinated MLT rather than tritiated MLT as a tracer, costly and cumbersome scintillation counting can be avoided (Leung 1991). Also, because of its higher specific activity 2-[125I]-iodomelatonin allows a lower detection limit thus allowing the use of smaller amounts of sample. This is especially important in measurements not preceded by MLT purification. However, [125I] is more prone to nonspecific binding so that some determinations can be faulty (de Almeida, Di Mascio, Harumi, Warren Spence, et al. 2011).

Studies of antisera resulting from Mannich coupling of MLT to bovine serum albumin (BSA) have revealed that this approach leads to a highly specific MLT antiserum as shown by cross-reactivity studies in RIA (de Almeida, Di Mascio, Harumi, Warren Spence, et al. 2011; Lemaitre and Hartmann 1980; Pang et al. 1977; Bubenik, Brown, and Grota 1976; Yang et al. 2006).

However, Van Tassel et al. (2001) (Van Tassel et al. 2001) reported that the amount of MLT in tomato fruits determined by RIA was higher than that determined by GC–MS. Therefore, they claimed the RIA analysis is overestimated and creates the false positive findings (Feng et al. 2014). The hypothesis to explain this result is the presence of a compound or compounds which cross-react(s) with MLT RIA antiserum (Garcia-Parrilla, Cantos, and Troncoso 2009).

4.2. Chromatographic Techniques

It was proven that for RIA and EIA methods, a variety of molecules in samples may cross-react with the related antibodies and enzymes, and this can lead to potential overestimation of the actual melatonin values. For this reason, relatively new and rapid technologies were used for the determination of MLT. Those include high-performance liquid chromatography (HPLC) with UV/VIS photodiode array (PDA), fluorescence detector (FLD) or electrochemical detector (ECD), and coupled to mass spectrometry (LC-MS). These are powerful and precise methods. Especially, LC-MS methods have largely replaced the conventional techniques for the analysis of MLT (Huang and Mazza 2011). Also gas chromatography coupled to mass spectrometry (GC-MS) has been widely applied for the purpose.

The concerns about specificity and sensibility of the analytical methods led to comparing immunoassays with GC–MS. As commercially available antibodies lack specificity in the presence of related molecules, an LC–RIA procedure was developed (Simonin et al., 1999) (Simonin, Bru, Lelièvre, et al. 1999). The detection limit (15 pg/mL) was higher than that obtained with GC–MS (1 pg/mL) after automated solid phase extraction for plasma samples or liquid–liquid extraction for saliva samples. Therefore, chromatographic techniques proved to be a good alternative to immunoassay techniques (Garcia-Parrilla, Cantos, and Troncoso 2009).

Vitale et al. (1996) (Vitale et al. 1996) reported an HPLC method with fluorometric detection with enough sensitivity as to detect MLT in a single pineal cell. The achieved limit of detection was 3 pg/mg pineal tissue similar to the 4 pg/mg reported for RIA. In the case of vegetal matrices, HPLC coupled to fluorescence, coulometric or mass detectors have been used. MLT determination by mass spectrometry makes identity confirmation possible, thus increasing the validity of results. Liquid chromatography avoids the derivatisation step, which is an advantage. Therefore it has been applied for both biological samples and food matrices (Garcia-Parrilla, Cantos, and Troncoso 2009).

4.2.1. High-Performance Liquid Chromatography (HPLC)

High Performance Liquid Chromatography (HPLC) has its roots back into the early 20th century (Lynch, Chen, and Liu 2017). It is known as a powerful separation method, able to resolve mixtures with a large number of similar analytes, and providing directly both qualitative and quantitative information. In theory, the stationary phase requires very small particles and hence high pressures are essential for forcing the mobile phase through the column (Meyer 2013).

In many studies, RIA methodology has been replaced by HPLC with electrochemical and fluorescence detection for MLT evaluation, due to its greater sensitivity and specificity (Vitale et al. 1996). However, this procedure is more adequate for MLT alone, and not for mixtures of several indoles, such as serotonin and tryptamine among others, that can cause disruptions in the assay. For example, the serotonin/melatonin ratio is higher than 100 in rat pineal. This high ratio can cause disturbances in chromatographic separations that can make melatonin detection difficult, and thus requires a good procedure for MLT extraction. All the same, the avoidance of partial coelution with other indoles is mostly a matter of the art of chromatograph (de Almeida, Di Mascio, Harumi, Warren Spence, et al. 2011).

HPLC with isocratic mobile phases is a widely used technique for the qualitative and quantitative analyses of biogenic amines. Since the HPLC system allows high resolution and sensitive detection of indoleamines with sample preparation, many investigators used the system for analysis of MLT. Vitale et al. (Vitale et al. 1996) reported that HPLC with a normal-phase column and an organic non-polar mobile phase is a sensitive method for the MLT determination. However, many investigators recommended HPLC with reversed-phase columns (RP C18 or RP C8) using and aqueous mobile phase for the determination of MLT (Feng et al. 2014; Harumi and Matsushima 2000). Anderson and co-workers reported the use of HPLC mobile phases containing 35% methanol for MLT determinations (Harumi and Matsushima 2000; Anderson et al. 1981). On the other hand, Mills et al. (Mills, Finlay, and Haddad 1991) recommended the use of a mobile phase containing 16% acetonitrile for MLT determinations. Mefford and Barchas (Mefford and Barchas 1980) reported that the mobile phase with 25% methanol for MLT determinations is a good option (Harumi and Matsushima 2000).

When a polar or ionized solute is mixed with an ion charged opposite or counter ion, an ion pair is formed between the solute ion and the counter ion, and the polarity of the resulting complex is decreased. Since the polarity or hydrophobicity of samples is a major factor determining the retention time on RP-HPLC, addition of counter ions or ion paring agents into the RP-HPLC mobile phase results in a change of the retention time of the polar samples. Many kinds of ion paring agents including alkylamonium ions, alkylsulfonates, inorganic ions and surface-active ions were often utilized for the separation of polar samples by RP-HPLC (Harumi and Matsushima 2000). Raynaud and Pévet (Raynaud and Pevet 1991) used trimethylamine (TEA) and sodium 1-octanesulfonate (SOS) as ion pairing agents for the determination of MLT and other indoleamines. The authors report that the retention times of MLT are not affected by these ion paring agents. The effect of SOS concentrations in the HPLC mobile phase on the retention time was examined by Harumi et al. (Harumi and Matsushima 2000). The increase of the SOS concentration from 0 to 3 mM results in a slight decrease of the retention time of MLT (Harumi and Matsushima 2000).

After the chromatographic separation, either or several types of detection methods can be used, the most common being fluorometric detection (FLD), electrochemical detection (ECD), diode array detection (DAD), and mass detection (MS).

4.2.1.1. Fluorometric Detection (FLD)

In fluorescence, the molecular absorption of a photon triggers the emission of another photon with a longer wavelength. This difference in wavelengths (absorption vs. emission) provides more selectivity, and the fluorescent light is measured against a very low-light background, thus improving S/N ratio. Unfortunately, relatively few natural products fluoresce in a practical range of wavelengths. However, many substances can be made to fluoresce by forming appropriate derivatives (Wolfender 2009).

For this method of detection, the excitation and emission wavelengths are two important parameters. In most of the published HPLC-FLD methods used to determine MLT content in edible plants, the excitation and emission wavelengths used are 280 and 340–350 nm (Chen et al. 2003; Iriti, Rossoni, and Faoro 2006; Hernández-Ruiz, Cano, and Arnao 2005), and 245 and 380 nm (Pape and Lüning 2006), but it's not consensual. The mobile-phase composition is exceedingly varied in different published methods. The aqueous phase is generally adjusted to pH 2.8–5.5 with acetate or phosphate, and water without acidification is commonly used. Isocratic-, linear-, or step-gradient elution have been used for the separation of melatonin (Huang and Mazza 2011).

One advantage on the use of a FLD detector, is the fact that there is no limit for the concentration of organic solvents. This is a better point in the use of FLD because increasing of the concentrations of methanol or acetonitrile in the mobile phase results in shortening of the retention time of MLT (Harumi and Matsushima 2000). In relation to this, Kulczykowska and Iuvone (Kulczykowska and Iuvone 1998) reported the case of plasma MLT analysis by HPLC with FLD using a mobile phase containing 60% methanol. An additional advantage of the use of FLD is that the compounds can be recovered after detection (Harumi and Matsushima 2000). A review on the chromatographic conditions, column and mobile phases applied, as well as the limits of detection (LOD) and/or limits of quantification (LOQ) achieved for MLT is presented in Table 2.

In some cases in which MLT concentration is very low, derivatization is recommended to enhance the signal (Iinuma et al. 1999; de Almeida, Di Mascio, Harumi, Warren Spence, et al. 2011). An oxidation procedure that can enhance MLT fluorescence by 5.8 times (allowing its determination at attomole levels) has been described using biological samples (Tomita et al. 2003; de Almeida, Di Mascio, Harumi, Warren Spence, et al. 2011). Melatonin is usually derivatized under alkaline conditions in the presence of hydrogen peroxide (often oxidized to a new fluorescent compound with sodium carbonate and hydrogen peroxide). However, precautions should be taken when using this kind of approach, because other components in the biological sample may lead to the generation of fluorophores, which in turn could interfere with the determination of the correct level, thus impairing method specificity (Tomita et al. 2003; de Almeida, Di Mascio, Harumi, Warren Spence, et al. 2011). Iinuma et al. (1999) (Iinuma et al. 1999) conducted a study in order to optimize the derivatization procedure, the concentration of Na_2CO_3 and H_2O_2, reaction temperature and reaction time. The authors report that the peak height of MLT becomes higher with the increase in Na_2CO_3 concentration up to 300 mM (the value is the concentration of derivatizing reagent; the final concentration in the reaction mixture was 25 mM), and gradually decreased at concentrations higher than 300 mM. The effect of H_2O_2 concentration was also examined. The peak height of the MLT derivative reached a maximum when 5 mM H_2O_2 (final concentration was 0.42 mM) was used for the derivatization and did not change much when the concentration of H_2O_2 was increased. As a result, 300 mM Na_2CO_3 and 10 mM H_2O_2 were used for the derivatization of MLT. The reaction temperatures and reaction times were also investigated. The peak height of MLT increased with the increase of reaction temperature up to 90°C, and slightly decreased at 100°C. Thus, the time course of the peak height of the MLT derivative was examined at 90°C and 30 min was found to be the optimum time for the derivatization. The excitation and emission maxima of the melatonin derivative were 247 and 384 nm, respectively (Iinuma et al. 1999). Table 3 presents derivatization conditions applied to determine MLT in samples.

4.2.1.2. Electrochemical Detection (ECD)

ECD differs from other methods of detection because it alters the sample. Cells for HPLC-ECD usually consist of three electrodes (the working, counter, and reference electrodes) which can be aligned in several different geometries, although two main designs exist.

Table 2. Cromatographic conditions to determine melatonin by HPLC-FLD

Sample	Injection volume	Analytical Column	Column temperature	Mobile phase	Flow rate	Excitation (nm)	Emission (nm)	LOD	LOQ	Ref
Human samples										
Urine	20 µL	Capcell Pak C18 MG II column (250 mm length × 4.6 mm ID, with 5 µm beads)	n.s.	75 mM sodium acetate and acetonitrile (84:16, v/v)	1.0 mL/min	275	345	n.s.	0.2 ng/mL	(Minami et al. 2009)
Oral fluid	5 µL	CAPCELL PAK C18 MG S3 (75 mm × 1.0 mm i.d)	40°C	MeCN–TFA–water (10:0.05:90, v/v)	50 µL/min	247	392	n.s.	n.s.	(Hamase et al. 2004)
Extracts of ovary	n.s.	Superiorex ODS S-5 mm column (4.63150 mm, ID)	30°C	50 mM ammonium acetate buffer (pH 4.3) and 18% methanol (v/v)	1.0 mL/min	280	340	60 fmol	n.s.	(Itoh et al. 1999)
Plasma	25 µL	Hypersil 5-µm-particle ODS columns [100 mm x 4.6 mm	30°C	Solvent A (100 mL/L methanol) and Solvent B (900 mL/L methanol)	0.5 mL/min	286	352	6 pg on the column	n.s.	(Peniston-Bird et al. 1993)
Animal, organisms samples										
Fish plasma, bile and intestinal tissues	50 µL	Beckmann Ultrasphere ODS column (3 µm particles, 75mm×4.6mm i.d.)	room	85mM acetic acetate, 0.1mM EDTA-Na2, and acetonitrile 14% of final volume, pH adjusted to 4.7	1.0 mL/min.	285	360	8 pg	n.s.	(Muñoz et al. 2009)
Rat pineal gand	10 µL	CAPCELL PAK C18 MG (150 mm × 2.0 mm i.d)	40°C	MeCN–TFA–water (15:0.05:85, v/v)	0.2 mL/min.	247	392	n.s.	n.s.	(Hamase et al. 2004)

Table 2. (Continued)

Animal, organisms samples

Sample	Injection volume	Analytical Column	Column temperature	Mobile phase	Flow rate	Excitation (nm)	Emission (nm)	LOD	LOQ	Ref
Mice pineal gand	5 μL	CAPCELL PAK C18 MG S3 (1.0 mm i.d. × 75 mm; Shiseido)	n.s.	CF3COOH (TFA)/CH3CN/H2O =0.01/5/95 (v/v)	100 μL/min.	245	380	n.s.	200 amol	(Tomita et al. 2003)
Rat pineal gand	20 μL	TSKgel ODS-80Ts QA (150 x 4.6 mm)	40°C	100 mM Sodium phosphate buffer (pH 7.0) containing 10% of MeCN	1.0 mL/min.	245	380	n.s.	10 fmol	(Tomita et al. 2001)
Rats and mice pineal gland	20 μL	TSKgel ODS-80Ts QA (150 3 4.6 mm i.d.)	40°C	100 mM sodium-phosphate buffer (pH 7.0) containing 10% acetonitrile	1.0 mL/min.	245	380	1 fmol	n.s.	(Hamase et al. 2000)
Rat pineal gand	20 μL	J'sphere ODS-H80 (150×4.6 mm I.D.)	40°C	100 mM sodium phosphate buffer (pH 7.0) containing 12% acetonitrile	0.5 mL/min.	245	380	500 amol	500 amol	(Iinuma et al. 1999)
Planarians	25 - 50 μL	Superiorex ODS S-5 mm column 4.6x150 mm, i.d.	30°C	50 mM ammonium acetate buffer pH 4.3 and 22% methanol v/v.	1.0 mL/min.	280	340	n.s.	n.s.	(Itoh, Shinozawa, and Sumi 1999)
Flounder plasma	20 μL	Ultrasphere C1 8 column (250 × 4.6-mm i.d., 5-μm particle diameter, 80-Å pore size)	22°C	60% HPLC-grade methanol	0.6 mL/min.	286	352	3 pg/mL	n.s.	(Kulczykowska and Iuvone 1998)
Rat Ovary	30 - 50 μL	Superiox ODS S-5 μm column (4.6×150 mm, I.D)	30°C	50 mM ammonium acetate buffer (pH 4.3) and 20% methanol (v/v)	1.0 mL/min.	280	340	20 pg	n.s.	(Itoh et al. 1997)
Organs of adult female crickets of *Gryllus bimaculatus*	20 μL	Superiox ODS S-5 μm column (4.6 x 150 mm)	23-25°C	50 mM ammonium acetate and 30% methanol, adjusted to pH 4.25 with acetic acid	1.0 mL/min.	280	340	10 pg	n.s.	(Itoh et al. 1995)

Animal, organisms samples

Sample	Injection volume	Analytical Column	Column temperature	Mobile phase	Flow rate	Excitation (nm)	Emission (nm)	LOD	LOQ	Ref
Eyes and Cerebral Ganglia of *Aplysia californica*	n.s.	5-μm kromosyl C18 column	n.s.	33% methanol in 0.05 M ammonium acetate (pH 4.25)	1.0 mL/min.	245	350	n.s.	n.s.	(Abran, Anctil, and Ali 1994)
Rat Pineal Gland dialysate	50 μL	Reversed-phase C18 column (250 X 4.6 mm; Supelco)	30°C	Mixture of 10 mmol/L of disodium phosphate adjusted to a pH of 4.0 with concentrated phosphoric acid, 0.01 mmol/L of disodium EDTA, and 200 ml/L of acetonitrile.	1.0 mL/min.	280	345	5 fmol/inj	n.s.	(Drijfhout, Grol, and Westerink 1993)
dinoflagellate cells	n.s.	10-μm C18 reverse-phase column.	n.s.	20% methanol, 3.75 g NaH$_2$PO$_4$/L, 11.25 g citric acid/L, 320 mg octane sulfonic acid and 32 mg EDTA/L	0.75 mL/min.	285	360	n.s.	n.s.	(Poeggeler and Hardeland 1994)

Plants, fruits and food samples

Sample	Injection volume	Analytical Column	Column temperature	Mobile phase	Flow rate	Excitation (nm)	Emission (nm)	LOD	LOQ	Ref
Fruits of sweet cherry	30 μL	5 μm Symmetry (4.6 × 250 mm) column	n.s.	Mixture of acetonitrile: 50 mm Na$_2$HPO$_4$/H$_3$PO$_4$ buffer pH 4.5 (15:85)	1.0 mL/min.	280	348	n.s.	n.s.	(Zhao et al. 2013)

Table 2. (Continued)

Plants, fruits and food samples

Sample	Injection volume	Analytical Column	Column temperature	Mobile phase	Flow rate	Excitation (nm)	Emission (nm)	LOD	LOQ	Ref
Grapes, must, wine and grappa	n.s.	Microsorb MV-Rainin C8 reversed-phase column (250 × 4.6 mm I.D., 5 μm)	n.s.	Mixture of acetonitrile (20%, v/v) and pH 3.0, 40 mm aqueous phosphate buffer (80%, v/v)	0.8 mL/min.	298	386	0.02 ng/mL	0.05 ng/mL	(Mercolini, Mandrioli, and Raggi 2012)
Tissues of Coffea canephora	20 μL	C-18 Column (Waters, Atlantis dC18, 3.9 × 150 mm, particle size 3 μm)	n.s.	0.1 M sodium acetate, 0.1 M citric acid, 0.5 mM sodium octanylsulfonate, 0.15 M EDTA, adjusted to pH 3.7 and 5% methanol	1.0 mL/min.	n.s.	n.s.	n.s.	n.s.	(Ramakrishna et al. 2012)
Rice grains	10 μL	RP-18 Lichrospher Column (LiChroCART 250 × 4 (5 μm))	n.s.	A (2% acetic acid and 5% methanol in water) and B (2% acetic acid and 88% methanol in water). The gradient applied was as follows: (time, solvent B): 0 min., 0%; 5 min., 35%; 12 min., 40%; 15 min., 40%; 20 min., 45%; 25 min., 50%	0.5 mL/min.	290	330	n.s.	n.s.	(Setyaningsih, Palma, and Barroso 2012)
Beans of Coffea species	20 μL	C18 column, 300 × 3.9, 10 mm	n.s.	A, a solution of 0.2 m sodium acetate and 15 mm 1-octanesulphonic acid sodium salt, Adjusted to pH 4.9 with acetic acid, and B, acetonitrile.	0.6 mL/min.	340	445	n.s.	n.s.	(Ramakrishna et al. 2012)

Plants, fruits and food samples

Sample	Injection volume	Analytical Column	Column temperature	Mobile phase	Flow rate	Excitation (nm)	Emission (nm)	LOD	LOQ	Ref
Wine	10 µL	Phenomenex Luna C18 column (250 mm × 4.6 mm, 5 µm)	n.s.	Formic acid/water 0.1% (A) and methanol (B), with 40% of A and 60% of B	0.8 mL/min.	285	345	51.72 ng/mL	58 ng/mL	(Rodriguez-Naranjo et al. 2011)
Wine	n.s.	Zorbax C8 RP column (150 mm 64.6 mm id, 5 µm) equipped with a C8 cartridge precolumn.	n.s.	ACN (21%,v/v) and a pH 3,0, 50 mM aqueous phosphate buffer containing 0.2% triethylamine (79%, v/v)	1.5 mL/min.	298	386	0.01 ng/mL	0.03 ng/mL	(Mercolini et al. 2008)
Roots in etiolated hypocotyls of *Lupinus albus* L.	n.s.	RP-C18 column (ODS2)	n.s.	water:methanol (60:40)	0.7 mL/min.	280	348	n.s.	n.s.	(Arnao and Hernandez-Ruiz 2007)
Higher plants and other photoauto-trophic organisms	100 µL	TSKgel Super ODS, 4.6 mm ID, length 10 cm, particle size 2 µm	n.s.	12.66 g citric acid, 2.21 g NaH$_2$PO$_4$, 2.21 g Na$_2$HPO$_4$, 1.0 g octanesulfonic acid, 33 mg EDTA and 185.5 mL methanol adjusted with water to 1 L total volume. The pH was adjusted to 5.5 by use of (solid) NaOH.	1.0 mL/min.	245	380	n.s.	n.s.	(Pape and Lüning 2006)

Table 2. (Continued)

Plants, fruits and food samples

Sample	Injection volume	Analytical Column	Column temperature	Mobile phase	Flow rate	Excitation (nm)	Emission (nm)	LOD	LOQ	Ref
Grape skin	20 µL	Phenomenex Luna RP C18 column (250 × 4.6 mm i.d., particle size 5 µm)	room	Mixture of sodium acetate (10 mmol L−1, adjusted to pH 4.5 with concentrated acetic acid, Na₂EDTA (0.01 mmol L−1) and 200 mL L−1 acetonitrile	1.0 mL/min.	280	345	n.s.	10 pg/mL	(Iriti, Rossoni, and Faoro 2006)
Herbal pouder	20 µL	RP 5 µm Hypersil ODS (4.6×250 mm)	n.s.	Methanol: 50 mM Na₂HPO₄/H₃PO₄ buffer (40:60, v/v) pH 4.5	1.0 mL/min.	280	348	n.s.	n.s.	(Chen et al. 2003)

*n.s. (not specified); LOD (Limit of Detection); LOQ (Limit of Quantitation).

Table 3. Derivatization procedures to determine melatonin by HPLC-FLD

	Sample	Supernatant	Derivatizing agent	Temperature	Time	Ref
Human	Oral Fluid	Dried	40 µL of water, and 5 µL of aqueous 2 M Na₂CO₃ and 5 µL of aqueous 50 mM H₂O₂ solutions	100 °C	30 min.	(Hamase et al. 2004)
	Rat pineal gland	Dried	40 µL of water, and 5 µL of aqueous 2 M Na₂CO₃ and 5 µL of aqueous 50 mM H₂O₂ solutions	100 °C	30 min.	(Hamase et al. 2004)
	Mice pineal gland	Dried	40 µl of H₂O, 2.5 µl of aqueous 4 M Na₂CO₃ and 1 µl of aqueous 250 mM H₂O₂	101 °C	31 min.	(Tomita et al. 2003)
Animals	Rat pineal gland	50 µL	5µL 2M Na₂CO₃; 5µL 50 mM H₂O₂	100 °C	30 min.	(Tomita et al. 2001)
	Rats and mice pineal gland	Dried	80 µl of water, 10 µl of aqueous 2 M Na₂CO₃ and 10 µl of aqueous 50 mM H₂O₂	100 °C	31 min.	(Hamase et al. 2000)
	Rat pineal gland	100 µL	10 µl of 300 mM Na₂CO₃ in H₂O; 10 µl of 10 mM H₂O₂ in H₂O	90 °C	30 min.	(Iinuma et al. 1999)
Plants	Higher plants and other photoautotrophic organisms	200 µL	20 µL of 1 m Na₂CO₃ and 20 µL H₂O₂ (50 mM)	93 °C	90 min.	(Pape and Lüning 2006)

In coulometric systems, the eluent is directed through the electrode, while in amperometric systems, the eluent passes over it. ECD detection is usually performed by maintaining the potential of the working electrode at a fixed value relative to the potential of the electrolyte, which is measured by the reference electrode. The fixed potential difference applied between the working and the reference electrodes drives the electrochemical reaction, and the resulting current is measured as a function of the elution time. This allows for detection limits at the pmol level. The selectivity of ECD depends on the accessible potential range, the number of compounds that are active in this range, and the half-widths of the individual signals (Wolfender 2009).

HPLC-ECD has been used to quantify MLT levels in several tissues even though it is not a highly specific method. The problem is that these tissues may contain a variety of compounds with comparable oxidation potentials and retention times similar to MLT (Huang and Mazza 2011). However, the first finding of MLT in the brown alga (*Pterygophora californica Rupr*) was identified by HPLC with ECD by Fuhrberg et al., in 1996 (Fuhrberg et al. 1996) and MLT was well separated from the other constituents of the extract (Huang and Mazza 2011).

In ECD of indoleamines, the setting of the applied potential is important. Usually, an applied potential greater than 800 mV gives the higher sensitivity for MLT detection. However, a higher potential causes an elevation of the baseline noise. Therefore, the applied potential for MLT determination is usually set at between 700 mV and 900 mV (Harumi and Matsushima 2000).

Harumi et al. (Harumi, Akutsu, and Matsushima 1996) successfully determined MLT by HPLC with ECD, with very clear peak separation for different indoleamines among MLT. The authors reported the use of a higher potential, 900 mV, for good MLT signal with their graphite carbon working electrode, and even so they detected MLT at very low levels. However, the sensitivity of this procedure depends on the model of electrochemical cell. Amperometric-based electrochemical cells are generally less sensitive than coulometric cells, so that the adequate potential should be previously optimized by the construction of hydrodynamic voltammograms (de Almeida, Di Mascio, Harumi, Warren Spence, et al. 2011). De Almeida et al. (2011) (de Almeida, Di Mascio, Harumi, Warren Spence, et al. 2011) reported though that, using their coulometric electrochemical system, the best melatonin signal was obtained at 600 mV. Sensitivity could be also greater with coulometric electrochemical detectors such as the ESA coulochem III model (ESA, Bedford, MA, USA), which uses porous electrochemical cells that allow greater accuracy in MLT peak resolution (de Almeida, Di Mascio, Harumi, Warren Spence, et al. 2011).

Table 4. Cromatographic conditions to determine melatonin by HPLC-ECD

Human Samples

Sample	Injection volume	Column	Column temperature	Mobile phase	Flow rate	Analytical Cell 1	Analytical Cell 2	LOD	LOQ	Ref.
Human bone marrow	20 µL	Bischoff Chromatography stainless steel column (250×4 mm, C8 5 mm nucleosil)	n.s.	Acetonitrile (10%) and 10% methanol in 0.1 M citric acid, 0.1 M sodium acetate, pH 4	1.0 mL/min.	400 mV	800 mV	5 pg	n.s.	(Conti, Conconi, Hertens, Skwarlo-Sonta, et al. 2000)
Plasma	40 µL	3-µm Spherisorb C, reversed-phase analytical column (100 X 4.6 mm id.)	room	50 mM sodium acetate-100 mM acetic acid (pH 4.3), 0.1mM Na₂-EDTA, and acetonitrile 80:20 (v/v), mixed with acetonitrile (75:25, v/v)	0.8 mL/min.	350 mV	500 mV	1 pg	n.s.	(Vieira et al. 1992)
Plasma	40 µL	5-µm Spherisorb ODS-I reversed-phase analytical column (150 X 4.6 mm i.d.)	room	50 mM sodium acetate-100 mM acetic acid (pH 4.3), 0.1mM Na₂-EDTA, and acetonitrile 80:20 (v/v)	1.0 mL/min.	900 mV	n.s.	8.5 pg	n.s.	(Vieira et al. 1992)

Animal, organisms samples

Sample	Injection volume	Column	Column temperature	Mobile phase	Flow rate	Analytical Cell 1	Analytical Cell 2	LOD	LOQ	Ref.
Hamster pineal gland	20 µL	Eicompack CA-5ODS reversed-phase column (150 x 4.6 mm I.D., 5 µm particles)	25 °C	0.1 M sodium phosphate, 0.1 mM EDTA, 25% methanol (v/v)	1.0 mL/min.	n.s.	n.s.	5 pg	n.s.	(Harumi, Akutsu, and Matsushima 1996)
Mouse bone marrow	20 µL	Bischoff Chromatography stainless steel column (250×4 mm, C8 5 mm nucleosil)	n.s.	Acetonitrile (10%) and 10% methanol in 0.1 M citric acid, 0.1 M sodium acetate, pH 4	1.0 mL/min.	400 mV	800 mV	5 pg	n.s.	(Conti, Conconi, Hertens, Skwarlo-Sonta, et al. 2000)

Animal, organisms samples

Sample	Injection volume	Column	Column temperature	Mobile phase	Flow rate	Analytical Cell 1	Analytical Cell 2	LOD	LOQ	Ref.
Bile of mammals	30 μL	C18 reversed-phase column	n.s.	0.1 M potassium phosphate buffer (pH 4.5)	1.0 mL/min.	200 mV	900 mV	n.s.	n.s.	(Tan, Manchester, Reiter, Qi, Hanes, et al. 1999)
Slides of fresh bone marrow smears	30 μL	C18 reversed-phase column	n.s.	0.1 M potassium phosphate buffer (pH 4.5)	1.0 mL/min.	200 mV	900 mV	n.s.	n.s.	(Tan, Manchester, Reiter, Qi, Zhang, et al. 1999)
Dinoflagellate *Gonyaulax polyedra*	n.s.	Spherisorb ODS 1	40 °C	n.s.	n.s.	900 mV	n.s.	n.s.	n.s.	(Fuhrberg et al. 1997)
Mice pineal gland	20 μL	125 mm stainless steel column packed with 3 μm nucleosil reversed phase (Nucelosil 120-3 C 18)	n.s.	15% acetonitrile and 5% methanol in 0.1 M citric acid, 0.1 M sodium acetate, (pH 4.1)	0.9 mL/min.	750 mV	n.s.	n.s.	n.s.	(Conti and Maestroni 1996)
Dinoflagellate cells	n.s.	10-μm C18 reverse-phase column.	n.s.	20% methanol, 3.75 g NaH$_2$PO$_4$/L, 11.25 g citric acid/L, 320 mg octane sulfonic acid and 32 mg EDTA/L	0.75 mL/min.	990 mV	n.s.	n.s.	n.s.	(Poeggeler and Hardeland 1994)
Rat plasma	n.s.	Ultrasphere ODS (75 X 4.6 mm i.d. 3μm)	n.s.	Phosphate citrate buffer: Acetonitrile (11.5%)	1.5 mL/min.	200 mV	700 mV	n.s.	n.s.	(Raynaud et al. 1993)

Table 4. (Continued)

Animal, organisms samples

Sample	Injection volume	Column	Column temperature	Mobile phase	Flow rate	Analytical Cell 1	Analytical Cell 2	LOD	LOQ	Ref.
Rat pineal gland	50 µL	3-µm Spherisorb C, reversed-phase analytical column (100 X 4.6 mm i.d.)	room	50 mM sodium acetate-100 mM acetic acid (pH 4.3), 0.1mM Na₂-EDTA, and acetonitrile 80:20 (v/v), mixed with acetonitrile (75:25, v/v)	0.8 mL/min.	350 mV	500 mV	1 pg	n.s.	(Vieira et al. 1992)
Rat pineal gland	50 µL	5-µm Spherisorb ODS-I reversed-phase analytical column (150 X 4.6 mm i.d.)	room	50 mM sodium acetate-100 mM acetic acid (pH 4.3), 0.1mM Na₂-EDTA, and acetonitrile SO:20 (v/v)	1.0 mL/min.	900 mV	ns	8.5 pg	n.s.	(Vieira et al. 1992)
Micro-dialysates from rat pineal gland	n.s.	Eicom-Prepak column (AC-ODS, 4 mm x 5 mm I.D.)	28 °C	0.1 M KH2PO4, 0.05 M HsP04, pH 3.1	1.0 mL/min.	850 mV	n.s.	n.s.	5 pg	(Vieira et al. 1992)
Chicks and Rats serum, pineals and gut (upper duodenum)	n.s.	C18 reversed phase column (Phenomenex Kingsorb 250 x 4.6mm, 5mm)	n.s.	50mM phosphoric acid, 50mM citric acid, 60 mM EDTA, 8mM heptane sulphonic acid, 2mM sodium chloride, with the pH adjusted to 3.1 with potassium hydroxide, and completed by the addition of 5% methanol.	n.s.	750 mV	960 mV	n.s.	n.s.	(Vieira et al. 1992)

Plants, fruits and food samples

Sample	Injection volume	Column	Column temperature	Mobile phase	Flow rate	Analytical Cell 1	Analytical Cell 2	LOD	LOQ	Ref.
Cherry cultivars	20 µL	Zorbax 300SB-C18 (5 µm, 4.6 × 250 mm i.d.)	30 °C	Solution 1 corresponded to 85% of the total mobile phase with 75 mM NaH2PO4, 1.7 mM OSA, and 25 µM EDTA in deionized water, pH 3.5, while solution 2 (15%) was acetonitrile	1.0 mL/min.	600 mV	300 mV	25 ng/mL	25 ng/mL	(Rosado et al. 2017)

Plants, fruits and food samples

Sample	Injection volume	Column	Column temperature	Mobile phase	Flow rate	Analytical Cell 1	Analytical Cell 2	LOD	LOQ	Ref.
Raw walnuts	30 μL	C18 reverse-phase column	n.s.	0.1 mM potassium phosphate buffer (pH 4.5) with acetonitrile (20%)	1.0 mL/min.	200 mV	900 mV	n.s.	n.s.	(Reiter, Manchester, and Tan 2005)
Plant tissues	n.s.	RP-C18 column (ODS2)	n.s.	n.s.	n.s.	850 mV	n.s.	n.s.	n.s.	(Hernández-Ruiz, Cano, and Arnao 2005)
Seeds of lupin	10 - 60 μL	C18 RP-ODS-Spherisorb column (250 × 4.5 mm, 5 μm particle size)	30 °C	water:acetonitrile:glacial acetic acid (78:20:2) (pH 3.0)	0.7 mL/min.	850 mV	n.s.	n.s.	n.s.	(Hernandez-Ruiz, Cano, and Arnao 2004)
Herbal powder	n.s.	Nova-Pak C18 8 mm×100 mm Radial-Pak cartridge	n.s.	3.58 g/L tricine, 1.8 g/L NaCl, and 40 mL/L acetone. The pH was adjusted to 7.0 with NaOH	n.s.	n.s.	n.s.	n.s.	n.s.	(Van Tassel et al. 2001)
Montmorency and Balaton tart cherries	30 μL	YMC-BD (4.6 mm x 250 mm, Partisil 5 μm ODS3	n.s.	0.1 M potassium phosphate buffer with acetonile (20%)(pH 4.5)	1.0 mL/min.	800 mV	n.s.	20 pg/inj	n.s.	(Burkhardt et al. 2001)
In vitro cultured plants of St. John's wort	n.s.	Catecholamin.e column (3.9″ 150 mm; Waters)	room	buffer consisting of 0.1 M sodium acetate, 0.1 M citric acid, 0.5 mM sodium octanyl sulfonate, 0.15 mM EDTA, pH 3.7 and 5% methanol	0.8 mL/min.	850 mV	n.s.	n.s.	n.s.	(Murch, Campbell, and Saxena 2001)
Seeds	30 μL	C18 reversed-phase column	n.s.	0.1 M potassium phosphate buffer (pH 4.5) with acetonitrile (20%)	1.0 mL/min.	200 mV	900 mV	n.s.	n.s.	(Manchester et al. 2000)

Table 4. (Continued)

Plants, fruits and food samples

Sample	Injection volume	Column	Column temperature	Mobile phase	Flow rate	Analytical Cell 1	Analytical Cell 2	LOD	LOQ	Ref.
St. John's wort	200 μL	Catecholamine column (3.9" 150 mm; Waters)	room	buffer consisting of 0.1 M sodium acetate, 0.1 M citric acid, 0.5 mM sodium octanyl sulfonate, 0.15 mM EDTA, pH 3.7 and 5% methanol	0.8 mL/min.	850 mV	n.s.	n.s.	n.s.	(Murch, KrishnaRaj, and Saxena 2000)

*n.s. (not specified); LOD (Limit of Detection); LOQ (Limit of Quantitation).

One disadvantage of the use of ECD is that the high concentration of organic solvent such as methanol or acetonitrile in HPLC mobile phases shortens the life time of the working electrode in the detector. Therefore, the concentration of organic solvent in the mobile phase for ED is usually held at less than 30% (v/v) (Harumi and Matsushima 2000). An additional disadvantage of the use of HPLC-ECD is that it is a destructive method and the analytes detected by HPLC-ECD cannot be recovered after detection (Huang and Mazza 2011). Table 4 reviews the chromatographic conditions, column and mobile phases applied, as well as LODs and LOQs achieved for MLT when ECD is used.

4.2.1.3. Ultraviolet Detection (UV)

In UV detection, the relationship between the intensity of light transmitted through the detector cell and the solute concentration is given by Beer's Law. The two factors that control detector sensitivity are the magnitude of the extinction coefficient of the analyte of interest at a given wavelength and the path length of the light passing through the UV cell. The sensitivity will increase with increasing the path length, but a compromise must be found with the cell volume in order to avoid peak dispersion (Wolfender 2009).

In HPLC with UV-detection methods for MLT, the used wavelength is generally 280 nm. The mobile phases containing acidic water adjusted with weak acid (trifluoroacetic acid or acetic acid) and methanol or acetronitrile are most commonly used. Isocratic elution is used in the determinations of MLT separately. Gradient elution is used in the determinations of MLT and other analytes simultaneously (Huang and Mazza 2011).

The present detector was commonly used in edible plant analysis. The determination of melatonin in the seeds, leafs, stem, and root tissues of Chinese licorice (*Glycyrrhiza uralensis Fisch.*) plants was performed using an HPLC-UV/VIS photodiode array (PDA) method reaching a detection limit of 10 pg/µL (Huang and Mazza 2011; Afreen, Zobayed, and Kozai 2006). Also, the use of HPLC-UV method allowed the determination of melatonin in purple wheat by Hosseinian et al. (2008) with a limit of detection of 0.1 ng/g (Huang and Mazza 2011; Hosseinian, Li, and Beta 2008). However, its application has been extended to other types of samples, the conditions of which are reported in Table 5.

4.2.1.4. Mass Spectrometry Detection (MS)

The application of LC-MS in the analysis and characterisation of natural products has been recognised as a major breakthrough. Indeed, while it is expensive, the use of a mass spectrometer as an LC detector provides important structural information, such as MW, molecular formula, and diagnostic fragments, which are crucial for dereplication and rapid characterisation (Wolfender 2009). The LC–MS methods offer high sensitivity and excellent detection specificity, thus increasing the validity of results and allowing qualitative and quantitative determination of melatonin (Feng et al. 2014).

Nevertheless, analytical conditions have to be optimized using MLT standards. The acidic aqueous mobile phase and gradient elution can be used to enhance the LC separation, to reduce the co-eluting compounds, and to improve the ionization efficiency. Different types of mass analyzers have been used for MLT LC-MS determination. Quadrupole is the most common and versatile of all the mass analyzers and has a long applications record and an extensive history of use with mass-spectral libraries. Quadrupole is used in two ways. The first is the single quadrupole mode, which generates significant fragments. The structure is elucidated from the observed masses and intensities. The most sensitive operating mode of single quadrupole is known as SIM (Selected Ion Monitoring). The second is the triple quadrupole mode, which allows analysts to fragment their ions within the quadrupole mass analyzer (MS/MS). Significant improvement is made in specificity and selectivity with the SRM (Selected Reaction Monitoring) and MRM (Multiple Reaction Monitoring) modes. Single and triple quadrupoles are particularly well-suited for quantitative applications in LC-MS. In the quantification of MLT in edible plants by LC-MS, the triple quadrupole was used more than the single quadrupole (Huang and Mazza 2011).

Only one determination operated in the negative ion mode (Hosseinian, Li, and Beta 2008), while the others used the positive ion mode for melatonin (Huang and Mazza 2011). The positive ion mode is more sensitive than the negative ion mode for the determination of MLT (Huang and Mazza 2011).

Melatonin was detected and quantified in eight different sweet cherry cultivars (*Prunus avium L.*) using LC-MS by Gonzalez-Gomez et al., 2009 (González-Gómez et al. 2009). The chromatographic separations were carried out on an Agilent 1100 series instrument equipped with a MS-ESI-quadrupole detector. The optimized voltage was determined to monitor the precursor ion for melatonin (*m/z* 233) achieving a limit of detection of 4.3 pg/µL (Huang and Mazza 2011).

LC-MS is in general more sensitive than HPLC-ECD, FLD, or UV. However, the so-called "matrix effect" is often observed during LC-MS detection. This effect usually reduces the ionization efficiency and consequently changes the peak area of the analytes due to co-eluting compounds or the chemicals or labware used in the analyses. The possibility of matrix effects should therefore be taken into account while developing LC-MS methods for the determination of melatonin in samples. These effects could be minimized or eliminated by a more selective extraction, a more efficient chromatographic separation, or optimized ionization parameters (Huang and Mazza 2011).

Table 5. Cromatographic conditions to determine melatonin by HPLC-UV

Human samples

Sample	Injection volume	Column	Column temperature	Mobile phase	Flow rate	Wavelength (nm)	LOD	LOQ	Ref
Prostate normal amd cancer cell lines	50 μL	Tracer Extrasil ODS1 column (250 mm × 0.46 mm, 5 μm)	35°C	Sodium acetate 50 mmol L^{-1} (pH 4.1) in 30% acetonitrile	1.0 mL/min.	279	> 1.83 nmol L-1/cell	> 4.03 nmol L^{-1}/cell	(Hevia et al. 2010)
Skin	20 μL	LUNA 54, C18, 4.6 × 150 mm	room	methanol:water (50:50)	0.5 mL/min.	230	n.s.	n.s.	(Dubey, Mishra, and Jain 2007)
keratinocytes and cell-free systems	20 μL	C18 Nova-pak™ reverse-phase column (4 μm particle size; 10 cm x 5 mm id)	n.s.	a gradient (5–15% over 40 min.) of HPLC-grade acetonitrile in phosphate buffer (0.01 M; pH 7.2)	1.0 mL/min.	223	n.s.	n.s.	(Fischer et al. 2006)
Skin	20 μL	LUNA 54, C18, 4.6×150 mm	n.s.	methanol–water (50:50)	0.5 mL/min.	223	n.s.	n.s.	(Dubey et al. 2006)
Skin fibroblasts *in vitro*	20 μL	Microsorb™ C18 column (5 M, 4.6 mm, 25 cm)	n.s.	acetonitrile–water (85:15,v/v)	1.0 mL/min.	242	n.s.	n.s.	(Lee et al. 2003)

Animal, organisms samples

Sample	Injection volume	Column	Column temperature	Mobile phase	Flow rate	Wavelength (nm)	LOD	LOQ	Ref
Rat, dog, and monkey plasma samples	n.s.	Narrow-bore Hypersil C18 (2.1×200 mm) 5μ particle	n.s.	21% acetonitrile	0.3 mL/min.	225	n.s.	12.5 ng	(Yeleswaram et al. 1997)
Hamster skin culture	100 μL	RP-HPLC column (Versapack C18 1OU	n.s.	0.1 N acetic acid (pH 4.0); methanol	1.0 mL/min.	280	n.s.	n.s.	(Slominski et al. 1996)
Murine B16 melanoma cells	n.s.	Beckman C18 (5×100 mm)	n.s.	5–50% acetonitrile in methanol over 30 min.	1.0 mL/min.	254	n.s.	n.s.	(Helton et al. 1993)

Table 5. (Continued)

Plants, fruits and food samples

Sample	Injection volume	Column	Column temperature	Mobile phase	Flow rate	Wavelength (nm)	LOD	LOQ	Ref
Alcoholic and hot water extracts of plants	n.s.	Nucleosil-100 C18 column (25cm × 4.6 mm ID, particle size 5 μm)	n.s.	0.1 M potassium phosphate buffer (pH 4.5) with acetonitrile (20%)	1.0 mL/min.	280	0.5 μg/mL	0.5 μg/mL	(Ansari et al. 2010)
Grape skin	20 μL	Phenomenex Luna RP C18 column (250 × 4.6 mm i.d., particle size 5 μm)	room	Mixture of sodium acetate (10 mmol L−1), adjusted to pH 4.5 with concentrated acetic acid, Na$_2$EDTA (0.01 mmol L^{-1}) and 200 mL L^{-1} acetonitrile	1.0 mL/min.	310	n.s.	n.s.	(Iriti, Rossoni, and Faoro 2006)
Glycyrrhiza uralensis	20 μL	Phenomenex Hypersil C18 column (3.0 μm; 4.6 · 100 mm)	n.s.	0.1 M Na$_2$HPO$_4$ and acetonitrile (65:35, v:v)	1.0 mL/min.	220	n.s.	n.s.	(Afreen, Zobayed, and Kozai 2006)
In vitro cultured plants of St. John's wort	200 μL	Catecholamine column (3.9 x 150 mm; Waters)	room	buffer [0.1 M sodium acetate, 0.1 M citric acid, 0.5 mM sodium octanyl sulfonate, 0.15 mM EDTA] pH 3.7 and 5% methanol	0.8 mL/min.	278	n.s.	n.s.	(Murch, Campbell, and Saxena 2001)
St. John's wort	200 μL	Catecholamine column (3.9 " 150 mm; Waters)	room	buffer consisting of 0.1 M sodium acetate, 0.1 M citric acid, 0.5 mM sodium octanyl sulfonate, 0.15 mM EDTA, pH 3.7 and 5% methanol	0.8 mL/min.	278	n.s.	n.s.	(Murch, KrishnaRaj, and Saxena 2000)

Commercial formulations samples

Sample	Injection volume	Column	Column temperature	Mobile phase	Flow rate	Wavelength (nm)	LOD	LOQ	Ref
Transdermal Patches	n.s.	C18 (ODS-AQTM S5 mm 120 A 3.0 150 mm)	n.s.	methanol and water (50:50)	0.5 mL/min.	223	n.s.	n.s.	(Kanikkannan et al. 2004)
Tablets	n.s.	5-μm C-18 Luna Phenomenex (150×2.0 mm i.d.)	n.s.	water-methanol 80:20 (v/v)	n.s.	225, 275, 350	n.s.	n.s.	(Andrisano et al. 2000)
Lipidic microparticles (lipospheres)	n.s.	5-μm Luna C18 column (150mm×4.6mm i.d.)	n.s.	sodium acetate buffer (pH 4.0; 0.05 M)–methanol (50:50, v/v)	1 mL/min.	280	n.s.	n.s.	(Tursilli et al. 2006)

*n.s. (not specified); LOD (Limit of Detection); LOQ (Limit of Quantitation).

An internal standard is usually mandatory. Yang and co-workers (Yang et al. 2002) reported a methodology by using acetyltryptamine as internal standard; however, this is not the ideal situation. The most appropriate is the employment of a labeled internal standard whose structure is the same as the analyte with slight difference in the masss. The addition of an isotopically labeled internal standard prior to analysis improves the method's confidence level. One analytical method enabled the determination of endogenous MLT in human saliva, by using column-switching semi-microcolumn liquid chromatography/mass spectrometry and selected ion monitoring (SIM). Melatonin was monitored based on its fragment ion at *m/z* 174 by in-source dissociation and using deuterated melatonin (N-[2-(5-methoxy-1H indol-3-yl)tetradeuteroethyl] trideutero-acetamide, 7-D-melatonin) as internal standard (Martinez et al. 2005). Table 6 reports several chromatographic and detection conditions applied to determine MLT in several samples (using either SIM or SRM/MRM modes).

4.2.2. Gas Chromatography – Mass Spectrometry Detection (GC-MS)

Gas chromatography (GC) is one specific form of the more general separation process of chromatography. In GC the mobile phase is an inert carrier gas and the stationary phase is a high molecular weight liquid which is deposited either on the surface of finely divided particles or on the walls of a long capillary tubing (Karasek and Clement 2012). Usually helium, hydrogen or nitrogen gas compressed in cylinders is used as the carrier gas. Regarding MLT, helium is the most reported.

The GC column is attached to the injection port and samples are introduced into the gas stream at a temperature sufficient to ensure vaporization of all components (Karasek and Clement 2012).Typically, the sample is introduced with a microliter syringe in a volume substantially lower when compared to the previously described techniques.

The GC-MS technique is very sensitive and offers more specificity than HPLC-ECD or HPLC-FLD, however, a difficulty with this technique is the need of derivatization, and thus it has been gradually substituted by LC-MS procedures (de Almeida, Di Mascio, Harumi, Warren Spence, et al. 2011) (Martinez et al. 2005).

Appropriate derivatization of MLT gives this indoleamine an adequate vapour pressure for gas chromatography. MLT and biogenic samples were performed with silanizing agent to form trimethylsilyl melatonin (TMS-MLT) (Nunez-Vergara et al. 2001), with heptafluorobutyrylimidazole to form diheptafluorobutyryl melatonin (HFB-MLT) (Aboul-Enein, Doneanu, and Covaci 1999), or with pentafluoropropionic anhydride to form pentafluoropropionyl melatonin (PFP-MLT) (Fourtillan et al. 1994). TMS-MLT gives rise to two major fragment ions at *m/z* 232 and 245 due to the β and γ cleavages of side chains to the pyrrole ring. The fragment ions at *m/z* 232 and 245 derived from TMS-MLT are much higher than the molecular ion at *m/z* 304.

Table 6. Chromatographic conditions to determine melatonin by LC-MS (SIM and SRM/MRM mode)

	Sample	Injection volume	Column	Column temperature	Mobile phase	Flow rate	Ions (m/z)	LOD	LOQ	Ref.
					SIM mode					
Human	Serum	50 µL	C8 reversed-phase column (Symmetry C8; 5 µm; 150×2.1 mm I.D.)	35°C	40% methanol and 10 mM ammonium formate (pH 3.6)	250 µL/min.	174	0.5 pmoL	n.s.	(Rolčík et al. 2002)
Animals	Slides of fresh bone marrow smears	30 µL	C18 reversed-phase column	n.s.	0.1 M potassium phosphate buffer (pH 4.5)	1.0 mL/min.	160; 232;173; 145; 117	n.s.	n.s.	(Tan, Manchester, Reiter, Qi, Zhang, et al. 1999)
	Tissues of *Coffea canephora*	20 µL	n.s.	n.s.	n.s.	n.s.	233	n.s.	n.s.	(Ramakrishna et al. 2012)
Plants	Purple wheat	10 µL	Luna 3u C18 column (150 × 3 mm i.d., 3 µm)	35°C	Solvent A: 0.1% acetic acid in double deionized water and solvent B: 0.1% acetic acid in acetonitrile. The gradient conditions were as follows: solvent B: 0 min, 10%; 5 min., 10%; 10–40 min., 40%; 41–50 min., 10%	0.5 mL/min.	231;175	100 ng/mg	200 ng/mg	(Hosseinian, Li, and Beta 2008)

Table 6. (Continued)

	Sample	Injection volume	Column	Column temperature	Mobile phase	Flow rate	Ions (m/z)	LOD	LOQ	Ref.
Human	Oral fluid	100 μL	150 mm×3.9 mm C8 Symmetry column (Waters)	n.s.	Methanol and water with 0.1% formic acid. A linear gradient from 40% methanol to 95% methanol was run over 3 min. and 95% methanol was then maintained for 2 min. A linear gradient back to 40% methanol was run during 0.5 min. followed by an equilibration for 1 min.	0.8 mL/min.	233> 174	1.05 pg/mL	3.0 pg/mL	(Eriksson, Östin, and Levin 2003)
	Serum	n.s.	n.s	n.s	Solvent A: 5 mM ammonium formate at pH 4.0. Solvent B: 95% acetonitrile+5% 5 mM ammonium formate at pH 4.0.	1 mL/min.	233> 174	0.1 ng/mL	0.5 ng/mL	(Yang et al. 2002)
Animals	Rat, dog, and monkey plasma samples	n.s.	Hypersil BDS column (50 x 2.1 mm id; 3 μ particle size)	60°C	Solvent A 10 mM ammonium acetate in 25:75 (v/v) methanol/water adjusted to pH 5.5 with acetic acid and Solvent B 100% acetonitrile.	0.3 mL/min.	233> 174	n.s.	3.9 ng/mL	(Yeleswaram et al. 1997)

SRM/MRM mode

SRM/MRM mode

	Sample	Injection volume	Column	Column temperature	Mobile phase	Flow rate	Ions (m/z)	LOD	LOQ	Ref.
Plants	Wine	3 µL	Agilent 150 mm × 0.5 mm i.d., 5-µm Zorbax Sb-18 column	n.s.	Binary gradient consisting of (A) water and (B) methanol both containing 0.1% formic acid (v/v). The elution profile was: 40% B (2 min.), 85% B (4 min.), 90% B (9 min.)	10 µL/min.	233> 174; 233> 216	n.s.	n.s.	(Rodriguez-Naranjo et al. 2011)
	Roots in etiolated hypocotyls of *Lupinus albus* L.	n.s.	n.s.	n.s.	n.s.	n.s.	233> 174	n.s.	n.s.	(Arnao and Hernandez-Ruiz 2007)
	Seeds of lupin	n.s.	n.s	n.s	Solvent A: 5 mM ammonium formate at pH 4.0. Solvent B: 95% acetonitrile +5% 5 mM ammonium formate at pH 4.0	1 mL/min.	233> 174	n.s.	n.s.	(Hernandez-Ruiz, Cano, and Arnao 2004)
	Seeds of *Chenopodium rubrum* L.	20 µL	(0.3 mm I.D. × 150 mm, Hypersil C18 BDS, 3µmol/L particle size, LCPackings)	n.s	Solvent A (0.1% HCOOH, 1% CH3CN, 98.9% H2O); Solvent B (0.1% HCOOH, 30% CH3CN, 69.9% H2O)	35 µL/min.	233> 174; 233> 216	n.s.	n.s.	(Wolf et al. 2001)

Table 6. (Continued)

				SRM/MRM mode					
	Sample	Injection volume	Column	Column temperature	Mobile phase	Flow rate	Ions (m/z)	LOD	LOQ	Ref.
Plants	St. John's wort	100 μL	Prodigy 5 ODS-2 column (150 " 3.2 mm i.d. with a 5 mm particle size	n.s	90% pure water containing 0.1% trifluoroacetic acid (TFA) and 10% acetonitrile, delivered using a quaternary step gradient program [10% CH$_3$CN in 90% acidified water (0.1% TFA) for 5 min.; 50% CH^{3}CN for 10 min.]	n.s	233>174	n.s.	n.s.	(Murch, KrishnaRaj, and Saxena 2000)
	Chenopodium rubrum	10 μL	Separon RP-C-18 column (7 μm, i.d. 8 mm)	40°C	Solvent A (30% MeOH) and Solvent B (70% MeOH). A gradient was run: 0-30 rain 30% B, 30-31 min. 80% B, 31-41 min. 30% B	1 mL/min.	233>174	5 pg/tube	n.s.	(Kolář et al. 1997)
	Chenopodium rubrum	10 μL	C8 reversed phase column (Merck, Lichrosphere 60 RP Select B, 5 μm, 125 × 4 mm)	n.s	MeCN-H$_2$0 (9:11)	0.8 mL/min.	233>174	5 pg/tube	n.s.	(Kolář et al. 1997)

*n.s. (not specified); LOD (Limit of Detection); LOQ (Limit of Quantitation).

Table 7. Chromatographic conditions to determine melatonin by GC-MS

Human samples

Sample	Injection volume	Injector temperature	Column	Oven temperature	Carrier Gas	Ion source temperature	Ion (m/z)	LOD	LOQ	Ref.
Plasma, Oral fluid	2 μL	60°C	methyl silicone (12.5 m×0.32 mm); second analytical column (25 m×0.22 mm)	60–230°C with a heating rate of 20°C min^{-1}	Helium	250°C	320; 324	n.s.	1 pg/ mL	(Simonin, Bru, Lelievre, et al. 1999)
Plasma	1 μL	275°C	DB5-ms (30 m x 0.25 mm i.d. x 0.25 μm film thickness)	90°C (1 min.) then 15°C/ min. to 275°C (maintained 10 min.)	Helium	280°C	360	n.s.	50 pg/ mL	(Covaci, Doneanu, Aboul-Enein, et al. 1999)
Plasma	2 μL	300 °C	CP Sil5 CB non-polar (25mm x 0.25 mm i.d. x 0.25 μm)	160°C to 300°C with a programme rate of 20°C min.$^{-1}$	Helium	200°C	320	n.s.	0.5 pg/ mL	(Fourtillan et al. 1994)
Plasma	5 μL	240 °C	glass column 10% OV17 (1 m x 6 mm x 3 mm)	220 °C	Helium	150°C	360;32 0; 310	n.s.	5 pg on column	(Skene et al. 1983)

Animal samples

Sample	Injection volume	Injector temperature	Column	Oven temperature	Carrier Gas	Ion source temperature	Ion (m/z)	LOD	LOQ	Ref.
Mice pineal gland	1 μL	n.s.	n.s.	n.s.	n.s.	n.s.	320, 324	n.s.	2.5 pg/ mL	(Vivien-Roels et al. 1998)
Hamster skin culture	1 μL	n.s.	DB-5ms (15 m x 0.25 mm i.d. x 0.25μm)	100–300°C, 20°C/min.; with a final hold of 5 min. was used along with a column temperature of 250°C	Helium 1.0 mL/min.	275°C	n.s.	n.s.	n.s.	(Slominski et al. 1996)

Table 7. (Continued)

Plants and food samples

Sample	Injection volume	Injector temperature	Column	Oven temperature	Carrier Gas	Ion source temperature	Ion (m/z)	LOD	LOQ	Ref.
Herbal powder	n.s.	n.s.	n.s.	n.s.	n.s.	n.s.	340, 320	n.s.	n.s.	(Van Tassel et al. 2001)
Fruits or vegetables	1 µL	240 °C	HP-ultra 1 (12 m x 0.2 mm i.d.x 0.33 µm)	100 °C/3 min., rate 30°C/min. to 310°C/5 min.	Helium 0.5 mL/min.	280°C	186, 213, 241, 360	n.s.	n.s.	(Dubbels et al. 1995)

Commercial formulations samples

Sample	Injection volume	Injector temperature	Column	Oven temperature	Carrier Gas	Ion source temperature	Ion (m/z)	LOD	LOQ	Ref.
tablets	2 µL	250 °C	HP-ultra 1 (25 m×0.2 mm i.d.×0.11 µm)	140 to 315°C (2 min.) at 15°C min.$^{-1}$	Helium 0.3 mL/min.	300°C	232	5 ng/mL	10 ng/mL	(Nunez-Vergara et al. 2001)
tablets	1 µL	n.s.	DB-5Ms (30 m x 25-mm i.d. x 0.25 µm)	60°C to 280°C with a programme rate of 15°C min.	n.s.	n.s.	232	n.s.	n.s.	(Aboul-Enein, Doneanu, and Covaci 1999)
tablets	1 µL	275 °C	DB5-ms (30 m x 0.25 mm i.d. x 0.25 µm)	90°C (1 min.) then 15°C/ min. to 275°C (maintained 10 min.)	Helium	280°C	360	n.s.	5 ng on column	(Covaci, Doneanu, Aboul-Enein, et al. 1999)

*n.s. (not specified); LOD (Limit of Detection); LOQ (Limit of Quantitation).

Table 8. Derivatization procedures to determine melatonin by GC-MS

	Sample	Supernatant volume	Derivatizing agent	Temperature	Time	Ref
Human samples	Plasma, Oral fluid	200 µL	25 µL pentafluoropropionic anhydride	60°C	15 min.	(Simonin, Bru, Lelievre, et al. 1999)
	Plasma, Oral fluid	Dried	40 µL pentafluoropropionic anhydride	n.s.	n.s.	(Fourtillan et al. 1994)
	Plasma	Dried	100 µL ethyl acetate added. The solution was vortexed for 10 s and 100 µL pentafluoropropionic anhydride then added	60°C	10 min.	(Skene et al. 1983)
Animal samples	Mice pineal gland	200 µL	10 µL pentafluoropropionic anhydride	60°C	15 min.	(Vivien-Roels et al. 1998)
Plant samples	Herbal powder	n.s.	pentafluoropropionic anhydride, dissolved in iso-octane	n.s.	n.s.	(Van Tassel et al. 2001)
	Fruits or vegetables	200 µL	200 µL pentafluoropropionic anhydride	60°C	10 min.	(Dubbels et al. 1995)
Commercial formulations samples	Tablets	Dried	100 µL of the reagent mixture composed of N-methyl-N-trimethylsilyl-trifluoroacetamide plus 1% trimethylchlorosilane (MSTFA/TMSCl 100:1)	75°C	30 min.	(Nunez-Vergara et al. 2001)
	Tablets	Dried	200 µL of N-methyl-N-trimethylsilyl-heptafluorobutyramide (MTSHFBA)	60°C	15 min.	(Aboul-Enein, Doneanu, and Covaci 1999)

*n.s. (not specified).

The peak height of fragment ions of HFB-MLT appear at *m/z* 159, 356 and 369 are also higher than the height of the molecular ion at *m/z* 582. On the other hand, PFP-MLT is a spirocyclic derivate. The molecular ion of this derivate at *m/z* 360 is the most abundant ion on the mass spectrum, and its fragment ions at *m/z* 186 and 213 appear as smaller peaks. Therefore, the peak height of the molecular ion at *m/z* 360 is utilized for the quantitative analysis using PFP-MLT (Harumi and Matsushima 2000). Table 7 presents several chromatographic conditions to determine MLT by GC-MS with the selected *m/z*, while table 8 reports derivatization conditions applied to enhance MLT determination.

Conclusion

This chapter had the main goal to focus on both pre-analytical and analytical concerns that occur whenever MLT monitoring is necessary. It's a fact that, nowadays, the highly sensitive and robust equipment allows an unequivocal determination of a wide range of molecules. The evolution on sample extraction procedures also tends to fasten up and simplify this critical step.

The big number of reports on MLT determination is due to the physiological properties attributed to this indoleamine, and also to the wide diversity of samples used for its determination.

The correct sample homogenization will have a critical influence on MLT recovery or degradation. Several conditions applied have been summarized with the results subsequently obtained. The pre-concentration procedures can also dictate the limits of detection achieved for MLT determination. Even in the most complex specimens, the proper choice of solvent, sorbent and sample volume have to be carefully studied in order to improve analytical response and minimize the many possible interferences.

Chromatographic techniques seem to be more selective to determine MLT, and HPLC is the most reported. The chromatographic and detection conditions can also influence the correct identification of MLT, hence a great importance was given to this matter. Herein a comprehensive review on the latest applications of chromatography to determine MLT is presented.

Acknowledgments

The authors acknowledge the European Investment Funds by FEDER/COMPETE/POCI– Operational Competitiveness and Internationalization Programme (Project POCI-01-0145-FEDER-007491), and National Funds by FCT - Portuguese Foundation for Science and Technology, under the project UID/Multi /00709/2013.

REFERENCES

Aboul-Enein, Hassan Y, Catalin Doneanu, and Adrian Covaci. 1999. Capillary GCMS determination of melatonin in several pharmaceutical tablet formulations. *Biomedical Chromatography* 13 (1):24-26.

Abran, D., M. Anctil, and M. A. Ali. 1994. Melatonin activity rhythms in eyes and cerebral ganglia of Aplysia californica. *General and Comparative Endocrinology* 96 (2):215-22.

Afreen, F, S. M. A. Zobayed, and T. Kozai. 2006. Melatonin in *Glycyrrhiza uralensis*: response of plant roots to spectral quality of light and UV-B radiation. *Journal of pineal research* 41 (2):108-115.

Anderson, George M., J. Gerald Young, David K. Batter, Simon N. Young, Donald J. Cohen, and Bennett A. Shaywitz. 1981. Determination of indoles and catechols in rat brain and pineal using liquid chromatography with fluorometric and amperometric detection. *Journal of Chromatography B: Biomedical Sciences and Applications* 223 (2):315-320.

Andrisano, V., C. Bertucci, A. Battaglia, and V. Cavrini. 2000. Photostability of drugs: photodegradation of melatonin and its determination in commercial formulations. *Journal of pharmaceutical and biomedical analysis* 23 (1):15-23.

Ansari, M., Kh. Rafiee, N. Yasa, S. Vardasbi, S. M. Naimi, and A. Nowrouzi. 2010. Measurement of melatonin in alcoholic and hot water extracts of Tanacetum parthenium, Tripleurospermum disciforme and Viola odorata. *DARU Journal of Pharmaceutical Sciences* 18 (3):173.

Arnao, M. B., and J. Hernandez-Ruiz. 2007. Melatonin promotes adventitious- and lateral root regeneration in etiolated hypocotyls of Lupinus albus L. *Journal of Pineal Research* 42 (2):147-52.

Repeated Author. 2014. Melatonin: plant growth regulator and/or biostimulator during stress? *Trends in Plant Sciences* 19 (12):789-97.

Arnao, Marino B., and Josefa Hernández-Ruiz. 2007. Melatonin promotes adventitious- and lateral root regeneration in etiolated hypocotyls of Lupinus albus L. *Journal of pineal research* 42 (2):147-152.

Azekawa, Takaharu, Atsuko Sano, Kazuhiro Aoi, Hiroyoshi Sei, and Yusuke Morita. 1990. Concurrent on-line sampling of melatonin in pineal microdialysates from conscious rat and its analysis by high-performance liquid chromatography with electrochemical detection. *Journal of Chromatography B: Biomedical Sciences and Applications* 530:47-55.

Benloucif, Susan, Helen J. Burgess, Elizabeth B. Klerman, Alfred J. Lewy, Benita Middleton, Patricia J. Murphy, Barbara L. Parry, and Victoria L. Revell. 2008. Measuring melatonin in humans. *Journal of clinical sleep medicine: JCSM: official publication of the American Academy of Sleep Medicine* 4 (1):66.

Berrueta, L. A., B. Gallo, and F. Vicente. 1995. A review of solid phase extraction: Basic principles and new developments. *Chromatographia* 40 (7):474-483.

Bubenik, G. A. 2002. Gastrointestinal melatonin: localization, function, and clinical relevance. *Digestive Diseases and Sciences* 47 (10):2336-48.

Bubenik, G. A., G. M. Brown, and L. J. Grota. 1976. Immunohistochemical localization of melatonin in the rat Harderian gland. *Journal of Histochemistry & Cytochemistry* 24 (11):1173-1177.

Burkhardt, Susanne, Dun Xian Tan, Lucien C. Manchester, Rüdiger Hardeland, and Russel J. Reiter. 2001. Detection and quantification of the antioxidant melatonin in Montmorency and Balaton tart cherries (Prunus cerasus). *Journal of Agricultural and Food Chemistry* 49 (10):4898-4902.

Caniato, Rosamaria, Raffaella Filippini, Anna Piovan, Lucia Puricelli, Anna Borsarini, and Elsa M. Cappelletti. 2003. Melatonin in plants. In *Developments in Tryptophan and Serotonin Metabolism*: Springer.

Cardinali, D. P., and J. M. Rosner. 1971. Metabolism of serotonin by the rat retina *in vitro*. *Journal of neurochemistry* 18 (9):1769-1770.

Carrillo-Vico, A., J. R. Calvo, P. Abreu, P. J. Lardone, S. Garcia-Maurino, R. J. Reiter, and J. M. Guerrero. 2004. Evidence of melatonin synthesis by human lymphocytes and its physiological significance: possible role as intracrine, autocrine, and/or paracrine substance. *Faseb j* 18 (3):537-9.

Champier, J., B. Claustrat, R. Besancon, C. Eymin, C. Killer, A. Jouvet, G. Chamba, and M. Fevre-Montange. 1997. Evidence for tryptophan hydroxylase and hydroxy-indol-O-methyl-transferase mRNAs in human blood platelets. *Life sciences* 60 (24):2191-7.

Chegini, Soheil, Birgit Ehrhart-Hofmann, Alexandra Kaider, and Franz Waldhauser. 1995. Direct enzyme-linked immunosorbent assay and a radioimmunoassay for melatonin compared. *Clinical chemistry* 41 (3):381-386.

Chen, Guofang, Yushu Huo, Dun-Xian Tan, Zhen Liang, Weibing Zhang, and Yukui Zhang. 2003. Melatonin in Chinese medicinal herbs. *Life sciences* 73 (1):19-26.

Conti, A., S. Conconi, E. Hertens, K. Skwarlo-Sonta, M. Markowska, and J. M. Maestroni. 2000. Evidence for melatonin synthesis in mouse and human bone marrow cells. *Journal of Pineal Research* 28 (4):193-202.

Conti, Ario, Stefano Conconi, Elisabeth Hertens, Krystyna Skwarlo-Sonta, Magda Markowska, and Georges J. M. Maestroni. 2000. Evidence for melatonin synthesis in mouse and human bone marrow cells. *Journal of pineal research* 28 (4):193-202.

Conti, Ario, and Georges J. M. Maestroni. 1996. HPLC validation of a circadian melatonin rhythm in the pineal gland of inbred mice. *Journal of pineal research* 20 (3):138-144.

Covaci, Adrian, Catalin Doneanu, Hassan Y Aboul-Enein, and Paul Schepens. 1999. Determination of melatonin in pharmaceutical formulations and human plasma by gas chromatography-electron impact mass spectrometry. *Biomedical Chromatography* 13 (6):431-436.

de Almeida, Eduardo Alves, Paolo Di Mascio, Tatsuo Harumi, D. Warren Spence, Adam Moscovitch, Rüdiger Hardeland, Daniel P. Cardinali, Gregory M. Brown, and S. R. Pandi-Perumal. 2011. Measurement of melatonin in body fluids: Standards, protocols and procedures. *Child's Nervous System* 27 (6):879-891.

de la Puerta, Cristina, María P. Carrascosa-Salmoral, Pedro P. García-Luna, Patricia J. Lardone, Juan L. Herrera, Rafael Fernández-Montesinos, Juan M. Guerrero, and David Pozo. 2007. Melatonin is a phytochemical in olive oil. *Food Chemistry* 104 (2):609-612.

Drijfhout, W. J., C. J. Grol, and B. H. Westerink. 1993. Microdialysis of melatonin in the rat pineal gland: methodology and pharmacological applications. *Journal of Neurochemistry* 61 (3):936-42.

Dubbels, R, R. J. Reiter, E. Klenke, A. Goebel, E. Schnakenberg, C. Ehlers, H. W. Schiwara, and W. Schloot. 1995. Melatonin in edible plants identified by radioimmunoassay and by high performance liquid chromatography-mass spectrometry. *Journal of pineal research* 18 (1):28-31.

Dubey, Vaibhav, Dinesh Mishra, Abhay Asthana, and Narendra Kumar Jain. 2006. Transdermal delivery of a pineal hormone: melatonin via elastic liposomes. *Biomaterials* 27 (18):3491-3496.

Dubey, Vaibhav, Dinesh Mishra, and N. K. Jain. 2007. Melatonin loaded ethanolic liposomes: physicochemical characterization and enhanced transdermal delivery. *European Journal of Pharmaceutics and Biopharmaceutics* 67 (2):398-405.

Eriksson, Kåre, Anders Östin, and Jan-Olof Levin. 2003. Quantification of melatonin in human saliva by liquid chromatography–tandem mass spectrometry using stable isotope dilution. *Journal of Chromatography B* 794 (1):115-123.

Erland, Lauren A. E., Abhishek Chattopadhyay, Andrew Maxwell P. Jones, and Praveen K. Saxena. 2016. Melatonin in Plants and Plant Culture Systems: Variability, Stability and Efficient Quantification. *Frontiers in Plant Science* 7:1721.

Erland, Lauren A. E., Abhishek Chattopadhyay, Andrew Maxwell P. Jones, and Praveen K. Saxena. 2016. Melatonin in Plants and Plant Culture Systems: Variability, Stability and Efficient Quantification. *Frontiers in plant science* 7.

Feng, Xiaoyuan, Meng Wang, Yanyun Zhao, Ping Han, and Ying Dai. 2014. Melatonin from different fruit sources, functional roles, and analytical methods. *Trends in Food Science & Technology* 37 (1):21-31.

Ferreira, Mônica Siqueira, Diogo Noin de Oliveira, Caroline Costa Mesquita, Ana Paula de Lima Barbosa, Gabriel Forato Anhê, and Rodrigo Ramos Catharino. 2016. MALDI-MSI: a fast and reliable method for direct melatonin quantification in biological fluids. *Journal of Analytical Science and Technology* 7 (1):28.

Fischer, Tobias W, Trevor W. Sweatman, Igor Semak, Robert M. Sayre, Jacobo Wortsman, and Andrzej Slominski. 2006. Constitutive and UV-induced metabolism of melatonin in keratinocytes and cell-free systems. *The FASEB journal* 20 (9):1564-1566.

Fourtillan, J. B., P. Gobin, B. Faye, and J. Girault. 1994. A highly sensitive assay of melatonin at the femtogram level in human plasma by gas chromatography/negative ion chemical ionization mass spectrometry. *Biological mass spectrometry* 23 (8):499-509.

Fraser, S, P. Cowen, M. Franklin, C. Franey, and J. Arendt. 1983. Direct radioimmunoassay for melatonin in plasma. *Clinical chemistry* 29 (2):396-397.

Fuhrberg, Birgit, Ivonne Balzer, Rüdiger Hardeland, Astrid Werner, and Klaus Lüning. 1996. The vertebrate pineal hormone melatonin is produced by the brown alga Pterygophora californica and mimics dark effects on growth rate in the light. *Planta* 200 (1):125-131.

Fuhrberg, Birgit, Rüdiger Hardeland, Burkhard Poeggeler, and C. Behrmann. 1997. Dramatic rises of melatonin and 5-methoxytryptamine in Gonyaulax exposed to decreased temperature. *Biological Rhythm Research* 28 (1):144-150.

Gallardo, Eugenia, and J. A. Queiroz. 2008. The role of alternative specimens in toxicological analysis. *Biomedical Chromatography* 22 (8):795-821.

Garcia-Parrilla, M. Carmen, Emma Cantos, and Ana M. Troncoso. 2009. Analysis of melatonin in foods. *Journal of Food Composition and Analysis* 22 (3):177-183.

Gern, William A, David W. Owens, and Charles L. Ralph. 1978. Plasma melatonin in the trout: Day-night change demonstrated by radioimmunoassay. *General and comparative endocrinology* 34 (4):453-458.

González-Gómez, D, M. Lozano, M. F. Fernández-León, M. C. Ayuso, M. J. Bernalte, and A. B. Rodríguez. 2009. Detection and quantification of melatonin and serotonin in eight Sweet Cherry cultivars (Prunus avium L.). *European Food Research and Technology* 229 (2):223-229.

Grota, L. J., V. Snieckus, S. O. de Silva, and G. M. Brown. 1983. Antibodies to indolealkylamines II: site of conjugation of melatonin to protein using formaldehyde. *Canadian Journal of Biochemistry and Cell Biology* 61 (10):1096-1101.

Guerrero, Juan M, Francisco Martínez-Cruz, and Félix L. Elorza. 2008. WITHDRAWN: Significant amounts of melatonin in red wine: Its consumption increases blood melatonin levels in humans. *Food Chemistry*.

Hamase, Kenji, Junzo Hirano, Yuki Kosai, Tatsunosuke Tomita, and Kiyoshi Zaitsu. 2004. A sensitive internal standard method for the determination of melatonin in mammals using precolumn oxidation reversed-phase high-performance liquid chromatography. *Journal of chromatography B* 811 (2):237-241.

Hamase, Kenji, Tatsunosuke Tomita, Ayako Kiyomizu, and Kiyoshi Zaitsu. 2000. Determination of pineal melatonin by precolumn derivatization reversed-phase high-performance liquid chromatography and its application to the study of circadian rhythm in rats and mice. *Analytical biochemistry* 279 (1):106-110.

Hardeland, R., J. A. Madrid, D. X. Tan, and R. J. Reiter. 2012. Melatonin, the circadian multioscillator system and health: the need for detailed analyses of peripheral melatonin signaling. *Journal of Pineal Research* 52 (2):139-66.

Harumi, T., and S. Matsushima. 2000. Separation and assay methods for melatonin and its precursors. *Journal of Chromatography B: Biomedical Sciences and Applications* 747 (1-2):95-110.

Harumi, Tatsuo, Hiroaki Akutsu, and Shoji Matsushima. 1996. Simultaneous determination of serotonin, N-acetylserotonin and melatonin in the pineal gland of the juvenile golden hamster by high-performance liquid chromatography with electrochemical detection. *Journal of Chromatography B: Biomedical Sciences and Applications* 675 (1):152-156.

Hattori, Atsuhiko, Hiro Migitaka, Masayuki Iigo, Masanori Itoh, Koji Yamamoto, Ritsuko Ohtani-Kaneko, Masayuki Hara, Takuro Suzuki, and Russel J. Reiter. 1995. Identification of melatonin in plants and its effects on plasma melatonin levels and binding to melatonin receptors in vertebrates. *Biochemistry and molecular biology international* 35 (3):627-634.

Helton, R. A., W. A. Harrison, K. Kelley, and M. A. Kane. 1993. Melatonin interactions with cultured murine B16 melanoma cells. *Melanoma research* 3 (6):403-414.

Hemwimon, Surasak, Prasert Pavasant, and Artiwan Shotipruk. 2007. Microwave-assisted extraction of antioxidative anthraquinones from roots of Morinda citrifolia. *Separation and Purification Technology* 54 (1):44-50.

Hernandez-Ruiz, Josefa, Antonio Cano, and Marino B. Arnao. 2004. Melatonin: a growth-stimulating compound present in lupin tissues. *Planta* 220 (1):140-144.

Hernández-Ruiz, Josefa, Antonio Cano, and Marino B. Arnao. 2005. Melatonin acts as a growth-stimulating compound in some monocot species. *Journal of pineal research* 39 (2):137-142.

Hevia, D, J. C. Mayo, I. Quiros, C. Gomez-Cordoves, and R. M. Sainz. 2010. Monitoring intracellular melatonin levels in human prostate normal and cancer cells by HPLC. *Analytical and bioanalytical chemistry* 397 (3):1235-1244.

Hickie, Ian B., and Naomi L. Rogers. 2011. Novel melatonin-based therapies: potential advances in the treatment of major depression. *The Lancet* 378 (9791):621-631.

Hosseinian, Farah S, Wende Li, and Trust Beta. 2008. Measurement of anthocyanins and other phytochemicals in purple wheat. *Food Chemistry* 109 (4):916-924.

Huang, X., and G. Mazza. 2011. Application of LC and LC-MS to the analysis of melatonin and serotonin in edible plants. *Critical Reviews in Food Science and Nutrition* 51 (4):269-84.

Huether, Gerald, Burkhard Poeggeler, Andreas Reimer, and Annette George. 1992. Effect of tryptophan administration on circulating melatonin levels in chicks and rats:

evidence for stimulation of melatonin synthesis and release in the gastrointestinal tract. *Life sciences* 51 (12):945-953.

Iinuma, Fumio, Kenji Hamase, Sayaka Matsubayashi, Masaki Takahashi, Mitsuo Watanabe, and Kiyoshi Zaitsu. 1999. Sensitive determination of melatonin by precolumn derivatization and reversed-phase high-performance liquid chromatography. *Journal of Chromatography A* 835 (1):67-72.

Iriti, Marcello, Mara Rossoni, and Franco Faoro. 2006. Melatonin content in grape: myth or panacea? *Journal of the Science of Food and Agriculture* 86 (10):1432-1438.

Iriti, Marcello, and Ileana Vigentini. 2015. Tryptophan-ethylester, the false (unveiled) melatonin isomer in red wine. *International journal of tryptophan research: IJTR* 8:27.

Ishizaki, Atsushi, Akiko Uemura, and Hiroyuki Kataoka. 2017. A sensitive method to determine melatonin in saliva by automated online in-tube solid-phase microextraction coupled with stable isotope-dilution liquid chromatography-tandem mass spectrometry. *Analytical Methods* 9 (21):3134-3140.

Itoh, Masanori T, Atsuhiko Hattori, Yawara Sumi, and Takuro Suzuki. 1995. Day-night changes in melatonin levels in different organs of the cricket (Gryllus bimaculatus). *Journal of pineal research* 18 (3):165-169.

Itoh, Masanori T, Bunpei Ishizuka, Yoshiko Kudo, Sigeyoshi Fusama, Akira Amemiya, and Yawara Sumi. 1997. Detection of melatonin and serotonin N-acetyltransferase and hydroxyindole-O-methyltransferase activities in rat ovary. *Molecular and cellular endocrinology* 136 (1):7-13.

Itoh, Masanori T, Bunpei Ishizuka, Yasushi Kuribayashi, Akira Amemiya, and Yawara Sumi. 1999. Melatonin, its precursors, and synthesizing enzyme activities in the human ovary. *Molecular human reproduction* 5 (5):402-408.

Itoh, Masanori T, Takao Shinozawa, and Yawara Sumi. 1999. Circadian rhythms of melatonin-synthesizing enzyme activities and melatonin levels in planarians. *Brain research* 830 (1):165-173.

Johns, Nutjaree Pratheepawanit, Jeffrey Johns, Supatra Porasuphatana, Preeyaporn Plaimee, and Manit Sae-Teaw. 2013. Dietary intake of melatonin from tropical fruit altered urinary excretion of 6-sulfatoxymelatonin in healthy volunteers. *Journal of agricultural and food chemistry* 61 (4):913-919.

Kanikkannan, N, S. Andega, S. Burton, R. J. Babu, and Mandip Singh. 2004. Formulation and *in vitro* evaluation of transdermal patches of melatonin. *Drug development and industrial pharmacy* 30 (2):205-212.

Karasek, Francis W, and Ray E. Clement. 2012. *Basic gas chromatography-mass spectrometry: principles and techniques*: Elsevier.

Kennaway, D., R. Frith, G. Phillipou, C. Matthews, and R. Seamark. 1977. A specific radioimmunoassay for melatonin in biological tissue and fluids and its validation by gas chromatography-mass spectrometry. *Endocrinology* 101.

Kolář, Jan, and Ivana Macháčková. 2005. Melatonin in higher plants: occurrence and possible functions. *Journal of pineal research* 39 (4):333-341.

Kolář, Jan, Ivana Macháčková, Josef Eder, Els Prinsen, Walter Van Dongen, Harry Van Onckelen, and Helena Illnerová. 1997. Melatonin: occurrence and daily rhythm in Chenopodium rubrum. *Phytochemistry* 44 (8):1407-1413.

Kollmann, M. T, M. Locher, F. Hirche, K. Eder, H. H. D. Meyer, and R. M. Bruckmaier. 2008. Effects of tryptophan supplementation on plasma tryptophan and related hormone levels in heifers and dairy cows. *Domestic animal endocrinology* 34 (1):14-24.

Kulczykowska, Ewa, and P. Michael Iuvone. 1998. Highly sensitive and specific assay of plasma melatonin using high-performance liquid chromatography with fluorescence detection preceded by solid-phase extraction. *Journal of chromatographic science* 36 (4):175-178.

Lee, Kyu-Suk, Won-Suk Lee, Seong-Il Suh, Sang-Pyo Kim, Sung-Ryong Lee, Young-Wook Ryoo, and Byung-Chun Kim. 2003. Melatonin reduces ultraviolet-B induced cell damages and polyamine levels in human skin fibroblasts in culture. *Experimental & molecular medicine* 35 (4):263.

Lemaitre, Béatrice J, and Lucien Hartmann. 1980. Preparation of anti-melatonin antibodies and antigenic properties of the molecule. *Journal of immunological methods* 32 (4):339-347.

Lerner, Aaron B, James D. Case, Yoshiyata Takahashi, Teh H. Lee, and Wataru Mori. 1958. Isolation of melatonin, the pineal gland factor that lightens melanocyteS1. *Journal of the American Chemical Society* 80 (10):2587-2587.

Leung, Frederick C. 1991. Circadian rhythms of melatonin release from chicken pineal *in vitro*: modified melatonin radioimmunoassay. *Proceedings of the Society for Experimental Biology and medicine* 198 (3):826-832.

Li, Ye, and Vincent M. Cassone. 2015. A simple, specific high-throughput enzyme-linked immunosorbent assay (ELISA) for quantitative determination of melatonin in cell culture medium. *International Immunopharmacology* 28 (1):230-234.

Liu, C., C. Fukuhara, J. H. Wessel, 3rd, P. M. Iuvone, and G. Tosini. 2004. Localization of Aa-nat mRNA in the rat retina by fluorescence in situ hybridization and laser capture microdissection. *Cell and Tissue Research* 315 (2):197-201.

Lynch, Kyle B., Apeng Chen, and Shaorong Liu. 2017. Miniaturized High-Performance Liquid Chromatography Instrumentation. *Talanta* in press, DOI [10.1016/j.talanta.2017.09.016]

Manchester, Lucien C, Dun-Xian Tan, Russel J. Reiter, Won Park, Kanishka Monis, and Wenbo Qi. 2000. High levels of melatonin in the seeds of edible plants: possible function in germ tissue protection. *Life sciences* 67 (25):3023-3029.

Martinez, G. R., E. A. Almeida, C. F. Klitzke, J. Onuki, F. M. Prado, M. H. Medeiros, and P. Di Mascio. 2005. Measurement of melatonin and its metabolites: importance for the evaluation of their biological roles. *Endocrine* 27 (2):111-8.

Martins, Leiziani Gnatkowski, Najeh Maissar khalil, and Rubiana Mara Mainardes. 2017. Application of a validated HPLC-PDA method for the determination of melatonin content and its release from poly(lactic acid) nanoparticles. *Journal of Pharmaceutical Analysis* in press, DOI [/10.1016/j.jpha.2017.05.007].

Mefford, Ivan N, and Jack D. Barchas. 1980. Determination of tryptophan and metabolites in rat brain and pineal tissue by reversed-phase high-performance liquid chromatography with electrochemical detection. *Journal of Chromatography B: Biomedical Sciences and Applications* 181 (2):187-193.

Mercolini, L., M. Addolorata Saracino, F. Bugamelli, A. Ferranti, M. Malaguti, S. Hrelia, and M. A. Raggi. 2008. HPLC-F analysis of melatonin and resveratrol isomers in wine using an SPE procedure. *J Sep Sci* 31 (6-7):1007-14.

Mercolini, Laura, Maria Addolorata Saracino, Francesca Bugamelli, Anna Ferranti, Marco Malaguti, Silvana Hrelia, and Maria Augusta Raggi. 2008. HPLC-F analysis of melatonin and resveratrol isomers in wine using an SPE procedure. *Journal of separation science* 31 (6-7):1007-1014.

Mercolini, Laura, Roberto Mandrioli, and Maria Augusta Raggi. 2012. Content of melatonin and other antioxidants in grape-related foodstuffs: measurement using a MEPS-HPLC-F method. *Journal of pineal research* 53 (1):21-28.

Meyer, Veronika R. 2013. *Practical high-performance liquid chromatography*: John Wiley & Sons.

Mills, Malcolm H, David C. Finlay, and Paul R. Haddad. 1991. Determination of melatonin and monoamines in rat pineal using reversed-phase ion-interaction chromatography with fluorescence detection. *Journal of Chromatography B: Biomedical Sciences and Applications* 564 (1):93-102.

Minami, Masayasu, Hideyo Takahashi, Hirofumi Inagaki, Yuko Yamano, Sakura Onoue, Shun Matsumoto, Tsukasa Sasaki, and Kazuhiro Sakai. 2009. Novel tryptamine-related substances, 5-sulphatoxydiacetyltryptamine, 5-hydroxydiacetyltryptamine, and reduced melatonin in human urine and the determination of those compounds, 6-sulphatoxymelatonin, and melatonin with fluorometric HPLC. *Journal of Chromatography B* 877 (8):814-822.

Muñoz, José L. P., Rosa M. Ceinos, José L. Soengas, and Jesús M. Míguez. 2009. A simple and sensitive method for determination of melatonin in plasma, bile and intestinal tissues by high performance liquid chromatography with fluorescence detection. *Journal of Chromatography B* 877 (22):2173-2177.

Murch, S. J, S. KrishnaRaj, and P. K. Saxena. 2000. Tryptophan is a precursor for melatonin and serotonin biosynthesis in *in vitro* regenerated St. John's wort (Hypericum perforatum L. cv. Anthos) plants. *Plant Cell Reports* 19 (7):698-704.

Murch, Susan J, Skye S. B. Campbell, and Praveen K. Saxena. 2001. The role of serotonin and melatonin in plant morphogenesis: regulation of auxin-induced root organogenesis in *in vitro*-cultured explants of St. John's wort (Hypericum perforatum L.). *In vitro Cellular & Developmental Biology-Plant* 37 (6):786-793.

Nawaz, Muhammad A., Yuan Huang, Zhilong Bie, Waqar Ahmed, Russel J. Reiter, Mengliang Niu, and Saba Hameed. 2015. Melatonin: Current Status and Future Perspectives in Plant Science. *Frontiers in Plant Science* 6:1230.

Nunez-Vergara, Luis J, J. A. Squella, J. C. Sturm, H. Baez, and Cristián Camargo. 2001. Simultaneous determination of melatonin and pyridoxine in tablets by gas chromatography-mass spectrometry. *Journal of pharmaceutical and biomedical analysis* 26 (5):929-938.

Pang, S. F, G. M. Brown, L. J. Grota, J. W. Chambers, and R. L. Rodman. 1977. Determination of N-acetylserotonin and melatonin activities in the pineal gland, retina, Harderian gland, brain and serum of rats and chickens. *Neuroendocrinology* 23 (1):1-13.

Pape, Carsten, and Klaus Lüning. 2006. Quantification of melatonin in phototrophic organisms. *Journal of pineal research* 41 (2):157-165.

Paredes, Sergio D, Ahmet Korkmaz, Lucien C Manchester, Dun-Xian Tan, and Russel J. Reiter. 2008. Phytomelatonin: a review. *Journal of Experimental Botany* 60 (1):57-69.

Peniston-Bird, J. F, W. L. Di, Catherine A. Street, ABAN Kadva, Maria A. Stalteri, and Robert E Silman. 1993. HPLC assay of melatonin in plasma with fluorescence detection. *Clinical chemistry* 39 (11):2242-2247.

Plebani, Mario, Maurizio Masiero, Alessandro P. Burlina, Maria Laura Chiozza, Massimo Scanarini, and Angelo Burlina. 1990. Measurement of melatonin in blood by radioimmunoassay. *Child's Nervous System* 6 (4):220-221.

Poeggeler, Burkhard, and Rüdiger Hardeland. 1994. Detection and quantification of melatonin in a dinoflagellate, Gonyaulax polyedra: Solutions to the problem of methoxyindole destruction in non-vertebrate material. *Journal of pineal research* 17 (1):1-10.

Posmyk, Małgorzata M, and Krystyna M. Janas. 2009. Melatonin in plants. *Acta Physiologiae Plantarum* 31 (1):1.

Psillakis, E., and N. Kalogerakis. 2003. Developments in liquid-phase microextraction. *TrAC Trends in Analytical Chemistry* 22 (9):565-574.

Pucci, Vincenzo, Anna Ferranti, Roberto Mandrioli, and Maria Augusta Raggi. 2003. Determination of melatonin in commercial preparations by micellar electrokinetic chromatography and spectrofluorimetry. *Analytica Chimica Acta* 488 (1):97-105.

Radi, A., and G. E. Bekhiet. 1998. Voltammetry of melatonin at carbon electrodes and determination in capsules. *Bioelectrochemistry and Bioenergetics* 45 (2):275-279.

Razgi, Maria Augusta, Francesca Bugamelli, and Vincenzo Pucci. 2002. Determination of melatonin in galenic preparations by LC and voltammetry. *Journal of Pharmaceutical and Biomedical Analysis* 29 (1):283-289.

Ramakrishna, Akula, Parvatam Giridhar, Kadimi Udaya Sankar, and Gokare Aswathanarayana Ravishankar. 2012. Endogenous profiles of indoleamines: serotonin and melatonin in different tissues of Coffeacanephora P ex Fr. as analyzed by HPLC and LC-MS-ESI. *Acta physiologiae plantarum* 34 (1):393-396.

Repeated Author. 2012. Melatonin and serotonin profiles in beans of Coffea species. *Journal of pineal research* 52 (4):470-476.

Raynaud, F, F. Mauviard, M. Geoffriau, B. Claustrat, and P. Pevet. 1993. Plasma 6-hydroxymelatonin, 6-sulfatoxymelatonin and melatonin kinetics after melatonin administration to rats. *Neurosignals* 2 (6):359-366.

Raynaud, F, and P. Pevet. 1991. Determination of 5-methoxyindoles in pineal gland and plasma samples by high-performance liquid chromatography with electrochemical detection. *Journal of Chromatography B: Biomedical Sciences and Applications* 564 (1):103-113.

Reiter, R. J., D. X. Tan, L. C. Manchester, A. P. Simopoulos, M. D. Maldonado, L. J. Flores, and M. P. Terron. 2007. Melatonin in edible plants (phytomelatonin): Identification, concentrations, bioavailability and proposed functions. *World Review of Nutrition and Dietetics* 97:211-30.

Reiter, Russel J, L. C. Manchester, and Dun-xian Tan. 2005. Melatonin in walnuts: influence on levels of melatonin and total antioxidant capacity of blood. *Nutrition* 21 (9):920-924.

Reiter, Russel J, Dun Xian Tan, Susanne Burkhardt, and Lucien C. Manchester. 2001. Melatonin in plants. *Nutrition Reviews* 59 (9):286-290.

Rizzo, Vittoria, Camillo Porta, Mauro Moroni, Enrico Scoglio, and Remigio Moratti. 2002. Determination of free and total (free plus protein-bound) melatonin in plasma and cerebrospinal fluid by high-performance liquid chromatography with fluorescence detection. *Journal of Chromatography B* 774 (1):17-24.

Rodriguez-Naranjo, M. Isabel, Angel Gil-Izquierdo, Ana M. Troncoso, Emma Cantos, and M. Carmen Garcia-Parrilla. 2011. Melatonin: a new bioactive compound in wine. *Journal of Food Composition and Analysis* 24 (4):603-608.

Rolčík, Jakub, René Lenobel, Věra Siglerová, and Miroslav Strnad. 2002. Isolation of melatonin by immunoaffinity chromatography. *Journal of Chromatography B* 775 (1):9-15.

Rollag, M. D, and G. D. Niswender. 1976. Radioimmunoassay of serum concentrations of melatonin in sheep exposed to different lighting regimens. *Endocrinology* 98 (2):482-489.

Rosado, T., I. Henriques, Eugenia Gallardo, and A. P. Duarte. 2017. Determination of melatonin levels in different cherry cultivars by high-performance liquid chromatography coupled to electrochemical detection. *European Food Research and Technology* 243 (10):1749-1757.

Setyaningsih, W, M. Palma, and C. G. Barroso. 2012. A new microwave-assisted extraction method for melatonin determination in rice grains. *Journal of cereal science* 56 (2):340-346.

Silvestre, Cristina I. C., João L. M. Santos, José L. F. C. Lima, and Elias A. G. Zagatto. 2009. Liquid–liquid extraction in flow analysis: A critical review. *Analytica Chimica Acta* 652 (1):54-65.

Simonin, Gilles, Laurence Bru, E. Lelievre, Jean-Philippe Jeanniot, Norbert Bromet, Bernard Walther, and Claire Boursier-Neyret. 1999. Determination of melatonin in biological fluids in the presence of the melatonin agonist S 20098: comparison of immunological techniques and GC-MS methods. *Journal of pharmaceutical and biomedical analysis* 21 (3):591-601.

Simonin, Gilles, Laurence Bru, E. Lelièvre, Jean-Philippe Jeanniot, Norbert Bromet, Bernard Walther, and Claire Boursier-Neyret. 1999. Determination of melatonin in biological fluids in the presence of the melatonin agonist S 20098: comparison of immunological techniques and GC-MS methods. *Journal of Pharmaceutical and Biomedical Analysis* 21 (3):591-601.

Skene, D. J, R. M. Leone, I. M. Young, and R. E. Silman. 1983. The assessment of a plasma melatonin assay using gas chromatography negative ion chemical ionization mass spectrometry. *Biological Mass Spectrometry* 10 (12):655-659.

Slominski, A., T. W. Fischer, M. A. Zmijewski, J. Wortsman, I. Semak, B. Zbytek, R. M. Slominski, and D. J. Tobin. 2005. On the role of melatonin in skin physiology and pathology. *Endocrine* 27 (2):137-48.

Slominski, Andrzej, James Baker, Thomas G. Rosano, Lawrence W. Guisti, Gennady Ermak, Melissa Grande, and Stephen J. Gaudet. 1996. Metabolism of serotonin to N-acetylserotonin, melatonin, and 5-methoxytryptamine in hamster skin culture. *Journal of Biological Chemistry* 271 (21):12281-12286.

Stege, Patricia W, Lorena L. Sombra, Germán Messina, Luis D. Martinez, and María F. Silva. 2010. Determination of melatonin in wine and plant extracts by capillary electrochromatography with immobilized carboxylic multi-walled carbon nanotubes as stationary phase. *Electrophoresis* 31 (13):2242-2248.

Talebianpoor, M. S., S. Khodadoust, A. Rozbehi, M. Akbartabar Toori, M. Zoladl, M. Ghaedi, R. Mohammadi, and A. S. Hosseinzadeh. 2014. Application of optimized dispersive liquid–liquid microextraction for determination of melatonin by HPLC–UV in plasma samples. *Journal of Chromatography B* 960 (Supplement C):1-7.

Tan, Dun-Xian. 2015. Melatonin and plants. *Journal of Experimental Botany* 66 (3):625-626.

Tan, Dun-Xian, Rudiger Hardeland, Lucien C. Manchester, Ahmet Korkmaz, Shuran Ma, Sergio Rosales-Corral, and Russel J. Reiter. 2011. Functional roles of melatonin in plants, and perspectives in nutritional and agricultural science. *Journal of experimental botany* 63 (2):577-597.

Tan, Dun-xian, Lucien C. Manchester, Russel J. Reiter, Wen-bo Qi, Ming Zhang, Susan T. Weintraub, Javier Cabrera, Rosa M. Sainz, and Juan C. Mayo. 1999. Identification of highly elevated levels of melatonin in bone marrow: its origin and significance. *Biochimica et Biophysica Acta (BBA)-General Subjects* 1472 (1):206-214.

Tan, Dun-xian, Lucien C. Manchester, Russel J. Reiter, Wenbo Qi, Martha A. Hanes, and Norma J. Farley. 1999. High physiological levels of melatonin in the bile of mammals. *Life sciences* 65 (23):2523-2529.

Tan, Dun-Xian, Lucien C. Manchester, Maria P. Terron, Luis J. Flores, and Russel J. Reiter. 2007. One molecule, many derivatives: A never-ending interaction of melatonin with reactive oxygen and nitrogen species? *Journal of pineal research* 42 (1):28-42.

Tomita, Tatsunosuke, Kenji Hamase, Hiromi Hayashi, Hiroko Fukuda, Junzo Hirano, and Kiyoshi Zaitsu. 2003. Determination of endogenous melatonin in the individual pineal glands of inbred mice using precolumn oxidation reversed-phase micro-high-performance liquid chromatography. *Analytical biochemistry* 316 (2):154-161.

Tomita, Tatsunosuke, Kenji Hamase, Hiromi Hayashi, and Kiyoshi Zaitsu. 2001. Attomole analysis of melatonin by precolumn derivatization reversed-phase micro-HPLC. *Chromatography-Tokyo-Society for Chromatographic Sciences* 22 (1):41-48.

Tosini, G., and M. Menaker. 1998. The clock in the mouse retina: melatonin synthesis and photoreceptor degeneration. *Brain Reseach* 789 (2):221-8.

Tursilli, Rosanna, Alberto Casolari, Valentina Iannuccelli, and Santo Scalia. 2006. Enhancement of melatonin photostability by encapsulation in liposheres. *Journal of pharmaceutical and biomedical analysis* 40 (4):910-914.

Vakkuri, O. 1985. Diurnal rhythm of melatonin in human saliva. *Acta Physiologica* 124 (3):409-412.

Vakkuri, Olli, Erkki Lämsä, Erkki Rahkamaa, Heikki Ruotsalainen, and Juhani Leppäluoto. 1984. Iodinated melatonin: preparation and characterization of the molecular structure by mass and 1H NMR spectroscopy. *Analytical biochemistry* 142 (2):284-289.

Van Tassel, D. L., and S. D. O'Neill. 2001. Putative regulatory molecules in plants: evaluating melatonin. *Journal of pineal research* 31 (1):1-7.

Van Tassel, David L, Nicholas Roberts, Alfred Lewy, and Sharman D O'neill. 2001. Melatonin in plant organs. *Journal of pineal research* 31 (1):8-15.

Vieira, Raúl, Jesús Míguez, Maria Lema, and Manuel Aldegunde. 1992. Pineal and plasma melatonin as determined by high-performance liquid chromatography with electrochemical detection. *Analytical biochemistry* 205 (2):300-305.

Vigentini, Ileana, Claudio Gardana, Daniela Fracassetti, Mario Gabrielli, Roberto Foschino, Paolo Simonetti, Antonio Tirelli, and Marcello Iriti. 2015. Yeast contribution to melatonin, melatonin isomers and tryptophan ethyl ester during alcoholic fermentation of grape musts. *Journal of pineal research* 58 (4):388-396.

Vitale, Arturo A, Carina C. Ferrari, Hernán Aldana, and Jorge M. Affanni. 1996. Highly sensitive method for the determination of melatonin by normal-phase high-performance liquid chromatography with fluorometric detection. *Journal of Chromatography B: Biomedical Sciences and Applications* 681 (2):381-384.

Vivien-Roels, Berthe, André Malan, Marie-Claire Rettori, Philippe Delagrange, Jean-Philippe Jeanniot, and Paul Pévet. 1998. Daily variations in pineal melatonin concentrations in inbred and outbred mice. *Journal of Biological Rhythms* 13 (5):403-409.

Webley, G. E, H. Mehl, and K. P. Willey. 1985. Validation of a sensitive direct assay for melatonin for investigation of circadian rhythms in different species. *Journal of Endocrinology* 106 (3):387-394.

Welp, André, Bernhard Manz, and Elmar Peschke. 2010. Development and validation of a high throughput direct radioimmunoassay for the quantitative determination of serum and plasma melatonin (N-acetyl-5-methoxytryptamine) in mice. *Journal of immunological methods* 358 (1):1-8.

Wolf, Karel, Jan Kolář, Erwin Witters, Walter van Dongen, Harry van Onckelen, and Ivana Macháčková. 2001. Daily profile of melatonin levels in Chenopodium rubrum L. depends on photoperiod. *Journal of plant physiology* 158 (11):1491-1493.

Wolfender, Jean-Luc. 2009. HPLC in natural product analysis: the detection issue. *Planta medica* 75 (07):719-734.

Yang, Shuming, Xiaohui Zheng, Yan Xu, and Xiang Zhou. 2002. Rapid determination of serum melatonin by ESI–MS–MS with direct sample injection. *Journal of Pharmaceutical and Biomedical Analysis* 30 (3):781-790.

Yang, Tangbin, Jianhe Wang, Lina Qu, Ping Zhong, and Yanhong Yuan. 2006. Preparation and identification of anti-melatonin monoclonal antibodies. *Journal of pineal research* 40 (4):350-354.

Ye, Tiantian, Yan-Hong Hao, Lei Yu, Haitao Shi, Russel J. Reiter, and Yu-Qi Feng. 2017. A Simple, Rapid Method for Determination of Melatonin in Plant Tissues by UPLC Coupled with High Resolution Orbitrap Mass Spectrometry. *Frontiers in Plant Science* 8:64.

Yeleswaram, Krishnaswamy, Lee G. McLaughlin, Jay O. Knipe, and David Schabdach. 1997. Pharmacokinetics and oral bioavailability of exogenous melatonin in preclinical animal models and clinical implications. *Journal of pineal research* 22 (1):45-51.

Zhao, Huimin, Yifei Wang, Yi Jin, Shu Liu, Haiyan Xu, and Xiumei Lu. 2016. Rapid and sensitive analysis of melatonin by LC-MS/MS and its application to pharmacokinetic study in dogs. *Asian Journal of Pharmaceutical Sciences* 11 (2):273-280.

Zhao, Yu, Dun-Xian Tan, Qiong Lei, Hao Chen, Lin Wang, Qing-tian Li, Yinan Gao, and Jin Kong. 2013. Melatonin and its potential biological functions in the fruits of sweet cherry. *Journal of Pineal Research* 55 (1):79-88.

In: Melatonin: Medical Uses and Role in Health and Disease ISBN: 978-1-53612-987-8
Editors: Lore Correia and Germaine Mayers © 2018 Nova Science Publishers, Inc.

Chapter 2

MELATONIN AS A THERAPEUTIC OPTION FOR OPTIC NEUROPATHIES

*María F. González Fleitas, Marcos L. Aranda,
Hernán H. Dieguez, Julián D. Devouassoux,
Georgia A. Milne, Mónica Chianelli,
María I. Keller Sarmiento, Pablo H. Sande,
Damián Dorfman and Ruth E. Rosenstein*[*]

Laboratorio de Neuroquímica Retiniana y Oftalmología Experimental, Departamento de Bioquímica Humana, Facultad de Medicina/CEFyBO, Universidad de Buenos Aires, CONICET, Buenos Aires, Argentina

ABSTRACT

Glaucoma is a leading cause of irreversible blindness worldwide, and provokes progressive visual impairment attributable to the dysfunction and loss of retinal ganglion cells and optic nerve axons. Currently, clinical interventions for glaucoma are mainly restricted to the reduction of intraocular pressure, one of the major risk factors for the disease. However, lowering intraocular pressure is often insufficient to halt or reverse the progress of visual loss, underlining the need for the development of alternative treatment strategies.

Optic neuritis, the most common optic neuropathy affecting young adults, is a condition involving primary inflammation, demyelination, and axonal injury in the optic nerve which leads to retinal ganglion cell death and visual dysfunction. Although corticosteroids are the current mainstays of therapy for the treatment of optic neuritis, they fail to achieve long-term benefit on functional recovery, and provoke considerable side effects. Despite several differences between these diseases, glaucoma and optic

---

[*] Corresponding Author: Dr. Ruth E. Rosenstein, Department of Human Biochemistry, School of Medicine, CEFyBO, University of Buenos Aires, CONICET, Paraguay 2155, 5th floor, (1121), Buenos Aires, ARGENTINA, Tel: 54-11-45083672 (ext 37), Fax: 54-11-45083672 (ext 317), Email: ruthr@fmed.uba.ar.

neuritis are the two major degenerative causes of optic nerve damage, sharing a common feature which is retinal ganglion cell loss and axonal damage. Although the current management of glaucoma is mainly directed at the control of intraocular pressure, and the current therapy for optic neuritis mainly target the autoimmune response by using anti-inflammatory and immunosuppressive agents, a neuroprotective therapy able to avoid the death of retinal ganglion cells and optic nerve fiber loss should be the main goal of treatment for both diseases. In this chapter, we will discuss evidence supporting the neuroprotective effect of melatonin, a very safe compound (even at high doses) for human use, which prevents and slows the progression of retinal ganglion cell and optic nerve axon damage induced by experimental glaucoma or optic neuritis in rodents, and as such, it could be considered as a new therapeutic resource for the management of the most prevalent forms of optic neuropathy.

ABBREVIATIONS

BDNF	brain-derived neurotrophic factor
CAT	cationic amino acid transporter
COX-2	cyclooxygenase-2
EAAT-1	excitatory amino acid transporter type 1
EAE	experimental autoimmune encephalomyelitis
GPX	glutathione peroxidase
GS	glutamine synthetase
HA	hyaluronic acid
iNOS	inducible NOS or NOS-2
IOP	intraocular pressure
LPS	bacterial lipopolysaccharide
MS	multiple sclerosis
NMDA	N-methyl D-aspartate
NOS	NO synthase
nNOS	neuronal NOS or NOS-1
ON	optic neuritis
ONH	optic nerve head
POAG	primary open-angle glaucoma
PLR	pupil light reflex
RGC	retinal ganglion cell
ROS	reactive oxygen species
SOD	superoxide dismutase
TNFα	tumor necrosis factor α
VEPs	visual evoked potentials

MELATONIN IN THE RETINA

Numerous studies have firmly established that melatonin synthesis occurs in the retina of vertebrates, including mammals (reviewed by Tosini and Fukuhara, 2003). Although available data indicate that photoreceptors synthesize melatonin independently from the rest of the retina (Cahill and Besharse, 1993), we have shown that melatonin could be also synthesized in chick retinal ganglion cells (RGCs) (Garbarino-Pico et al., 2004). In the vertebrate species studied so far, melatonin synthesis in the retina is elevated at night and reduced during the day in a fashion similar to events in the pineal gland. Melatonin synthesis in the retina is regulated by the photic stimulus (Faillace et al., 1995a, 1995b) and/or under the control of a circadian oscillator (Tosini and Menaker, 1996). Retinal biosynthesis of melatonin has been extensively studied and revised (Tosini and Fukuhara, 2003, Tosini et al., 2012). In contrast, although a mutual inhibitory relationship between melatonin and dopamine (Dubocovich, 1983; Jaliffa et al., 2000), as well as the involvement of melatonin in the regulation of photoreceptor disc shedding and phagocytosis (Besharse and Dunis, 1983), melanosome aggregation in pigment epithelium and cone photoreceptor retinomotor movements (Pierce and Besharse, 1985) were conclusively demonstrated, the full range of melatonin actions in the retina is far from being completely known. Retinal melatonin does not contribute to circulating levels, suggesting that it acts locally as a neurohormone and/or neuromodulator. Different subtypes of melatonin receptors are present on major types of retinal neurons, and the expression of these receptors is highly species- and neuron subtype-dependent. By activating different melatonin receptor subtypes, melatonin modulates the activity of retinal neurons (Huang et al., 2013). In addition, retinal melatonin may have clinical implications in the pathogenesis of a number of disorders of the eye, including keratopathies, corneal healing, age-related macular degeneration, and UVB/light-induced pathologies (Acuña-Castroviejo et al., 2014), as suggested by the different actions that melatonin may have in the retina, i.e., receptor mediated and non-receptor-mediated effects. Administration of exogenous melatonin may also benefit ocular health (Salido et al., 2013; Sande et al., 2008, 2014). In this chapter, we will discuss evidence supporting the therapeutic effect of melatonin for the treatment of glaucoma and optic neuritis, the most prevalent optic neuropathies.

GLAUCOMA

Glaucoma is currently recognized as a multifactorial, progressive and neurodegenerative disorder, characterized by the death of RGCs associated to the loss of axons that make up the optic nerve (Nucci et al., 2016). With more than 60 million people affected, and ~ 8 million suffering from bilateral blindness caused by this disease, glaucoma is recognized as a leading cause of irreversible blindness worldwide (Cedrone et al., 2012; Nucci et al., 2016; Quigley and Broman, 2006). It is estimated that half of those affected may be not aware of their condition because symptoms may not occur during the early stages of the disease. When vision loss appears, considerable and permanent damage has already occurred. Glaucomatous visual loss often starts in the periphery and advances to involve the central vision, with devastating consequences to the patient's quality of life (Almasieh et al., 2012; Cesareo et al., 2015). Several clinical studies demonstrated that increased intraocular pressure (IOP), over the levels considered physiological, is the main risk factor for the onset and progression of neuronal damage (Nucci et al., 2016). The Ocular Hypertensive Treatment Study (Kass et al., 2002) and the Early Manifest Glaucoma Trial (Heijl et al., 2002) show that reducing IOP by ocular medications is effective in preventing the onset or delaying the progression of the disease. In accordance to these data, glaucoma therapy is currently based on IOP reduction by medical, surgical or parasurgical treatments (Nucci et al., 2016). Notwithstanding, a considerable percentage of patients experiences disease progression, despite their IOP values do not differ from the normal range or are satisfactorily controlled by therapy (Leske et al., 2003). Moreover, although ocular hypertension is common among open-angle glaucoma patients, only a limited subset of individuals with ocular hypertension will develop the disease (Friedman et al., 2004). Thus, although the current management of glaucoma is mainly directed at the control of IOP, a therapy that prevents the death of RGCs should be the main goal of treatment. Unfortunately, the mechanisms that lead to RGC loss in glaucoma are still under debate. Being probably a complex and multifactorial disease, it is likely that several molecular pathways converge to induce RGC death. Signals that promote RGC death in glaucoma might be exacerbated by risks factors, tilting the neuron's fate toward dysfunction and demise. In recent years, there has been considerable progress in our understanding of multiple pathways that lead to RGC degeneration following optic nerve injury. In this vein, several factors such as a reduced antioxidant defense system activity (Aslan et al., 2008; Moreno et al., 2004; Tezel, 2006), an increase in the nitridergic pathway activity (Belforte et al., 2007; Neufeld, 1999; Neufeld et al., 1999), and glutamate excitotoxicity (Moreno et al., 2005a), among others, have been suggested as possible additional causes for early or advanced stages of glaucomatous damage.

Unraveling which are the most critical mechanisms involved in glaucoma is unlikely to be achieved in studies which are limited to the clinically observable changes to the retina and optic nerve head (ONH) that are seen in human glaucoma. Far more detailed and invasive studies are required, preferably in a readily available animal model. An experimental model system of pressure-induced optic nerve damage would greatly facilitate the understanding of the cellular events leading to RGC death, and how they are influenced by IOP and other risk factors associated to glaucoma. Several groups have developed various ways to increase IOP in the rat eye, generally by impeding the outflow of aqueous humor (Morrison et al., 1997; Shareef et al., 1995; Ueda et al., 1998). All of these models have both advantages and disadvantages. We have developed a model of glaucoma in rats through weekly intracameral injections of 1% hyaluronic acid (HA). Weekly injections of HA in the rat anterior chamber significantly increase IOP as compared with vehicle-injected contralateral eye (Benozzi et al., 2002; Moreno et al., 2005b). Although multiple injections of HA may be needed to obtain a sustained hypertension, we have shown that the injection procedure itself does not affect IOP and retinal function and histology. On the contrary, several advantages support our model: 1) a highly consistent hypertension is achieved, 2) it may have a reasonably long course, 3) daily variations in IOP persist in HA-injected eyes, 4) in contrast to other models, in all likelihood, HA does not impede the blood flow out of the eye, and 5) it is easy to perform. This experimental model that has been validated by several groups (Foureaux et al., 2015; Mayordomo-Febrer et al., 2015; Pirhan et al., 2016), may be employed in pharmacological studies, since the HA-induced hypertension is significantly reduced by the topic and acute application of therapeutically used hypotensive drugs (Benozzi et al., 2002). The chronic administration of HA significantly decreases the scotopic electroretinographic activity and provokes a significant loss of RGCs and optic nerve fibers (Moreno et al., 2005b). Based on both functional and histological evidence, these results indicate that intracameral injections of HA in the rat eye anterior chamber appear to mimic some key features of primary open-angle glaucoma (POAG), and therefore, it may be a useful tool to understand this ocular disease and to develop new therapeutic strategies.

OXIDATIVE DAMAGE IN GLAUCOMA AND MELATONIN AS A RETINAL ANTIOXIDANT

The retina is especially susceptible to oxidative stress because of its high oxygen consumption, its high proportion of polyunsaturated fatty acids, and its exposure to light. Among others, glutathione, and antioxidant enzymes such as superoxide dismutase (SOD), catalase, and glutathione peroxidase (GPX) provide a powerful antioxidant defense in the retina (Armstrong et al., 1981; Castorina et al., 1992; Ohta et al., 1996). SOD catalyzes the conversion of superoxide radicals to hydrogen peroxide (H_2O_2), which is the first step in the metabolic defense against cellular oxidative stress. Although H_2O_2 is not a free radical, it is highly reactive, membrane permeable, and can be converted to highly reactive metabolites of oxygen such as hydroxyl radical. Under normal conditions, most of H_2O_2 molecules generated by SOD are further metabolized to water by catalase and GPX. Thus, it is critical for the cellular survival that SOD activity should be coupled with similar GPX and catalase activities to safely detoxify H_2O_2. Despite having high levels of antioxidants, the retina is still susceptible to oxidative stress which has been observed in different retinal diseases (Rajesh et al., 2003; Tanito et al., 2002; Wu et al., 1997). Several investigations have supported a role of oxidative stress in the pathogenesis of glaucoma (Izzotti et al., 2006), showing lower levels of antioxidants (Ferreira et al., 2004; Goyal et al., 2014), elevated oxidative stress markers in the aqueous humor of glaucomatous eyes (Goyal et al., 2014), antibodies against glutathione-S-transferase (Yang et al., 2001), and decreased plasma levels of glutathione (Gherghel et al., 2005). Moreover, it has been reported that the level of lipid peroxidation products increases more than two fold and that the ocular antioxidant defense mechanism decreases in the anterior chamber of patients with advanced glaucoma (Kurysheva et al., 1996). We have demonstrated that SOD and catalase activities significantly decrease in eyes with ocular hypertension induced by intracameral HA injections, which seems to be compatible with a significant increase in lipid peroxidation observed in the retina from hypertensive eyes (Moreno et al., 2004).

The possibility that melatonin could detoxify highly reactive oxygen species was originally suggested by Ianas et al. (1991). Three years later, Reiter and coworkers (Reiter et al., 1994) using spin trapping and electron resonance spectroscopy, demonstrated that melatonin has capacity to directly scavenge highly reactive hydroxyl radicals. Since then, many reports have shown that melatonin acts as a free radical scavenger and an efficient antioxidant (Hardeland et al., 1995; Pandi-Perumal et al., 2006; Reiter, 1998; Reiter et al., 1997, 2000; Turjanski et al., 1998;). Not only melatonin, but also several of its metabolites generated during its free radical scavenging action may act as antioxidants (Tan et al., 2007), greatly increasing melatonin's efficacy. Melatonin has been shown to scavenge free radicals generated in mitochondria, reduce electron leakage from the respiratory complexes and improve ATP synthesis (Acuña-Castroviejo et al., 2003; Leon et al., 2005). Moreover, melatonin preserves mitochondrial glutathione levels, thereby enhancing its antioxidant potential (Leon et al., 2004). It was demonstrated that melatonin significantly increases SOD activity and reduced glutathione (GSH) levels, whereas it decreases retinal lipid peroxidation in the rat retina (Belforte et al., 2010). By scavenging free radicals, increasing the antioxidant defense system activity, and improving the electron transport chain at mitochondrial level, melatonin is able to protect ocular tissues from oxidative damage (Lundmark et al., 2006; Siu et al., 2005). In the experimental model of glaucoma induced by injections of HA, we have demonstrated a significant decrease in retinal melatonin levels (Moreno et al., 2004). Taking into account the conclusive evidence on the role of melatonin as antioxidant, together with the decrease in retinal melatonin levels, and in the antioxidant defense system activity in hypertensive eyes, it is tempting to speculate about a causal relationship between these latter phenomena. Moreover, based on these results, it seems possible that a decrease in some of these anti-oxidant enzymatic activities may overcome the capacity of RGCs to resist oxidative damage, further supporting the involvement of oxidative stress in glaucomatous damage. Thus, manipulation of intracellular redox status using antioxidants such as melatonin may be a new therapeutic strategy to prevent glaucomatous cell death.

NITROSATIVE STRESS IN GLAUCOMA AND ANTI-NITRIDERGIC EFFECTS OF MELATONIN

NO is a ubiquitous signaling molecule that participates in a variety of cellular functions. However, in concert with reactive oxygen species, NO can be transformed into a highly potent and effective cytotoxic entity of pathophysiological significance. In fact, NO modulates the activity of various proteins that contribute to apoptosis (Melino et al., 1997). Furthermore, it was demonstrated that an extracellular proteolytic pathway in the retina contributes to RGC death via NO-activated metalloproteinase-9 (Manabe et al., 2005).

Several lines of evidence support a link between NO and glaucoma. In that vein, an increased presence of neuronal NOS (NOS-1 or nNOS) and inducible NOS (NOS-2 or iNOS), was reported in astrocytes of the lamina cribrosa and ONH of patients with glaucoma (Liu and Neufeld, 2000, 2003; Neufeld et al., 1997). In rats whose extraocular veins were cauterized to produce chronic ocular hypertension and retinal damage, expression of NOS-2 but not NOS-1 increases in ONH astrocytes (Neufeld and Liu, 2003; Shareef et al., 1999). Moreover, elevation of hydrostatic pressure *in vitro* upregulates the expression of NOS-2 in human astrocytes derived from the ONH (Liu and Neufeld, 2001). Most important, inhibition of NOS-2 by aminoguanidine or L-N[6]-[1-iminoethyl]lysine 5-tetrazole amide protects against RGC loss in a rat glaucoma model (Neufeld, 2004; Neufeld et al., 1999). These data support that activation of NOS, especially NOS-2, may play a significant role in glaucomatous optic neuropathy. However, later on, Pang et al. (2005) showed that chronically elevated IOP in rats induced by episcleral injection of hypertonic saline does not increase NOS-2 immunoreactivity in the optic nerve, ONH, or ganglion cell layer. Moreover, retinal and ONH NOS-2 mRNA levels did not correlate with either IOP level or severity of optic nerve injury. In addition, there was no difference in NOS-2 immunoreactivity in the optic nerve or ONH between POAG and non-glaucomatous eyes (Pang et al., 2005), and aminoguanidine treatment did not affect the development of pressure-induced optic neuropathy in rats (Pang et al., 2005). A significant activation of the retinal nitridergic pathway was described in the experimental model of glaucoma induced by intracameral injections of HA (Belforte et al., 2007). Despite that other studies previously addressed the issue of NO involvement in glaucoma (mostly based on Western blotting or immunohistochemical analysis), they did not assessed changes in the functional capacity of the retinal nitridergic pathway. Although no changes in the levels of NOS isoforms were observed in HA-treated eyes, a significant increase of the retinal arginine to citrulline conversion was demonstrated in HA-injected eyes (Belforte et al., 2007). The intracellular events triggered by ocular hypertension that could explain the increase in retinal NOS activity, remain to be established. However, since glutamate acting through N-methyl D-aspartate (NMDA) receptors is one of the most conspicuous activators of NOS-1 activity, the raise in glutamate synaptic levels in HA-treated eyes (as discussed below) could account for it. In this sense, it was shown that RGCs in the nNOS-deficient mouse were relatively resistant to NMDA, while damage in the retina of the endothelial NOS-deficient mouse was not distinguishable from that observed in control animals (Vorwerk et al., 1997). Moreover, it was demonstrated that intravitreal injection of NMDA in rats induces accumulation of nitrite/nitrate (El-Remessy et al., 2003).

A significant increase in the retinal uptake of L-arginine (a NOS substrate) was demonstrated in HA-treated eyes. Purified NOS from different sources has been reported to have a low half-saturating L-arginine concentration (EC50) ~10 µM. Since high levels of intracellular L-arginine ranging from 0.1 - 1 mM have been measured in many systems (Block et al., 1995), it is expected that endogenous L-arginine would support maximal activation of NOS. However, a number of *in vivo* and *in vitro* studies indicate that NO production under physiological conditions can be increased by extracellular L-arginine, despite saturating intracellular L-arginine concentrations. This has been termed "the arginine paradox" (Kurz and Harrison, 1997). One possible explanation could be that intracellular L-arginine is sequestered in one or more pools that are poorly, if at all, accessible to NOS, whereas extracellular L-arginine transported into the cells is preferentially delivered to NO biosynthesis (Kurz and Harrison, 1997). Accordingly, it was demonstrated that L-arginine availability controls NMDA-induced NO synthesis in the rat central nervous system (Grima et al., 1998). Therefore, it seems likely that to induce the activation of NOS, an obligatory influx of L-arginine is required. The coordination between NOS activity and L-arginine uptake has been demonstrated in several systems (do Carmo et al., 1998; Stevens et al., 1996). A similar coordination between NO biosynthesis and intracellular L-arginine availability seems to occur in hypertensive retinas. It was demonstrated that activation of NMDA receptors in cultured retinal cells promotes an increase of the intracellular L-arginine pool available for NO synthesis (Cossenza et al., 2006). This way, the increase in both NOS activity and L-arginine influx, could be triggered by higher levels of synaptic glutamate levels.

It was demonstrated that the uptake of L-arginine in retinas from rats and hamsters occurs through a transporter resembling the y+ system (Carmo et al., 1999; Saenz et al., 2002a). This transport system encompasses three homologous proteins (named cationic amino acid transporter (CAT)-1, CAT-2, and CAT-3) that have been characterized in several tissues. RT-PCR analysis demonstrated an increase of mRNAs for both CAT-1 and CAT-2 in retinas from hypertensive eyes, suggesting that ocular hypertension could induce an upregulation of L-arginine transporters (Belforte et al., 2007; Carmo et al., 1999).

Melatonin inhibits the nitridergic pathway activity in the golden hamster (Saenz et al., 2002b) and rat retina (Belforte et al. 2010). Melatonin significantly decreases retinal NOS activity and L-arginine uptake, and inhibits the accumulation of cGMP induced by both L-arginine and a NO donor. The inhibitory effect of melatonin on retinal NOS activity is consistent with the previously described effect of melatonin on this enzyme from other neural structures (Bettahi et al., 1996; Leon et al., 1998; Pozo et al., 1997). However, while the effect of melatonin in those tissues was evident up to 1 nM, a much higher sensitivity to the methoxyindole is evident in the hamster retina, since it is effective even at 1 pM, suggesting that the retinal nitridergic pathway is regulated by physiological concentrations of melatonin (Sáenz et al., 2002b). In addition to inhibiting NOS activity, melatonin is able to directly scavenge NO, generating at least one stable product, i.e., N-nitrosomelatonin (Turjanski et al., 2000). Moreover, melatonin reduces NO-induced lipid peroxidation in rat retinal homogenates and ileum tissue sections (Cuzzocrea et al., 2000; Siu et al., 1999). Taken together, these results indicate that melatonin modulates the nitridergic pathway in an opposite way to that induced by ocular hypertension.

EXCITOTOXICITY IN GLAUCOMA AND THE EFFECT OF MELATONIN ON GLUTAMATE RETINAL SYNAPTIC LEVELS

Glutamate is the main excitatory neurotransmitter in the retina, but it is toxic when present in excessive amounts. Retinal tissue is in fact, an established paradigm for glutamate neurotoxicity for several reasons: insult leads to accumulation of relatively high levels of glutamate in the extracellular fluid (Louzada-Junior et al., 1992), administration of glutamate leads to neuronal cell death (David et al., 1988); and glutamate receptor antagonists can protect against neuronal degeneration (Mosinger et al., 1991). Thus, an appropriate clearance of synaptic glutamate is required for the normal function of retinal excitatory synapses and for prevention of neurotoxicity. Glial cells, mainly astrocytes and Müller glia, surround glutamatergic synapses, and express glutamate transporters and the glutamate-metabolizing enzyme, glutamine synthetase (GS) (Riepe and Norenburg, 1977; Sarthy and Lam, 1978). Glutamate is transported into glial cells and amidated by GS to the non-toxic aminoacid glutamine. Glutamine is then released by the glial cells and taken up by neurons, where it is hydrolyzed by glutaminase to form glutamate again, completing the retinal glutamate/glutamine cycle (Sarthy and Lam, 1978; Thoreson and Witkovsky, 1999). In this way, the neurotransmitter pool is replenished and glutamate neurotoxicity is prevented.

Several studies suggested that excitotoxicity plays a key role in RGC degeneration associated with glaucoma through the overactivation of both NMDA and non-NMDA glutamate receptors (Nucci et al., 2005; Russo et al., 2009). Indeed, despite the doubts that have been raised regarding the effective role of glutamate in glaucoma (Salt and Cordeiro, 2006), several authors reported neuroprotection following treatment with NMDA and non-NMDA antagonists in experimental models of RGC death (Adachi et al., 1998; Nucci et al., 2005; Sucher et al., 1997). The involvement of glutamate in glaucoma is supported by the demonstration that vitreal glutamate levels are elevated in glaucomatous dogs (Brooks et al., 1997), and quail (Brooks et al., 1997; Dkhissi et al., 1999) with congenital glaucoma. In addition, high glutamine levels have been found in retinal Müller cells of glaucomatous rat eyes (Brooks et al., 1997; Shen et al., 2004). In contrast, other authors show no significant elevation of glutamate in the vitreous of patients with glaucoma (Honkanen et al., 2003), or in rats (Levkovitch-Verbin et al., 2002), and monkeys with experimental glaucoma (Carter-Dawson et al., 2002; Wamsley et al., 2005). In any case, it seems limited a viewpoint to assume that high levels of glutamate in the vitreous are a necessary condition for excitotoxicity to be involved in glaucomatous neuropathy. The local concentration of glutamate at the membrane receptors of RGCs is the important issue for toxicity. This could be very different from the level in vitreous samples. Vitreous humor must be removed for experimental measurement by a process that inevitably disturbs its state before removal. These manipulations could themselves alter the measured amount of glutamate. At present, there are no available tools to directly assess retinal glutamate synaptic concentrations *in vivo*. However, glutamate synaptic concentrations could be estimated by studying the retinal mechanisms that regulate glutamate clearance and recycling. We have demonstrated a significant alteration of the retinal glutamate/glutamine cycle activity in rats exposed to experimentally elevated IOP (Moreno et al., 2005a). Since no enzymes exist extracellularly that degrade glutamate, glutamate transporters are responsible for maintaining low synaptic glutamate concentrations. Retinal glutamate uptake significantly decreases in HA-treated eyes. In agreement, a significant reduction in the amount of the main retinal glutamate transporter (excitatory amino acid transporter type 1, (EAAT-1)) assessed by Western blot analysis in a rat glaucoma model (Martin et al., 2002), and a downregulation of this transporter in retinal Müller cells from glaucoma patients (Naskar et al., 2000) were demonstrated. While these studies did not assess changes in the functional capacity of glutamate transporters, our results demonstrate a removing glutamate disability in retinas from hypertensive eyes. The synaptically released glutamate is taken up into glial cells, where GS converts it into glutamine. Since Müller cells rapidly convert glutamate to glutamine, the driving force for glutamate uptake would be stronger in these cells than in neurons, which have much higher intracellular free glutamate concentrations (Pow and Robinson, 1994). In fact, although glutamate uptake is controlled by the expression and post-translational modifications, physiological measurements suggest that glutamate uptake may also depend on its metabolism (Gegelashvili and Schousboe, 1998; Tanaka, 2000). Indeed, an increase in internal glutamate concentrations significantly slows down the net transport of glutamate, and it was suggested that instantaneous intracellular glutamate metabolism may be needed for efficient glutamate clearance of the extracellular milieu (Attwell et al., 1993; Otis and Jahr, 1998). Thus, a decrease in GS activity could account for a decrease in glutamate uptake. Glutamine is released from Müller cells and could be a precursor for neuronal glutamate synthesis. The increase in the basal release and the uptake of glutamine in HA-treated eyes could provoke a raise in the availability of substratum for glutamate synthesis. Moreover, this increase in glutamate production could be further potentiated by the decrease in GS activity. Decreasing the expression levels of EAAT1 increases vitreal glutamate, and is toxic to RGCs (Vorwerk et al., 2000). Thus, the decrease in glutamate influx could provoke an increase in synaptic glutamate levels. In addition, a decrease of GS activity, as well as an increase in glutaminase activity in retinas form hypertensive eyes could contribute synergically and/or redundantly to an excessive increase in synaptic glutamate levels (Moreno et al., 2008), which could be involved in glaucomatous RGC loss.

Nanomolar concentrations of melatonin significantly modulate the glutamate/glutamine cycle activity in the golden hamster (Sáenz et al., 2004) and rat (Belforte et al., 2010) retina. In that sense, it was demonstrated that low concentrations of melatonin significantly increase retinal glutamate uptake and GS activity, and decrease glutaminase activity. This way, melatonin may contribute to the conversion of glutamate to glutamine through a possibly redundant mechanism. The physiological consequences of a modulation by melatonin of the retinal glutamate/glutamine cycle are yet to be determined, although this effect could provide new insights into the neuroprotective potential of melatonin. In that respect, it was demonstrated that an increase in GS provides neuroprotection in experimental models of neurodegeneration (Gorovits et al., 1997; Heidinger et al., 1999). Induction of GS *in vivo* or *in vitro* by glucocorticoids was clearly demonstrated in different tissues, including the retina (Patel et al., 1983; Sarkar and Chaudhury, 1983). Physiological levels of glucocorticoids regulate GS expression by stimulating the gene transcription. This effect of glucocorticoids has been associated to their ability to protect against neuronal degeneration (Gorovits et al., 1997), as shown in animal models of brain injury (Hall, 1985), as well as after retinal photic injury (Rosner et al., 1992). However, since induction of GS expression by glucocorticoids takes about 24 h, there are some potential weaknesses in glucocorticoid treatment. In contrast, since the effect of melatonin is much faster (in the range of minutes) a treatment with the methoxyindole may circumvent this obstacle. Furthermore, this beneficial effect of melatonin may be further improved by its effect on glutamate uptake, and glutaminase activity. In summary, these findings suggest that a treatment with melatonin could be considered as a new approach to handling glutamate-mediated neuronal degeneration, such as that induced by glaucoma.

EFFECT OF MELATONIN ON GLAUCOMATOUS DAMAGE

As already discussed in this chapter, previous results indicate that melatonin is able to reduce oxidative stress, decrease NO levels and impair retinal glutamate neurotoxicity, and that an increase in oxidative damage, NO and glutamate levels, could be involved in glaucomatous neuropathy. Thus, it seems likely that melatonin could have a beneficial effect against glaucoma. This hypothesis has been analyzed in the glaucoma model induced by HA injections (Belforte et al., 2010). For this purpose, a pellet of melatonin was implanted subcutaneously 24 h before the first injection of HA, and it was replaced every 15 days. Melatonin, which did not affect IOP in this experimental model, prevented the effect of ocular hypertension on retinal function (assessed by electroretinography), and diminished the vulnerability of RGCs to the deleterious effects of ocular hypertension (Belforte et al., 2010). Although melatonin conferred neuroprotection in the experimental model of glaucoma induced by HA, the translational relevance of this result is limited by the fact that melatonin was administered before the induction of ocular hypertension (e.g., 24 h before the first injection of HA). To analyze the therapeutic effect of melatonin, a pellet of the methoxyindole was implanted at 6 weeks of ocular hypertension, a time point in which functional alterations are already evident. The results indicate that the delayed treatment with melatonin of eyes with ocular hypertension resulted in similar protection when compared with eyes treated from the onset of ocular hypertension. We do not have any clear explanation for these results. We have shown that an increase in synaptic glutamate concentrations (Moreno et al., 2005a) and NO production occur mostly prior to 6 weeks of ocular hypertension induced by HA. Based on these results, it seems possible that alterations in glutamate and NO may trigger an initial insult responsible for initiation of damage that is followed by a slower secondary degeneration that ultimately results in cell death. In that sense, we showed that oxidative stress is a longer lasting phenomenon which can be observed even at 10 weeks of ocular hypertension (Moreno et al., 2004). In this scenario, the preventive effect of melatonin (shown by the administration of melatonin before the first injection of HA) could be explained by the decrease in glutamate and NO levels, while its therapeutic effect (shown by the administration of melatonin at 6 weeks of ocular hypertension) can be explained essentially by its antioxidant effect. In this context, the fact that melatonin was similarly effective in the chronically treated animals and the delayed treatment, could support the hypothesis that in both cases melatonin is able to reverse oxidative damage which could be a key factor in glaucomatous dysfunction and cell death.

Besides the mechanisms already described, there are other beneficial mechanisms of melatonin which can support its usefulness for glaucoma treatment. Several lines of evidence support that the obstruction of retrograde transport at the ONH results in the deprivation of neurotrophic support to RGCs, leading to apoptotic cell death in glaucoma (Tang et al., 2006). An important corollary to this concept is the implication that appropriate enhancement of neurotrophic support will prolong the survival of injured RGCs. Of particular importance is the fact that brain-derived neurotrophic factor (BDNF) not only promotes RGC survival following damage to the optic nerve, but also helps to preserve the structural integrity of the surviving neurons, which in turn results in enhanced visual function (Weber et al., 2008). As for the link between melatonin and neurotrophins, it has been suggested that melatonin may participate in neurodevelopment and in the regulation of neurotrophic factors (Jimenez-Jorge et al., 2007; Niles et al., 2004). *In vitro*, melatonin promotes the viability and neuronal differentiation of neural stem cells and increases their production of BDNF (Kong et al., 2008). Moreover, ramelteon (a melatonin receptor agonist) is capable of increasing BDNF protein in primary cultures of cerebellar granule cells (Imbesi et al., 2008). Finally, while the cellular mechanisms involved in RGC loss observed in glaucomatous neuropathy are based on apoptosis phenomenon, melatonin was shown to have anti-apoptotic properties acting through several mechanisms, such as reduction of caspases, cytochrome c release, and modulation of Bcl-2 and Bax genes, among others (Hoijman et al., 2004; Zhao et al., 2015, 2017).

Melatonin has been involved in IOP regulation. In this line, it has been shown that a dysfunction of melatonin signaling triggers elevation of nocturnal IOP and loss of RGCs (Tosini et al., 2013). Moreover, exogenous melatonin and its analogues lower IOP in mammals, including humans (Crooke et al., 2012a; Ismail and Mowafi, 2009; Martínez-Águila et al., 2013; Pescosolido et al., 2015; Pintor et al., 2001, 2003; Samples et al., 1988; Serle et al., 2004). This ocular hypotensive effect seems to be mediated by ciliary body receptors and an unidentified receptor (classically named MT3) and final inhibition of ciliary chloride secretion (Alarma-Estrany et al., 2008, 2009; Huete-Toral et al., 2015; Vincent et al., 2010). Furthermore, melatonergic compounds provide a sustained reduction in IOP through regulation of ciliary genes expression and potentiate the hypotensive action of classic anti-glaucomatous drugs such as timolol and brimonidine (Crooke et al., 2011, 2012b, 2013).

Neuroprotection in glaucoma implies the use of drugs or chemicals to slow down whatever causes loss of vision (RGC death), without influencing IOP. In order to be effective, a neuroprotectant must reach the ONH and/or RGCs and will therefore probably have to be taken orally (Osborne et al., 1999). Because it will reach other parts of the body, any side-effect of an appropriate neuroprotectant must be reduced to a minimum. Melatonin, a very safe compound for human use, is highly lipophilic and readily diffuses into tissues. In fact, subcutaneously administered, melatonin reaches the retina (Sande et al., 2008), increasing the local levels of the methoxyindole. RGCs are induced to die by different triggers in glaucoma, suggesting that neuroprotectants with multiple modes of actions such as melatonin, are likely be effective in the therapeutic management of glaucoma (Marcic et al., 2003). Taking together, these data indicate that melatonin seems to fulfill all the requirements to be considered a promissory neuroprotectant for glaucoma treatment. Alone or combined with an ocular hypotensive therapy, a treatment with melatonin could be a new therapeutic tool helping the challenge faced by ophthalmologists treating glaucoma.

OPTIC NEURITIS

Optic neuritis (ON), the most common optic neuropathy affecting young adults, is a condition involving primary inflammation, demyelination, and axonal injury in the optic nerve which leads to RGC loss and visual dysfunction (reviewed by Toosy et al., 2014). Typically, ON affects young adults ranging from 18 to 45 years of age, and also children as young as 4 years (Wilejto et al., 2006). The annual incidence of ON is approximately 5 in 100,000, with a prevalence estimated to be 115 in 100,000 (Martínez-Lapiscina et al., 2014). Clinical features of ON include peri- or retro-ocular pain accentuated by eye movement, abnormal visual acuity and visual field, reduced color vision, a relative afferent pupillary defect, and abnormal visual evoked potentials (VEPs) (Hickman et al., 2002; Toosy et al., 2014). The fundoscopy may appear normal or demonstrate edema of the optic nerve head (papillitis) (Costello, 2013; Kaufman et al., 2000). The visual deficit of ON may worsen over 1 to 2 weeks, and usually begins improving over the next months. However, although the majority of patients recover after several weeks, in ~ 40% of these patients, vision loss remains permanent (Beck et al., 1992; Horstmann et al., 2013). Even if visual acuity fully improves, most patients have some residual visual function deficits such as disturbances of visual acuity, contrast sensitivity, color vision, visual field, stereopsis, pupillary contraction, and VEPs (Kupersmith et al., 1983; Kaufman et al., 2000). Some studies indicate that the loss of vision after an episode of ON correlates with RGC death (Costello et al., 2006; Trip et al., 2005).

ON has many causes; it may be associated to a range of autoimmune or infectious diseases (Horwitz et al., 2014; Toosy et al., 2014), and it is closely linked to multiple sclerosis (MS). In fact, ON is the initial symptom of MS in 25% of cases and may occur during the disease in about 70%, usually in the relapsing-remitting phase (Toosy et al., 2014). In addition, a broad range of conditions can give rise to or mimic ON. Several infectious and systemic diseases such as sarcoidosis (Constantino et al., 2000), systemic lupus erythematosus (Frigui et al., 2011), and inflammatory bowel disease (Felekis et al., 2009), as well as HIV (Kallenbach and Frederiksen, 2008), and Lyme disease (Blanc et al., 2010), among others (Eggenberger, 2001; Hickman et al., 2002) have been reported to cause ON. On the other hand, acute ON often occurs as an isolated clinical event without contributory systemic abnormalities, and it is retrospectively diagnosed as idiopathic (or primary) ON (Hickman et al., 2002; Optic Neuritis Study Group, 2008). Based on the clinical association between MS and ON, the most commonly used experimental model for ON studies is experimental autoimmune encephalomyelitis

(EAE), an established model of human MS, which involves an immune-mediated demyelination process. EAE-ON can be induced by immunization with different myelin antigens, such as myelin basic protein, proteolipid protein, and myelin oligodendrocyte glycoprotein (Furlan et al., 2009; Rangachari and Kuchroo, 2013; Simmons et al., 2013). In fact, when it is associated with MS, ON occurs as an inflammatory demyelinating disorder of the optic nerve (Toosy et al., 2014). However, since the optic nerve lesions in EAE are always associated with inflammation and demyelination of the brain and spinal cord, this model does not mimic the primary form of ON. Moreover, the optic nerve damage in EAE is relatively unpredictable, with different animals developing varying severities of ON (Bettelli et al., 2003; Shindler et al., 2006). Experimental ON was also induced by the injection of lysolecithin, a major component of oxidized low-density lipoproteins, which has a deterrent effect on myelin and myelinating cells. Lysolecithin has been successfully used to induce optic nerve demyelination in primates and rats (Lachapelle et al., 2005; You et al., 2011). However, although the lysolecithin model reproduces the focal demyelination in the optic nerve, this ON model lacks the primary inflammatory component, which is characteristic of human ON. Experimentally, inflammation can be induced in most tissues by the local injection of bacterial lipopolysaccharide (LPS). LPS affects microglia and astrocytes (reviewed by Cunningham, 2013), and produces the activation of glial cells *in vivo* after its intracerebral injection (Bourdiol et al., 1991; Herrera et al., 2000). Moreover, a substantial demyelination resulting from the focal inflammatory lesion caused by the injection of LPS directly into the rat dorsal funiculus was demonstrated (Felts et al., 2005). Based on these antecedents, we have developed a new experimental model of primary ON through the microinjection of LPS into the rat optic nerve (Aranda et al., 2015). LPS induces a significant and persistent decrease in VEP amplitude and pupil light reflex (PLR) (classical signs of human ON), without changes in the electroretinogram. In addition, LPS induces a deficit in anterograde transport, and an early inflammatory response consisting in an increased cellularity, and Iba-1 and ED1-immunoreactivity in the optic nerve, which are followed by changes in axonal density, astrocytosis, demyelination, and axon and RGC loss. These results suggest that the microinjection of LPS into the optic nerve may serve as a new experimental model of primary ON.

CURRENT THERAPY FOR OPTIC NEURITIS

Corticosteroids are the current mainstays of therapy for the treatment of ON. Results of the North American Optic Neuritis Treatment Trial, showed no improvement in visual acuity at 6 months after 3 days of high-dose (1 g per day) intravenous methylprednisolone followed by 11 days of low-dose oral prednisolone versus placebo, although visual recovery was faster (Beck et al., 1992, 2008), and mild benefits were noted for some secondary outcomes (visual fields, contrast sensitivity and color vision). Patients taking standard-dose (1 mg per kg) oral prednisolone did not differ from those taking placebo in visual outcomes, but there was an unexpected increased risk of ON recurrence for reasons that are still unclear. Intravenous methylprednisolone also delayed the onset of clinically definite MS at 2 years (Beck et al., 1993), but this difference decreased over time (Optic Neuritis Study Group, 2008). In a Cochrane review, no long-term benefit was reported for corticosteroid treatment for visual acuity, visual fields, or contrast sensitivity (Gal et al., 2012). In a rat model of MS, corticosteroids given before induction of experimental allergic encephalomyelitis reduce the incidence of ON and preserve RGCs after ON (Dutt et al., 2010). However, in another experimental model of ON, methylprednisolone even increases RGC degeneration (Diem et al., 2008). In humans, poor evidence for the efficacy of corticosteroids exists, and only results of uncontrolled studies examining very early treatment (within days of symptoms) have been published (Toosy et al., 2014). For other therapeutic approaches such as plasmaferesis or intravenous immunoglobulins, the evidence for a beneficial effect is still weak (Toosy et al., 2014). In summary, currently there are no therapeutic strategies able to improve the visual outcome in ON, and the development of therapies with the potential to prevent neuroaxonal loss following ON remains a significant unmet clinical need.

MELATONIN IN OPTIC NEURITIS

Recently, we have analyzed the effect of melatonin on the optic nerve and retinal alterations in LPS-induced ON. For this purpose, vehicle or LPS were injected into the optic nerve from adult male *Wistar* rats. One group of animals received a subcutaneous pellet of 20 mg melatonin at 24 h before vehicle or LPS injection, and another group was submitted to a sham procedure (Aranda et al., 2016). The results indicate that melatonin could improve functional signs of ON (VEPs and PRL). The functional protection was consistent with improved histopathologic outcomes: reduced microglial/macrophage reactivity, astrocytosis, demyelination, and axon and RGC loss (Figure 1). A defective axonal transport was described in mice with EAE-ON (Lin et al., 2014) and LPS-induced ON (Aranda et al., 2015), whereas melatonin avoids the misconnection between retina and its main synaptic target in rodents (the superior colliculus).

Microglial/macrophage activation is a key component of the inflammatory response. Increasing evidence indicates that an early inflammatory response contributes to the late stages of brain injury that result in neurological function loss (Kaushik et al., 2013), and it has been demonstrated that microglial activation and their progression towards a phagocytic state can lead to progressive and cumulative neuronal cell loss (Neher et al., 2013). Iba-1 labels both quiescent and activated microglia/macrophage (providing an index of microglial/macrophage density), whereas abundance of the lysosomal antigen ED1 offers a measure of microglial phagocytic activity (Ebneter et al., 2010). Although no changes in Iba-1(+) area were detected in LPS-injected optic nerves, melatonin decreased phagocytic microglial activation (as shown by a decrease in ED-1 immunoreactivity) induced by experimental ON. In addition, melatonin significantly prevents the reactive gliosis and the occurrence of demyelination in the optic nerve induced by LPS injection. Although we could not ascertain whether the protection induced by melatonin primarily occurs in axons or glial elements, our results suggest that alterations in the crosstalk between axons and glial cells induced by experimental ON can be positively affected by melatonin. In addition, since axonal damage often results in axonal degeneration and permanent loss of the cell body, the preservation of RGC number is consistent with the prevention of axon loss induced by melatonin. Improper upregulation of inflammatory signals such as tumor necrosis factor α (TNFα), cyclooxygenase-2 (COX-2) and NOS-2 has been associated with the pathophysiology of EAE-ON (Das et al., 2013). Moreover, oxidative stress is known to induce neuronal damage in experimental ON (Qi et al., 2007). Melatonin prevents the increase in NOS-2, COX-2, and TNFα levels, which could be consistent with the reduced microglial activation, and optic nerve lipid peroxidation, which is consistent with its well-known antioxidant activity. Thus, it seems likely that melatonin may attenuate the severity of experimental ON by exerting anti-inflammatory and antioxidant effects. Established strategies for the treatment of EAE-ON mainly target the autoimmune response by using anti-inflammatory, immunomodulatory and immunosuppressive agents, but none of them have clear neuroprotective properties. Since a significant RGC loss seems to be the main cause of permanent visual function damage in both experimental models of ON and patients with ON, an optimal therapeutic strategy may require a combination of a immunosuppressive and anti-inflammatory treatment combined with a neuroprotective agent (Li et al., 2014). Our results support that melatonin could be able to achieve these requirements, since it suppress visual impairment, prevents optic nerve structural alterations and RGC loss, and decreases inflammatory signals and oxidative stress induced by experimental ON. Since ON patients do not have symptoms until inflammation begins, the therapeutic benefit of the preventive effect of melatonin in LPS-induced ON has limited translational relevance. Thus, we analyzed whether melatonin also preserves visual function when the treatment is initiated after the onset of ON. A delayed treatment with melatonin (administered at 4 days post-injection of LPS) significantly reduces the visual dysfunction (VEP and PLR) induced by experimental ON, supporting that melatonin not only prevents, but also reduces the progression of optic nerve functional damage (Aranda et al., 2016). Overall, these results demonstrate an ability of melatonin to significantly preserve visual functions in experimental ON, with reduction of inflammation, astrocytosis, demyelination, and axon and RGC loss. These neuroprotective effects may involve anti-inflammatory and/or antioxidant effects; and by either mechanism, melatonin, a very safe compound for human use, could represent a promising potential therapy for primary ON.

CONCLUSION

Glaucoma and ON are the two major degenerative optic neuropathies, sharing a common feature which is RGC loss and axonal damage. RGCs bear the sole responsibility of propagating visual stimuli to the brain. Their axons, which make up the optic nerve, project from the retina to the brain through the lamina cribrosa and in rodents, decussate almost entirely at the optic chiasm before reaching central synaptic targets. For degenerative ocular conditions, particularly glaucoma and ON, the dysfunction and/or loss of RGC seems to be the primary determinant of irreversible visual loss. Glaucoma and ON, prevalent visual dysfunctions with potentially blinding sequels remain a challenging field to ophthalmologists, as both diseases cause significant morbidity and the use of traditional forms of treatment is restricted by limited effectiveness and/or considerable side effects. Glaucoma and ON differ in several aspects, including causes, risk factors, and cell types involved, among many others. However, several lines of evidence support that both diseases share some of pathogenic mechanisms such as oxidative damage, and increased NO levels. Melatonin by itself exhibits antioxidant, antinitridergic and anti-inflammatory properties, and has the ability to reduce glutamate synaptic levels. The results obtained using experimental models of glaucoma and ON in rats strongly support that melatonin, which lacks of adverse collateral effects even at high doses, could be a promissory resource in the management of both ocular neuropathies. Moreover, as already mentioned, a post-treatment with melatonin significantly attenuates ocular alterations induce by ocular hypertension or optic nerve inflammation, supporting that melatonin not only prevents but also reduces the progression of glaucoma and ON. In summary, melatonin appears as a potentially useful drug, particularly in ophthalmology, and as an alternative or eventually as complement to traditional treatments, since it resulted to be highly effective and it does not present side effects. Further investigations of the mechanisms by which melatonin affects ocular targets may unveil new concepts for its inclusion in ophthalmological therapy.

REFERENCES

Acuña-Castroviejo, D., Escames, G., León, J., Carazo, A. & Khaldy, H. (2003). Mitochondrial regulation by melatonin and its metabolites. *Adv Exp Med Biol, 527*, 549-557.

Acuña-Castroviejo, D., Escames, G., Venegas, C., Díaz-Casado, M. E., Lima-Cabello, E., López, L. C., Rosales-Corral, S., Tan, D. X. & Reiter, R. J. (2014). Extrapineal melatonin: sources, regulation, and potential functions. *Cell Mol Life Sci, 71*(16), 2997-3025. doi: 10.1007/s00018-014-1579-2.

Adachi, K., Kashii, S., Masai, H., Ueda, M., Morizane, C., Kaneda, K., Kume, T., Akaike, A. & Honda, Y. (1998). Mechanism of the pathogenesis of glutamate neuro toxicity in retinal ischemia. *Graefes Arch Clin Exp Ophthalmol*, *236*(10), 766-774.

Alarma-Estrany, P., Crooke, A., Mediero, A., Peláez, T. & Pintor, J. (2008). Sympathetic nervous system modulates the ocular hypotensive action of MT2-melatonin receptors in normotensive rabbits. *J Pineal Res*, *45*(4), 468-475. doi: 10.1111/j.1600-079X.2008.00618.x.

Alarma-Estrany, P., Crooke, A. & Pintor, J. (2009). 5-MCA-NAT does not act through NQO2 to reduce intraocular pressure in New-Zealand white rabbit. *J Pineal Res*, *47*(2), 201-209. doi: 10.1111/j.1600-079X.2009.00702.x.

Almasieh, M., Wilson, A. M., Morquette, B., Cueva Vargas, J. L. & Di Polo, A. (2012). The molecular basis of retinal ganglion cell death in glaucoma. *Prog Retin Eye Res*, *31*(2), 152-81. doi: 10.1016/ j.preteyeres.2011.11.002.

Aranda, M. L., Dorfman, D., Sande, P. H. & Rosenstein, R. E. (2015). Experimental optic neuritis induced by the microinjection of lipopolysaccharide into the optic nerve. *Exp Neurol*, *266*, 30-41. doi: 10.1016/j.expneurol.2015.01.010.

Aranda, M. L., González Fleitas, M. F., De Laurentiis, A., Keller Sarmiento, M. I., Chianelli, M., Sande, P. H., Dorfman, D. & Rosenstein, R. E. (2016). Neuroprotective effect of melatonin in experimental optic neuritis in rats. *J Pineal Res*, *60*(3), 360-372. doi: 10.1111/jpi.12318.

Armstrong, D., Santangelo, G. & Connole, E. (1981). The distribution of peroxide regulating enzymes in the canine eye. *Curr Eye Res*, *1*(4), 225-242.

Aslan, M., Cort, A. & Yucel, I. (2008). Oxidative and nitrative stress markers in glaucoma. *Free Radic Biol Med*, *45*(4), 367-376. doi: 10.1016/ j.freeradbiomed.2008.04.026.

Attwell, D., Barbour, B. & Szatkowski, M. (1993). Nonvesicular release of neurotransmitter. *Neuron*, *11*(3), 401-407.

Beck, R. W., Cleary, P. A., Anderson, M. M. Jr, Keltner, J. L., Shults, W. T., Kaufman, D. I., Buckley, E. G., Corbett, J. J., Kupersmith, M. J., Miller, N. R., Savino, P. J., Guy, J. R., Trobe, J. D., McCrary, J. A. III., Smith, C. H., Chrousos, G. A., Thompson, H. S., Katz, B. J., Brodsky, M. C., Goodwin, J. A. & Atwell, C. W. (1992). The Optic Neuritis Study Group. A randomized, controlled trial of corticosteroids in the treatment of acute optic neuritis. *N Engl J Med*, *326*(9), 581-588.

Beck, R. W., Cleary, P. A., Trobe, J. D., Kaufman, D. I., Kupersmith, M. J., Paty, D. W. & Brown, C. H. (1993). The Optic Neuritis Study Group. The effect of corticosteroids for acute optic neuritis on the subsequent development of multiple sclerosis. *N Engl J Med*, *329*(24), 1764-1769.

Beck, R. W. & Gal, R. L. (2008). Treatment of acute optic neuritis: a summary of findings from the optic neuritis treatment trial. *Arch Ophthalmol*, *126*(7), 994-995. doi: 10.1001/archopht.126.7.994.

Belforte, N., Moreno, M. C., Cymeryng, C., Bordone, M., Keller Sarmiento, M. I. & Rosenstein, R. E. (2007). Effect of ocular hypertension on retinal nitridergic pathway activity. *Invest Ophthalmol Vis Sci*, *48*(5), 2127-2133. doi: 10.1167/iovs.06-1229.

Belforte, N. A., Moreno, M. C., de Zavalía, N., Sande, P. H., Chianelli, M. S., Keller Sarmiento, M. I. & Rosenstein, R. E. (2010). Melatonin: a novel neuroprotectant for the treatment of glaucoma. *J Pineal Res*, *48*(4), 353-364. doi: 10.1111/j.1600-079X.2010.00762.x.

Benozzi, J., Nahum, L. P., Campanelli, J. L. & Rosenstein, R. E. (2002). Effect of hyaluronic acid on intraocular pressure in rats. *Invest Ophthalmol Vis Sci*, *43*(7), 2196-2200.

Besharse, J. C. & Dunis, D. A. (1983). Methoxyindoles and photoreceptor metabolism: activation of rod shedding. *Science*, *219*(4590), 1341-1343.

Bettahi, I., Pozo, D., Osuna, C., Reiter, R. J., Acuña-Castroviejo, D. & Guerrero, J. M. (1996). Melatonin reduces nitric oxide synthase activity in rat hypothalamus. *J Pineal Res*, *20*(4), 205-210.

Bettelli, E., Pagany, M., Weiner, H., Linington, C., Sobel, R. & Kuchroo, V. (2003). Myelin oligodendrocyte glycoprotein-specific T cell receptor transgenic mice develop spontaneous autoimmune optic neuritis. *J Exp Med*, *197*(9), 1073-1081. doi: 10.1084/jem.20021603.

Blanc, F., Ballonzoli, L., Marcel, C., De, M. S., Jaulhac, B. & De, S. J. (2010). Lyme optic neuritis. *J Neurol Sci*, *295*(1-2), 117-119. doi: 10.1016/j.jns.2010.05.009.

Block, E. R., Herrera, H. & Couch, M. (1995). Hypoxia inhibits L-arginine uptake by pulmonary artery endothelial cells. *Am J Physiol*, *269*(5Pt1), L574-L580.

Brooks, D. E., Garcia, G. A., Dreyer, E. B., Zurakowski, D. & Franco-Bourland, R. E. (1997). Vitreous body glutamate concentration in dogs with glaucoma. *Am J Vet Res*, *58*(8), 864-867.

Cahill, G. M. & Besharse, J. C. (1993). Circadian clock functions localized in xenopus retinal photoreceptors. *Neuron*, *10*(4), 573-577.

Carmo, A., Cunha-Vaz, J. G., Carvalho, A. P. & Lopes, M. C. (1999). L-arginine transport in retinas from streptozotocin diabetic rats: correlation with the level of IL-1 beta and NO synthase activity. *Vision Res*, *39*(23), 3817-3823.

Carter-Dawson, L., Crawford, M. L., Harwerth, R. S., Smith, E. L., 3rd, Feldman, R. & Shen, F. F. (2002). Vitreal glutamate concentration in monkeys with experimental glaucoma. *Invest Ophthalmol Vis Sci*, *43*(8), 2633-2637.

Castorina, C., Campisi, A., Di Giacomo, C., Sorrenti, V., Russo, A. & Vanella, A. (1992). Lipid peroxidation and antioxidant enzymatic systems in rat retina as a function of age. *Neurochem Res*, *17*(6), 599-604.

Cedrone, C., Mancino, R., Ricci, F., Cerulli, A., Culasso, F. & Nucci, C. (2012). The 12-year incidence of glaucoma and glaucoma-related visual field loss in Italy: the Ponza eye study. *J Glaucoma*, *21*(1), 1-6. doi: 10.1097/IJG.0b013e3182027796.

Cesareo, M., Ciuffoletti, E., Ricci, F., Missiroli, F., Giuliano, M. A., Mancino, R. & Nucci, C. (2015). Visual disability and quality of life in glaucoma patients. *Prog Brain Res*, *221*, 359-74. doi: 10.1016/ bs. pbr.2015.07.003.

Constantino, T., Digre, K. & Zimmerman, P. (2000). Neuro-ophthalmic complications of sarcoidosis. *Semin Neurol*, *20*(1), 123-137.

Cossenza, M., Cadilhe, D. V., Coutinho, R. N. & Paes-de-Carvalho, R. (2006). Inhibition of protein synthesis by activation of NMDA receptors in cultured retinal cells: a new mechanism for the regulation of nitric oxide production. *J Neurochem*, *97*(5), 1481-1493. doi:10. 1111/j.1471-4159.2006.03843.x.

Costello, F. (2013). The afferent visual pathway: designing a structural–functional paradigm of multiple sclerosis. *ISRN Neurol*, 2013, 134858. doi: 10.1155/2013/134858.

Costello, F., Coupland, S., Hodge, W., Lorello, G. R., Koroluk, J., Pan, Y. I., Freedman, M. S., Zackon, D. H. & Kardon, R. H. (2006). Quantifying axonal loss after optic neuritiswith optical coherence tomography. *Ann Neurol*, *59*(6), 963-969. doi:10.1002/ana.20851.

Crooke, A., Huete-Toral, F., Martínez-Águila, A., Alarma-Estrany, P. & Pintor, J. (2011). Regulation of ocular adrenoceptor genes expression by 5-MCA-NAT: implications for glaucoma treatment. *Pharmacogenet Genomics*, *21*(9), 587-589. doi: 10.1097/ FPC. 0b013e32834910d1.

Crooke, A., Colligris, B. & Pintor, J. (2012a). Update in glaucoma medicinal chemistry: emerging evidence for the importance of melatonin analogues. *Curr Med Chem*, *19*(21), 3508-3522.

Crooke, A., Huete-Toral, F., Martínez-Águila, A., Martín-Gil, A. & Pintor, J. (2012b). Involvement of carbonic anhydrases in the ocular hypotensive effect of melatonin analogue 5-MCA-NAT. *J Pineal Res*, *52*(3), 265-270. doi: 10.1111/j.1600-079X.2011.00938.x.

Crooke, A., Huete-Toral, F., Martínez-Águila, A., Martín-Gil, A. & Pintor, J. (2013). Melatonin and its analog 5-methoxycarbonylamino-N-acetyltryptamine potentiate adrenergic receptor-mediated ocular hypotensive effects in rabbits: significance for combination therapy in glaucoma. *J Pharmacol Exp Ther*, *346*(1), 138-145. doi: 10.1124/ jpet.112.202036.

Cunningham, C. (2013). Microglia and neurodegeneration: the role of systemic inflammation. *Glia*, *61*(1), 71-90. doi: 10.1002/glia.22350.

Cuzzocrea, S., Costantino, G., Mazzon, E., Micali, A., De Sarro, A. & Caputi, A. P. (2000). Beneficial effects of melatonin in a rat model of splanchnic artery occlusion and reperfusion. *J Pineal Res*, *28*(1), 52-63.

Das, A., Guyton, M. K., Smith, A., Wallace, G. 4th., McDowell, M. L., Matzelle, D. D., Ray, S. K. & Banik, N. L. (2013). Calpain inhibitor attenuated optic nerve damage in acute optic neuritis in rats. *J Neurochem*, *124*(1), 133-146. doi: 10.1111/jnc.12064.

David, P., Lusky, M. & Teichberg, V. I. (1988). Involvement of excitatory neurotransmitters in the damage produced in chick embryo retinas by anoxia and extracellular high potassium. *Exp Eye Res*, *46*(5), 657-662.

Diem, R., Demmer, I., Boretius, S., Merkler, D., Schmelting, B., Williams, S. K., Sättler, M. B., Bähr, M., Michaelis, T., Frahm, J., Brück, W. & Fuchs, E. (2008). Autoimmune optic neuritis in the common marmoset monkey: comparison of visual evoked potentials with MRI and histopathology. *Invest Ophthalmol Vis Sci*, *49*(8), 3707-3714. doi: 10.1167/iovs.08-1896.

Dkhissi, O., Chanut, E., Wasowicz, M., Savoldelli M., Nguyen-Legros, J., Minvielle, F. & Versaux-Botteri, C. (1999). Retinal TUNEL-positive cells and high glutamate levels in vitreous humor of mutant quail with a glaucoma-like disorder. *Invest Ophthalmol Vis Sci*, *409*(5), 90-95.

do Carmo, A., Lopes, C., Santos, M., Proenca, R., Cunha-Vaz, J. & Carvalho, A. P. (1998). Nitric oxide synthase activity and L-arginine metabolism in the retinas from streptozotocin-induced diabetic rats. *Gen Pharmacol*, *30*(3), 319-324.

Dubocovich, M. L. (1983). Melatonin is a potent modulator of dopamine release in the retina. *Nature*, *306*(5945), 782-784.

Dutt, M., Tabuena, P., Ventura, E., Rostami, A. & Shindler, K. S. (2010). Timing of corticosteroid therapy is critical to prevent retinal ganglion cell loss in experimental optic neuritis. *Invest Ophthalmol Vis Sci*, *51*(3), 1439-1445. doi: 10.1167/iovs.09-4009.

Ebneter, A., Casson, R. J., Wood, J. P. & Chidlow, G. (2010). Microglial activation in the visual pathway in experimental glaucoma: spatiotemporal characterization and correlation with axonal injury. *Invest Ophthalmol Vis Sci*, *51*(12), 6448-6460. doi: 10.1167/iovs.10-5284.

Eggenberger, E. R. (2001). Inflammatory optic neuropathies. *Ophthalmol Clin North Am*, *14*(1), 73-82.

El-Remessy, A. B., Khalil, I. E., Matragoon, S., Abou-Mohamed, G., Tsai, N. J., Roon, P., Caldwell, R. B., Caldwell, R. W., Green, K. & Liou, G. I. (2003). Neuroprotective effect of (-) Delta9-tetrahydrocannabinol and cannabidiol in N-methyl-D-aspartate-induced retinal neurotoxicity: involvement of peroxynitrite. *Am J Pathol*, *163*(5), 1997-2008.

Faillace, M. P., Cutrera, R., Sarmiento, M. I. & Rosenstein, R. E. (1995a). Evidence for local synthesis of melatonin in golden hamster retina. *Neuroreport*, *6*(15), 2093-2095.

Faillace, M. P., de las Heras, M. A., Sarmiento, M. I. & Rosenstein, R. E. (1995b). Daily variations in 2-[125I]melatonin specific binding in the golden hamster retina. *Neuroreport*, *7*(1), 141-144.

Felekis, T., Katsanos, K., Kitsanou, M., Trakos, N., Theopistos, V., Christodoulou, D., Asproudis, I. & Tsianos, E. V. (2009). Spectrumand frequency of ophthalmologic manifestations in patients with inflammatory bowel disease: a prospective single-center study. *Inflamm Bowel Dis*, *15*(1), 29-34. doi: 10.1002/ibd.20584.

Felts, P. A., Woolston, A. M., Fernando, H. B., Asquith, S., Gregson, N. A., Mizzi, O. J. & Smith, K. J. (2005). Inflammation and primary demyelination induced by the intraspinal injection of lipopolysaccharide. *Brain*, *128* (Pt 7), 1649-1666. doi:10.1093/brain/ awh516.

Ferreira, S. M., Lerner, S. F., Brunzini, R., Evelson, P. A. & Llesuy, S. F. (2004). Oxidative stress markers in aqueous humor of glaucoma patients. *Am J Ophthalmol*, *137*(1), 62-69.

Foureaux, G., Franca, J. R., Nogueira, J. C., Fulgêncio, Gde, O., Ribeiro, T. G., Castilho, R. O., Yoshida, M. I., Fuscaldi, L. L., Fernandes, S. O., Cardoso, V. N., Cronemberger, S., Faraco, A. A. & Ferreira, A. J. (2015). Ocular inserts for sustained release of the angiotensin-converting enzyme 2 activator, diminazene aceturate, to treat glaucoma in rats. *PLoS One*, *10*(7), e0133149. doi: 10.1371/journal.pone. 0133149.

Friedman, D. S., Wilson, M. R., Liebmann, J. M., Fechtner, R. D. & Weinreb, R. N. (2004). An evidence-based assessment of risk factors for the progression of ocular hypertension and glaucoma. *Am J Ophthalmol*, *138*(3 Suppl), S19-31. doi:10.1016/j.ajo.2004.04.058.

Frigui, M., Frikha, F., Sellemi, D., Chouayakh, F., Feki, J. & Bahloul, Z. (2011). Optic neuropathy as a presenting feature of systemic lupus erythematosus: two case reports and literature review. *Lupus*, *20*(11), 1214-1218. doi: 10.1177/0961203311403344.

Furlan, R., Cuomo, C. & Martino, G. (2009). Animal models of multiple sclerosis. *Methods Mol Biol*, *549*, 157-173. doi: 10.1007/978-1-60327-931-4_11.

Gal, R. L., Vedula, S. S. & Beck, R. (2012). Corticosteroids for treating optic neuritis. *Cochrane Database Syst Rev*, *4*, CD001430. doi: 10.1002/14651858.CD001430. pub3.

Garbarino-Pico, E., Carpentieri, A. R., Contin, M. A., Sarmiento, M. I., Brocco, M. A., Panzetta, P., Rosenstein, R. E., Caputto, B. L. & Guido, M. E. (2004). Retinal ganglion cells are autonomous circadian oscillators synthesizing N-acetylserotonin during the day. *J Biol Chem*, *279*(49), 51172-81. doi:10.1074/jbc.M309248200.

Gegelashvili, G. & Schousboe, A. (1998). Cellular distribution and kinetic properties of high affinity glutamate transporters. *Brain Res Bull*, *45*(3), 233-238.

Gherghel, D., Griffiths, H. R., Hilton, E. J., Cunliffe, I. A. & Hosking, S. L. (2005). Systemic reduction in glutathione levels occurs in patients with primary open-angle glaucoma. *Invest Ophthalmol Vis Sci*, *46*(3), 877-883.

Gorovits, R., Avidan, N., Avisar, N., Shaked, I. & Vardimon, L. (1997). Glutamine synthetase protects against neuronal degeneration in injured retinal tissue. *Proc Natl Acad Sci U S A*, *94*(13), 7024-7029.

Goyal, A., Srivastava, A., Sihota, R. & Kaur, J. (2014). Evaluation of oxidative stress markers in aqueous humor of primary open angle glaucoma and primary angle closure glaucoma patients. *Curr Eye Res*, *39*, 823-829. doi: 10.1007/s12262-013-0813-8.

Grima, G., Cuenod, M., Pfeiffer, S., Mayer, B. & Do, K. Q. (1998). Arginine availability controls the N-methyl-D-aspartate-induced nitric oxide synthesis: involvement of a glial-neuronal arginine transfer. *J Neurochem*, *71*(5), 2139-2144.

Hall, E. D. (1985). High-dose glucocorticoid treatment improves neurological recovery in head-injured mice. *J Neurosurg*, *62*(6), 882-887.

Hardeland, R., Balzer, I., Poeggeler, B., Fuhrberg, B., Uría, H., Behrmann, G., Wolf, R., Meyer, T. J. & Reiter, R. J. (1995). On the primary functions of melatonin in evolution: mediation of photoperiodic signals in a unicell, photooxidation, and scavenging of free radicals. *J Pineal Res*, *18*(2), 104-111.

Heidinger, V., Hicks, D., Sahel, J. & Dreyfus, H. (1999). Ability of retinal Müller glial cells to protect neurons against excitotoxicity *in vitro* depends upon maturation and neuron-glial interactions. *Glia*, *25*(3), 229-239.

Heijl, A., Leske, M. C., Bengtsson, B., Hyman, L., Bengtsson, B. & Hussein, M. (2002). Early Manifest Glaucoma Trial Group. Reduction of intraocular pressure and glaucoma progression: results from the Early Manifest Glaucoma Trial. *Arch Ophthalmol*, *120*(10), 1268-1279.

Herrera, A. J., Castaño, A., Venero, J. L., Cano, J. & Machado, A. (2000). The single intranigral injection of LPS as a new model for studying the selective effects of inflammatory reactions on dopaminergic system. *Neurobiol Dis*, *7*(4), 429-447. doi:10.1006/nbdi.2000.0289.

Hickman, S. J., Dalton, C. M., Miller, D. H. & Plant, G. T. (2002). Management of acute optic neuritis. *Lancet*, *360*(9349), 1953-1962.

Hoijman, E., Rocha Viegas, L., Keller Sarmiento, M. I., Rosenstein, R. E. & Pecci, A. (2004). Involvement of Bax protein in the prevention of glucocorticoid-induced thymocytes apoptosis by melatonin. *Endocrinology*, *145*(1), 418-425. doi: 10.1210/en.2003-0764.

Honkanen, R. A., Baruah, S., Zimmerman, M. B., Khanna, C. L., Weaver, Y. K., Narkiewicz, J., Waziri, R., Gehrs, K. M., Weingeist, T. A., Boldt, H. C., Folk, J. C., Russell, S. R. & Kwon, Y. H. (2003). Vitreous amino acid concentrations in patients with glaucoma undergoing vitrectomy. *Arch Ophthalmol*, *121*(2), 183-188.

Horstmann, L., Schmid, H., Heinen, A. P., Kurschus, F. C., Dick, H. B. & Joachim, S. C. (2013). Inflammatory demyelination induces glia alterations and ganglion cell loss in the retina of an experimental autoimmune encephalomyelitis model. *J Neuroinflammation*, *10*, 120. doi: 10.1186/1742-2094-10-120.

Horwitz, H., Friis, T., Modvig, S., Roed, H., Tsakiri, A., Laursen, B. & Frederiksen, J. L. (2014). Differential diagnoses to MS: experiences from an optic neuritis clinic. *J Neurol*, *261*(1), 98-105. doi: 10.1007/ s00415-013-7166-x.

Huang, H., Wang, Z., Weng, S. J., Sun, X. H. & Yang, X. L. (2013). Neuromodulatory role of melatonin in retinal information processing. *Prog Retin Eye Res*, *32*, 64-87. doi: 10.1016/j.preteyeres.2012.07.003.

Huete-Toral, F., Crooke, A., Martínez-Águila, A. & Pintor, J. (2015). Melatonin receptors trigger cAMP production and inhibit chloride movements in nonpigmented ciliary epithelial cells. *J Pharmacol Exp Ther*, *352*(1), 119-128. doi: 10.1124/ jpet.114.218263.

Ianăş, O., Olinescu, R. & Bădescu, I. (1991). Melatonin involvement in oxidative processes. *Endocrinologie*, *29*(3-4), 147-153.

Imbesi, M., Uz, T. & Manev, H. (2008). Role of melatonin receptors in the effects of melatonin on BDNF and neuroprotection in mouse cerebellar neurons. *J Neural Transm*, *115*(11), 1495-1499. doi: 10.1007/s00702-008-0066-z.

Ismail, S. A. & Mowafi, H. A. (2009). Melatonin provides anxiolysis, enhances analgesia, decreases intraocular pressure, and promotes better operating conditions during cataract surgery under topical anesthesia. *Anesth Analg*, *108*(4), 1146-1151. doi: 10.1213/ane.0b013e 3181907 ebe.

Izzotti, A., Bagnis, A. & Saccà, S. C. (2006). The role of oxidative stress in glaucoma. *Mutat Res*, *612*(2), 105-114. doi:10.1016/ j.mrrev.2005. 11.001.

Jaliffa, C. O., Lacoste, F. F., Llomovatte, D. W., Sarmiento, M. I. & Rosenstein, R. E. (2000). Dopamine decreases melatonin content in golden hamster retina. *J Pharmacol Exp Ther*, *293*(1), 91-95.

Jimenez-Jorge, S., Guerrero, J. M., Jimenez-Caliani, A. J., Naranjo, M. C., Lardone, P. J., Carrillo-Vico, A., Osuna, C. & Molinero, P. (2007). Evidence for melatonin synthesis in the rat brain during development. *J Pineal Res*, *42*(3), 240-246. doi: 10.1111/j.1600-079X.2006.00411.x.

Kallenbach, K. & Frederiksen, J. L. (2008). Unilateral optic neuritis as the presenting symptom of human immunodeficiency virus toxoplasmosis infection. *Acta Ophthalmol*, *86*(4), 459-460. doi: 10.1111/j.1600-0420. 2007.01034.x.

Kass, M. A., Heuer, D. K., Higginbotham, E. J., Johnson, C. A., Keltner, J. L., Miller, J. P., Parrish 2nd, R. K., Wilson, M. R. & Gordon, M. O. (2002). The Ocular Hypertension Treatment Study: a randomized trial determines that topical ocular hypotensive medication delays or prevents the onset of primary open-angle glaucoma. *Arch Ophthalmol*, *120*(6), 701-713.

Kaufman, D. I., Trobe, J. D., Eggenberger, E. R. & Whitaker, J. N. (2000). Practice parameter: the role of corticosteroids in the management of acute monosymptomatic optic neuritis. Report of the Quality Standards Subcommittee of the American Academy of Neurology. *Neurology*, *54*(11), 2039-2044.

Kaushik, D. K. & Basu, A. (2013). A friend in need may not be a friend indeed: role of microglia in neurodegenerative diseases. CNS Neurol. *Disord Drug Targets*, *12*(6), 726-740. doi: 10.2174/1871527311312 6660170.

Kong, X., Li, X., Cai, Z., Yang, N., Liu, Y., Shu, J., Pan, L. & Zuo, P. (2008). Melatonin regulates the viability and differentiation of rat midbrain neural stem cells. *Cell Mol Neurobiol*, *28*(4), 569-579. doi:10.1007/s10571-007-9212-7.

Kupersmith, M. J., Nelson, J. I., Seiple, W. H., Carr, R. E. & Weiss, P. A. (1983). The 20/20 eye in multiple sclerosis. *Neurology 33*(8), 1015-1020.

Kurysheva, N. I., Vinetskaia, M. I., Erichev, V. P., Demchuk, M. L. & Kuryshev, S. I. (1996). Contribution of free-radical reactions of chamber humor to the development of primary open-angle glaucoma. *Vestn Oftalmol*, *112*(4), 3-5.

Kurz, S. & Harrison, D. (1997). Insulin and the arginine paradox. *J Clin Invest*, *99*(3), 369-370. doi:10.1172/JCI119166.

Lachapelle, F., Bachelin, C., Moissonnier, P., Nait-Oumesmar, B., Hidalgo, A., Fontaine, D. & Baron-Van Evercooren, A. (2005). Failure of remyelination in the nonhuman primate optic nerve. *Brain Pathol*, *15*(3), 198-207.

Leon, J., Acuña-Castroviejo, D., Sainz, R. M., Mayo, J. C., Tan, D. X. & Reiter, R. J. (2004). Melatonin and mitochondrial function. *Life Sci*, *75*(7), 765-790. doi:10.1016/j.lfs.2004.03.003.

León, J., Acuña-Castroviejo, D., Escames, G., Tan, D. X. & Reiter, R. J. (2005). Melatonin mitigates mitochondrial malfunction. *J Pineal Res*, *38*(1), 1-9. doi:10.1111/j.1600-079X.2004.00181.x.

León, J., Vives, F., Crespo, E., Camacho, E., Espinosa, A., Gallo, M. A., Escames, G. & Acuña-Castroviejo, D. (1998). Modification of nitric oxide synthase activity and neuronal response in rat striatum by melatonin and kynurenine derivatives. *J Neuroendocrinol*, *10*(4), 297-302.

Leske, M. C., Heijl, A., Hussein, M., Bengtsson, B., Hyman, L. & Komaroff, E. (2003). Factors for glaucoma progression and the effect of treatment: the early manifest glaucoma trial. *Arch Ophthalmol*, *121*(1), 48-56.

Levkovitch-Verbin, H., Martin, K. R., Quigley, H. A., Baumrind, L. A., Pease, M. E. & Valenta, D. (2002). Measurement of amino acid levels in the vitreous humor of rats after chronic intraocular pressure elevation or optic nerve transection. *J Glaucoma*, *11*(5), 396-405.

Li, X., Du, Y., Fan, Q., Tang, C. Y. & He, J. F. (2014). Gypenosides might have neuroprotective and immunomodulatory effects on optic neuritis. *Med Hypotheses*, *82*(5), 636-638. doi: 10.1016/j.mehy.2014.02.030.

Lin, T. H., Kim, J. H., Perez-Torres, C., Chiang, C. W., Trinkaus, K., Cross, A. H. & Song, S. K. (2014). Axonal transport rate decreased at the onset of optic neuritis in EAE mice. *NeuroImage*, *100*, 244-253. doi: 10.1016/j.neuroimage.2014.06.009.

Liu, B. & Neufeld, A. H. (2000). Expression of nitric oxide synthase-2 (NOS-2) in reactive astrocytes of the human glaucomatous optic nerve head. *Glia*, *30*(2), 178-186.

Liu, B. & Neufeld, A. H. (2001). Nitric oxide synthase-2 in human optic nerve head astrocytes induced by elevated pressure *in vitro*. *Arch Ophthalmol*, *119*(2), 240-245.

Liu, B. & Neufeld, A. H. (2003). Activation of epidermal growth factor receptor signals induction of nitric oxide synthase-2 in human optic nerve head astrocytes in glaucomatous optic neuropathy. *Neurobiol Dis*, *13*(2), 109-123.

Louzada-Junior, P., Dias, J. J., Santos, W. F., Lachat, J. J., Bradford, H. F. & Coutinho-Netto, J. (1992). Glutamate release in experimental ischaemia of the retina: an approach using microdialysis. *J Neurochem*, *59*(1), 358-363.

Lundmark, P. O., Pandi-Perumal, S. R., Srinivasan, V. & Cardinali, D. P. (2006). Role of melatonin in the eye and ocular dysfunctions. *Vis Neurosci* 23(6), 853-862. doi:10.1017/S0952523806230189.

Manabe, S., Gu, Z. & Lipton, S. A. (2005). Activation of matrix metalloproteinase-9 via neuronal nitric oxide synthase contributes to NMDA-induced retinal ganglion cell death. *Invest Ophthalmol Vis Sci*, *46*(12), 4747-4753. doi:10.1167/iovs.05-0128.

Marcic, T. S., Belyea, D. A. & Katz, B. (2003). Neuroprotection in glaucoma: a model for neuroprotection in optic neuropathies. *Curr Opin Ophthalmol*, *14*(6), 353-356.

Martin, K. R., Levkovitch-Verbin, H., Valenta, D., Baumrind, L., Pease, M. E. & Quigley, H. A. (2002) Retinal glutamate transporter changes in experimental glaucoma and after optic nerve transection in the rat. *Invest Ophthalmol Vis Sci*, *43*(7), 2236-2243.

Martínez-Águila, A., Fonseca, B., Bergua, A. & Pintor, J. (2013). Melatonin analogue agomelatine reduces rabbit's intraocular pressure in normotensive and hypertensive conditions. *Eur J Pharmacol*, *701*(1-3), 213-217. doi: 10.1016/j.ejphar.2012.12.009.

Martínez-Águila, A., Fonseca, B., Pérez De Lara, M. J. & Pintor, J. (2016). Effect of melatonin and 5-methoxycarbonylamino-N-acetyltryptamine on the intraocular pressure of normal and glaucomatous mice. *J Pharmacol Exp Ther*, *357*(2), 293-299. doi: 10.1124/jpet.115.231456.

Martínez-Lapiscina, E. H., Fraga-Pumar, E., Pastor, X., Gómez, M., Conesa, A., Lozano-Rubí, R., Sánchez-Dalmau, B., Alonso, A. & Villoslada, P. (2014). Is the incidence of optic neuritis rising? Evidence from an epidemiological study in Barcelona (Spain), 2008-2012. *J Neurol*, *261*(4), 759-767. doi: 10.1007/s00415-014-7266-2.

Mayordomo-Febrer, A., López-Murcia, M., Morales-Tatay, J. M., Monleón-Salvado, D. & Pinazo-Durán, M. D. (2015). Metabolomics of the aqueous humor in the rat glaucoma model induced by a series of intracamerular sodium hyaluronate injection. *Exp Eye Res*, *131*, 84-92. doi: 10.1016/j.exer.2014.11.012.

Melino, G., Bernassola, F., Knight, R. A., Corasaniti, M. T., Nistico, G. & Finazzi-Agro, A. (1997). S-nitrosylation regulates apoptosis. *Nature*, *388*(6641), 432-433. doi:10.1038/41237.

Moreno, M. C., Campanelli, J., Sande, P., Sánez, D. A., Keller Sarmiento, M. I. & Rosenstein, R. E. (2004). Retinal oxidative stress induced by high intraocular pressure. *Free Radic Biol Med*, *37*(6), 803-812.

Moreno, M. C., Sande, P., Marcos, H. A., de Zavalía, N., Keller Sarmiento, M. I. & Rosenstein, R. E. (2005a). Effect of glaucoma on the retinal glutamate/glutamine cycle activity. *FASEB J*, *19*(9), 1161-1162. doi:10.1096/fj.04-3313fje.

Moreno, M. C., Marcos, H. J., Oscar Croxatto, J., Sande, P. H., Campanelli, J., Jaliffa, C. O., Benozzi, J. & Rosenstein, R. E. (2005b). A new experimental model of glaucoma in rats through intracameral injections of hyaluronic acid. *Exp Eye Res*, *81*(1), 71-80. doi:10.1016/ j.exer.2005.01.008.

Moreno, M. C., de Zavalía, N., Sande, P., Jaliffa, C. O., Fernandez, D. C., Keller Sarmiento, M. I. & Rosenstein, R. E. (2008). Effect of ocular hypertension on retinal GABAergic activity. *Neurochem Int*, *52*(4-5), 675-682. doi:10.1016/j.neuint. 2007.08.014.

Morrison, J. C., Moore, C. G., Deppmeier, L. M., Gold, B. G., Meshul, C. K. & Johnson, E. C. (1997). A rat model of chronic pressure-induced optic nerve damage. *Exp Eye Res*, *64*(1), 85-96. doi:10.1006/ exer.1996.0184.

Mosinger, J. L., Price, M. T., Bai, H. Y., Xiao, H., Wozniak, D. F. & Olney, J. W. (1991). Blockade of both NMDA and non-NMDA receptors is required for optimal protection against ischemic neuronal degeneration in the *in vivo* adult mammalian retina. *Exp Neurol*, *113*(1), 10-17.

Naskar, R., Vorwerk, C. K. & Dreyer, E. B. (2000). Concurrent down regulation of a glutamate transporter and receptor in glaucoma. *Invest Ophthalmol Vis Sci*, *41*(7), 1940-1944.

Neher, J. J., Emmrich, J. V., Fricker, M., Mander, P. K., Théry, C. & Brown, G. C. (2013). Phagocytosis executes delayed neuronal death after focal brain ischemia. *Proc Natl Acad Sci USA*, *110*(43), E4098-E4107. doi: 10.1073/pnas.1308679110.

Neufeld, A. H. (1999). Nitric oxide: a potential mediator of retinal ganglion cell damage in glaucoma. *Surv Ophthalmol*, *43* (Suppl 1), S129-135.

Neufeld, A. H. (2004). Pharmacologic neuroprotection with an inhibitor of nitric oxide synthase for the treatment of glaucoma. *Brain Res Bull*, *62*(6), 455-459. doi: 10.1016/j.brainresbull.2003.07.005.

Neufeld, A. H. & Liu, B. (2003). Glaucomatous optic neuropathy: when glia misbehave. *Neuroscientist*, *9*(6), 485-495. doi:10.1177/ 1073858403 253460.

Neufeld, A. H., Hernandez, M. R. & Gonzalez, M. (1997). Nitric oxide synthase in the human glaucomatous optic nerve head. *Arch Ophthalmol*, *115*(4), 497-503.

Neufeld, A. H., Sawada, A. & Becker, B. (1999). Inhibition of nitric-oxide synthase 2 by aminoguanidine provides neuroprotection of retinal ganglion cells in a rat model of chronic glaucoma. *Proc Natl Acad Sci U S A*, *96*(17), 9944-9948.

Niles, L. P., Armstrong, K. J., Rincon Castro, L. M., Dao, C. V., Sharma, R., McMillan, C. R., Doering, L. C. & Kirkham, D. L. (2004). Neural stem cells express melatonin receptors and neurotrophic factors: colocalization of the MT1 receptor with neuronal and glial markers. *BMC Neurosci*, *5*, 41. doi:10.1186/1471-2202-5-41.

Nucci, C., Tartaglione, R., Rombolà, L., Morrone, L. A., Fazzi, E. & Bagetta, G. (2005). Neurochemical evidence to implicate elevated glutamate in the mechanisms of high intraocular pressure (IOP)-induced retinal ganglion cell death in rat. *Neurotoxicology*, *26* (5), 935-941. doi: 10.1016/j.neuro.2005.06.002.

Nucci, C., Russo, R., Martucci, A., Giannini, C., Garaci, F., Floris, R., Bagetta, G. & Morrone, L. A. (2016). New strategies for neuroprotection in glaucoma, a disease that affects the central nervous system. *Eur J Pharmacol*, *787*, 119-126. doi: 10.1016/j.ejphar. 2016.04.030.

Ohta, Y., Yamasaki, T., Niwa, T., Niimi, K., Majima, Y. & Ishiguro, I. (1996). Role of catalase in retinal antioxidant defence system: its comparative study among rabbits, guinea pigs, and rats. *Ophthalmic Res*, *28*(6), 336- 342.

Optic Neuritis Study Group. (2008). Multiple sclerosis risk after optic neuritis: final optic neuritis treatment trial follow-up. *Arch Neurol*, *65*(6), 727-732. doi: 10.1001/archneur.65.6.727.

Osborne, N. N., Chidlow, G., Nash, M. S. & Wood, J. P. (1999). The potential of neuroprotection in glaucoma treatment. *Curr Opin Ophthalmol*, *10*(2), 82-92.

Otis, T. S. & Jahr, C. E. (1998). Anion currents and predicted glutamate flux through a neuronal glutamate transporter. *J Neurosci*, *18*(18), 7099-7110.

Pandi-Perumal, S. R., Srinivasan, V., Maestroni, G. J., Cardinali, D. P., Poeggeler, B. & Hardeland, R. (2006). Melatonin: Nature's most versatile biological signal? *FEBS J*, *273*(13), 2813-2838. doi:10. 1111/j.1742-4658.2006.05322.x

Pang, I. H., Johnson, E. C., Jia, L., Cepurna, W. O., Shepard, A. R., Hellberg, M. R., Clark, A. F. & Morrison, J. C. (2005). Evaluation of inducible nitric oxide synthase in glaucomatous optic neuropathy and pressure induced optic nerve damage. *Invest Ophthalmol Vis Sci*, *46*(4), 1313-1321. doi: 10.1167/iovs.04-0829.

Patel, A. J., Hunt, A. & Tahourdin, C. S. (1983). Regulation of *in vivo* glutamine synthetase activity by glucocorticoids in the developing rat brain. *Brain Res*, *312*(1), 83-91.

Pescosolido, N., Gatto, V., Stefanucci, A. & Rusciano, D. (2015). Oral treatment with the melatonin agonist agomelatine lowers the intraocular pressure of glaucoma patients. *Ophthalmic Physiol Opt*, *35*(2), 201-205. doi: 10.1111/opo.12189.

Pierce, M. E. & Besharse, J. C. (1985). Circadian regulation of retinomotor movements. I. Interaction of melatonin and dopamine in the control of cone length. *J Gen Physiol*, *86*(5), 671-689.

Pintor, J., Martin, L., Pelaez, T., Hoyle, C. H. & Peral, A. (2001). Involvement of melatonin MT(3) receptors in the regulation of intraocular pressure in rabbits. *Eur J Pharmacol*, *416*(3), 251-254.

Pintor, J., Peláez, T., Hoyle, C. H. & Peral, A. (2003). Ocular hypotensive effects of melatonin receptor agonists in the rabbit: further evidence for an MT3 receptor. *Br J Pharmacol*, *138*(5), 831-836. doi:10.1038/ sj.bjp.0705118.

Pirhan, D., Yüksel, N., Emre, E., Cengiz, A. & Kürşat Yıldız, D. (2016). Riluzole- and resveratrol-induced delay of retinal ganglion cell death in an experimental model of glaucoma. *Curr Eye Res*, *41*(1), 59-69. doi: 10.3109/02713683.2015.1004719.

Pow, D. V. & Robinson, S. R. (1994). Glutamate in some retinal neurons is derived solely from glia. *Neuroscience*, *60*(2), 355-366.

Pozo, D., Reiter, R. J., Calvo, J. R. & Guerrero, J. M. (1997). Inhibition of cerebellar nitric oxide synthase and cyclic GMP production by melatonin via complex formation with calmodulin. *J Cell Biochem*, *65*(3), 430-442.

Qi, X., Lewin, A. S., Sun, L., Hauswirth, W. W. & Guy, J. (2007). Suppression of mitochondrial oxidative stress provides long-term neuroprotection in experimental optic neuritis. *Invest Ophthalmol Vis Sci*, *48*(2), 681-691. doi:10.1167/iovs.06-0553.

Quigley, H. A. & Broman, A. T. (2006). The number of people with glaucoma worldwide in 2010 and 2020. *Br J Ophthalmol*, *90*(3), 262-726. doi:10.1136/bjo.2005.081224.

Rangachari, M. & Kuchroo, V. K. (2013). Using EAE to better understand principles of immune function and autoimmune pathology. *J Autoimmun*, *45*, 31-39. doi: 10.1016/j.jaut.2013.06.008.

Rajesh, M., Sulochana, K. N., Punitham, R., Biswas, J., Lakshmi, S. & Ramakrishnan, S. (2003). Involvement of oxidative and nitrosative stress in promoting retinal vasculitis in patients with Eales' disease. *Clin Biochem*, *36*(5), 377-385.

Reiter, R. J. (1998). Oxidative damage in the central nervous system: protection by melatonin. *Prog Neurobiol*, *56*(3), 359-384.

Reiter, R. J., Tan, D. X., Poeggeler, B., Menendez-Pelaez, A., Chen, L. D. & Saarela, S. (1994). Melatonin as a free radical scavenger: implications for aging and age-related diseases. *Ann NY Acad Sci*, *719*, 1-12.

Reiter, R. J., Guerrero, J. M., Escames, G., Pappolla, M. A. & Acuña-Castroviejo, D. (1997). Prophylactic actions of melatonin in oxidative neurotoxicity. *Ann NY Acad Sci*, *825*, 70-78.

Reiter, R. J., Tan, D. X., Osuna, C. & Gitto, E. (2000). Actions of melatonin in the reduction of oxidative stress. A review. *J Biomed Sci*, *7*(6), 444-458. doi: 25480.

Riepe, R. E. & Norenburg, M. D. (1977). Müller cell localization of glutamine synthetase in rat retina. *Nature*, *268*(5621), 654-655.

Rosner, M., Lam, T. T., Fu, J. & Tso, M. O. (1992). Methylprednisolone ameliorates retinal photic injury in rats. *Arch Ophthalmol*, *110*(6), 857-861.

Russo, R., Rotiroti, D., Tassorelli, C., Nucci, C., Bagetta, G., Bucci, M. G., Corasaniti, M. T. & Morrone, L. A. (2009). Identification of novel pharmacological targets to minimize excitotoxic retinal damage. *Int Rev Neurobiol*, *85*, 407-423. doi: 10.1016/S0074-7742(09)85028-9.

Sáenz, D. A., Cymeryng, C. B., De Nichilo, A., Sacca, G. B., Keller Sarmiento, M. I. & Rosenstein. R. E. (2002a). Photic regulation of L-arginine uptake in the golden hamster retina. *J Neurochem*, *80*(3), 512-519.

Sáenz, D. A., Turjanski, A. G., Sacca, G. B., Marti, M., Doctorovich, F., Sarmiento, M. I., Estrin, D. A. & Rosenstein, R. E. (2002b). Physiological concentrations of melatonin inhibit the nitridergic pathway in the Syrian hamster retina. *J Pineal Res*, *33*(1), 31-36.

Sáenz, D. A., Goldin, A. P., Minces, L., Chianelli, M., Keller Sarmiento, M. I. & Rosenstein, R. E. (2004). Effect of melatonin on the retinal glutamate/glutamine cycle in the golden hamster retina. *FASEB J*, *18*(15), 1912-3. doi:10.1096/fj.04-2062fje.

Salido, E. M., Bordone, M., De Laurentiis, A., Chianelli, M., Keller Sarmiento, M. I., Dorfman, D. & Rosenstein, R. E. (2013). Therapeutic efficacy of melatonin in reducing retinal damage in an experimental model of early type 2 diabetes in rats. *J Pineal Res*, *54*(2), 179-89. doi: 10.1111/jpi.12008.

Salt, T. E. & Cordeiro, M. F. (2006). Glutamate excitotoxicity in glaucoma: throwing the baby out with the bath water? *Eye*, *20*(6), 731-732, 730-1, author reply. doi:10.1038/sj.eye.6701967.

Samples, J. R., Krause, G. & Lewy, A. J. (1988). Effect of melatonin on intraocular pressure. *Curr Eye Res*, *7*(7), 649-653.

Sande, P. H., Fernandez, D. C., Aldana Marcos, H. J., Chianelli, M. S., Aisemberg, J., Silberman, D. M., Sáenz, D. A. & Rosenstein, R. E. (2008). Therapeutic effect of melatonin in experimental uveitis. *Am J Pathol*, *173*(6), 1702-1713. doi: 10.2353/ajpath.2008.080518.

Sande, P. H., Dorfman, D., Fernandez, D. C., Chianelli, M., Domínguez Rubio, A. P., Franchi, A. M., Silberman, D. M., Rosenstein, R. E. & Sáenz, D. A. (2014). Treatment with melatonin after onset of experimental uveitis attenuates ocular inflammation. *Br J Pharmacol*, *171*(24), 5696-5707. doi: 10.1111/bph.12873.

Sarkar, P. K. & Chaudhury, S. (1983). Messenger RNA for glutamine synthetase. *Mol Cell Biochem*, 53-54(1-2), 233-244.

Sarthy, P. V. & Lam, D. M. (1978). Biochemical studies of isolated glial (Müller) cells from the turtle retina. *J Cell Biol*, *78*(3), 675-684.

Serle, J. B., Wang, R. F., Peterson, W. M., Plourde, R. & Yerxa, B. R. (2004). Effect of 5-MCA-NAT, a putative melatonin MT3 receptor agonist, on intraocular pressure in glaucomatous monkey eyes. *J Glaucoma*, *13*(5), 385-388.

Shareef, S. R., Garcia-Valenzuela, E., Salierno, A., Walsh, J. & Sharma, S. C. (1995). Chronic ocular hypertension following episcleral venous occlusion in rats. *Exp Eye Res*, *61*(3), 379-382.

Shareef, S., Sawada, A. & Neufeld, A. H. (1999). Isoforms of nitric oxide synthase in the optic nerves of rat eyes with chronic moderately elevated intraocular pressure. *Invest Ophthalmol Vis Sci*, *40*(12), 2884-2891.

Shen, F., Chen, B., Danias, J., Lee, K. C., Lee, H., Su, Y., Podos, S. M. & Mittag, T. W. (2004) Glutamate-induced glutamine synthetase expression in retinal Muller cells after short term ocular hypertension in the rat. *Invest Ophthalmol Vis Sci*, *45*(9), 3107-3112. doi:10.1167/ iovs.03-0948.

Shindler, K. S., Guan, Y., Ventura, E., Bennett, J. & Rostami, A. (2006). Retinal ganglion cell loss induced by acute optic neuritis in a relapsing model of multiple sclerosis. *Mult Scler*, *12*(5), 526-532. doi: 10.1177/ 1352458506070629.

Simmons, S. B., Pierson, E. R., Lee, S. Y. & Goverman, J. M. (2013). Modeling the heterogeneity of multiple sclerosis in animals. *Trends Immunol*, *34*(8), 410-422. doi: 10.1016/j.it.2013.04.006.

Siu, A. W., Reiter, R. J. & To, C. H. (1999). Pineal indoleamines and vitamin E reduce nitric oxide-induced lipid peroxidation in rat retinal homogenates. *J Pineal Res*, *27*(2), 122-128.

Siu, A. W., Maldonado, M., Sanchez-Hidalgo, M., Tan, D. X. & Reiter, R. J. (2006). Protective effects of melatonin in experimental free radical-related ocular diseases *J Pineal Res*, *40*(2), 101-109. doi:10.1111/ j.1600-079X.2005.00304.x.

Stevens, B. R., Kakuda, D. K., Yu, K., Waters, M., Vo, C. B. & Raizada, M. K. (1996). Induced nitric oxide synthesis is dependent on induced alternatively spliced CAT-2 encoding L-arginine transport in brain astrocytes. *J Biol Chem*, *271*(39), 24017-24022.

Sucher, N. J., Lipton, S. A. & Dreyer, E. B. (1997). Molecular basis of glutamate toxicity in retinal ganglion cells. *Vis Res*, *37*(24), 3483-3493. doi: 10.1016/S0042-6989(97)00047-3.

Tan, D. X., Manchester, L. C., Terron, M. P., Flores, L. J. & Reiter, R. J. (2007). One molecule, many derivatives: a never-ending interaction of melatonin with reactive oxygen and nitrogen species? *J Pineal Res*, *42*(1), 28-42. doi: 10.1111/j.1600-079X.2006.00407.x.

Tanaka, K. (2000). Functions of glutamate transporters in the brain. *Neurosci Res*, *37*(1), 15-19.

Tang, Q., Hu, Y. & Cao, Y. (2006). Neuroprotective effect of melatonin on retinal ganglion cells in rats. *J Huazhong Univ Sci Technolog Med Sci*, *26*(2), 235-237, 253.

Tanito, M., Nishiyama, A., Tanaka, T., Masutani, H., Nakamura, H., Yodoi, J. & Ohira, A. (2002). Change of redox status and modulation by thiol replenishment in retinal photooxidative damage. *Invest Ophthalmol Vis Sci*, *43*(7), 2392-2400.

Tezel, G. (2006). Oxidative stress in glaucomatous neurodegeneration: mechanisms and consequences. *Prog Retin Eye Res*, *25*(5), 490-513. doi: 10.1016/j.preteyeres.2006.07.003.

Thoreson, W. B. & Witkovsky, P. (1999). Glutamate receptors and circuits in the vertebrate retina. *Prog Retin Eye Res*, *18*(6), 765-810.

Toosy, A. T., Mason, D. F. & Miller, D. H. (2014). Optic neuritis. *Lancet Neurol*, *13*(1), 83-99. doi: 10.1016/S1474-4422(13)70259-X.

Tosini, G. & Menaker, M. (1996). Circadian rhythms in cultured mammalian retina. *Science*, *272*(5260), 419-421.

Tosini, G. & Fukuhara, C. (2003). Photic and circadian regulation of retinal melatonin in mammals. *J Neuroendocrinol*, *15*(4), 364-369.

Tosini, G., Baba, K., Hwang, C. K. & Iuvone, P. M. (2012). Melatonin: an underappreciated player in retinal physiology and pathophysiology. *Exp Eye Res* 103: 82-89. doi:10.1016/j.exer.2012.08.009.

Tosini, G., Iuvone, M., Boatright, J. H. (2013). Is the melatonin receptor type 1 involved in the pathogenesis of glaucoma? *J Glaucoma*, Suppl 5, S49-50. doi: 10.1097/IJG.0b013e3182934bb4.

Trip, S. A., Schlottmann, P. G., Jones, S. J., Altmann, D. R., Garway-Heath, D. F., Thompson, A. J., Plant, G. T. & Miller, D. H. (2005). Retinal nerve fiber layer axonal loss and visual dysfunction in optic neuritis. *Ann Neurol*, *58*(3), 383-391. doi:10.1002/ana.20575.

Turjanski, A., Chaia, Z. D., Doctorovich, F., Estrin, D., Rosenstein, R. & Piro, O. E. (2000). N-nitrosomelatonin. *Acta Crystallogr C*, *56* (Pt 6), 682-683. doi: 10.1107/S010827010000456X.

Turjanski, A. G., Rosenstein, R. E. & Estrin, D. A. (1998). Reactions of melatonin and related indoles with free radicals: a computational study. *J Med Chem*, *41*(19), 3684-3689. doi:10.1021/jm980117m.

Ueda, J., Sawaguchi, S., Hanyu, T., Yaoeda, K., Fukuchi, T., Abe, H. & Ozawa, H. (1998). Experimental glaucoma model in the rat induced by laser trabecular photocoagulation after an intracameral injection of India ink. *Jpn J Ophthalmol*, *42*(5), 337-344.

Vincent, L., Cohen, W., Delagrange, P., Boutin, J. A. & Nosjean, O. (2010). Molecular and cellular pharmacological properties of 5-methoxycarbonylamino-N-acetyltryptamine (MCA-NAT): a nonspecific MT3 ligand. *J Pineal Res*, *48*(3), 222-229. doi: 10.1111/ j.1600-079X.2010.00746.x.

Vorwerk, C. K., Hyman, B. T., Miller, J. W., Husain, D., Zurakowski, D., Huang, P. L., Fishman, M. C. & Dreyer, E. B. (1997). The role of neuronal and endothelial nitric oxide synthase in retinal excitotoxicity. *Invest Ophthalmol Vis Sci, 38*(10), 2038-2044.

Vorwerk, C. K., Naskar, R., Schuettauf, F., Quinto, K., Zurakowski, D., Gochenauer, G., Robinson, M. B., Mackler, S. A. & Dreyer, E. B. (2000). Depression of retinal glutamate transporter function leads to elevated intravitreal glutamate levels and ganglion cell death. *Invest Ophthalmol Vis Sci, 41*(11), 3615-3621.

Wamsley, S., Gabelt, B. T., Dahl, D. B., Case, G. L., Sherwood, R. W., May, C. A., Hernandez, M. R. & Kaufman, P. L. (2005). Vitreous glutamate concentration and axon loss in monkeys with experimental glaucoma. *Arch Ophthalmol, 123*(1), 64-70. doi:10.1001/ archopht. 123.1.64.

Weber, A. J., Harman, C. D. & Viswanathan, S. (2008). Effects of optic nerve injury, glaucoma, and neuroprotection on the survival, structure, and function of ganglion cells in the mammalian retina. *J Physiol, 586*(18), 4393-4400. doi: 10. 1113/jphysiol.2008.156729.

Wilejto, M., Shroff, M., Buncic, J. R., Kennedy, J., Goia, C. & Banwell, B. (2006). The clinical features, MRI findings, and outcome of optic neuritis in children. *Neurology, 67*, 258-262. doi:10.1212/ 01.wnl. 0000224757.69746.fb.

Wu, G. S., Zhang, J. & Rao, N. A. (1997). Peroxynitrite and oxidative damage in experimental autoimmune uveitis. *Invest Ophthalmol Vis Sci, 38*(7), 1333-1339.

Yang, J., Tezel, G., Patil, R. V., Romano, C. & Wax, M. B. (2001). Serum autoantibody against glutathione S-transferase in patients with glaucoma. *Invest Ophthalmol Vis Sci, 42*(6), 1273-1276.

You, Y., Klistorner, A., Thie, J. & Graham, S. L. (2011). Latency delay of visual evoked potential is a real measurement of demyelination in a rat model of optic neuritis. *Invest Ophthalmol Vis Sci, 52*(9), 6911-6918. doi: 10.1167/iovs.11-7434.

Zhao, L., An, R., Yang, Y., Yang, X., Liu, H., Yue, L., Li, X., Lin, Y., Reiter, R. J. & Qu, Y. (2015). Melatonin alleviates brain injury in mice subjected to cecal ligation and puncture via attenuating inflammation, apoptosis, and oxidative stress: the role of SIRT1 signaling. *J Pineal Res, 59*(2), 230-239. doi: 10.1111/jpi.12254.

Zhao, L., Liu, H., Yue, L., Zhang, J., Li, X., Wang, B., Lin, Y. & Qu, Y. (2017). Melatonin attenuates early brain injury via the melatonin receptor/Sirt1/NF-κB signaling pathway following subarachnoid hemorrhage in mice. *Mol Neurobiol, 54*(3), 1612-1621. doi: 10.1007/ s12035-016-9776-7.

Chapter 3

THE PROTECTIVE ROLE OF EXOGEN MELATONIN ON THE PROSTATE GLAND UNDER EXPERIMENTAL MODELS OF METABOLIC DISEASES

Marina Guimarães Gobbo[1], Guilherme Henrique Tamarindo[2], Sebastião Roberto Taboga[1] and Rejane Maira Góes[1],*

[1]Department of Biology, Institute of Biosciences, Humanities and Exact Sciences, Universidade Estadual Paulista, UNESP, São José do Rio Preto, SP, Brazil

[2]Department of Functional and Structural Biology, Institute of Biology, University of Campinas, UNICAMP, Campinas, SP, Brazil

ABSTRACT

The prostate is an accessory gland of the male apparatus which is regulated by sexual steroids from its development phase. Several studies have reported that metabolic syndromes may also influence the prostate's physiology. Studies from our research group as well as other studies from the literature revealed that diabetes and high lipid intake are unfavorable to the prostatic homeostasis and may raise histophysiological alterations that can lead to cancer. Prostate stromal remodeling, imbalance on the gland proliferation/apoptotic rate, testosterone/estrogen ratio and tissue antioxidant system impairment are important features observed under metabolic disturbances and are often associated with carcinogenesis. In the last decade, the therapeutic potential of melatonin has been investigated due to it exerting antioxidant properties and influencing cell dynamics. As well as prostate cells expressing MTR1 and MTR2 receptors, the gland responses to melatonin treatment may also be triggered by several pathways, including the androgen pathway, antioxidant enzymes modulation and, as recently reported by our research group, changes in mitochondrial physiology. We have published articles describing conditions where tissue damage was induced, such as long-term diabetes,

---

* Corresponding Author Email: remagoes@ibilce.unesp.br.

when animals treated with low melatonin doses showed an effective restoration of their prostate homeostasis and functions and managed to recover cell proliferation. It is worth mentioning that melatonin also raised testosterone serum levels, which were impaired after long-term diabetes induction. Hyperglycemia is often associated to an increase in oxidative stress, which leads to damage and cell death. Melatonin supplementation, even at low doses, promoted the normalization of activities of antioxidant enzymes on the prostate, which can be associated with reduced apoptosis indexes. In addition, our experiments with human epithelial prostate cells with high proliferative potential proved that melatonin has antioxidant properties due to strongly reduced oxygen reactive species (ROS) generation associated to increased oxidative phosphorylation (OXPHOS). Although melatonin has been considered beneficial for the prostate gland under normal or diabetogenic situations, its effects on cell proliferation are still poorly understood. At pharmacological concentrations it may exert anti-clonogenic effects both under normal conditions and after a short period of pre-incubation of prostatic benign and cancerous cells in a hyperglycemic medium. On the other hand, with longer pre-incubation it may favor cell proliferation. Obesity is also considered a metabolic disorder and when associated with prostate aging had been reported to favor prostatic lesions. In this context, melatonin administration to middle-aged rats was able to recover prostatic morphology and strongly decrease amyloid bodies' deposition even after obesity exposure. These results were probably related to total glutathione *S-transferase* (GST) activity improvement. Interestingly, under pro-oxidant conditions, generated by docosahexaenoic acid incubation (omega-3 fatty acid), the indole not only reduced dramatically ROS generation and enhanced OXPHOS, but also restored cell capacity to survive stress situations independently of the sensitization of the membrane receptors. Furthermore, our studies also suggest melatonin supplementation as a benign prostate hyperplasia chemoprevention method due to the decreased cell proliferation it promotes, even in a hyperlipidemic medium. Our studies revealed that melatonin exerts a protective role on the prostate gland under several metabolic conditions in rodents, human benign prostatic hyperplasia models and cancer cells.

Keywords: melatonin, prostate, diabetes, obesity, oxidative stress

INTRODUCTION

The prostate is an accessory gland of the male reproductive system, which, together with the seminal vesicle, contributes to the production of nutrients for the seminal fluid and promotes the maintenance of the ionic gradient and adequate pH for spermatozoa survival (McNeal, 1983; Untergasser et al., 2005). The prostate is formed by branched tubule-alveolar secretory units with columnar pseudo-stratified secretory epithelium (McNeal, 1983; Schaklen and Van Leenders, 2003). In humans, the prostate exhibits a compact zonal morphology, which is differentiated into three regions: central, transitional and peripheral (McNeal, 1983). In rodents, this gland forms a complex composed of four distinct components: anterior or coagulating, dorsal, lateral and ventral lobes (Hayashi et al., 1991). The lobes are arranged around the base of the bladder, each exhibiting particularities regarding their branching patterns and proteins secretion (Sugimura et al., 1986).

The secretory epithelium of the prostate is composed of several interrelated cell types, such as stem cells, basal cells, transit-amplifying cells (TAC), intermediate cells and secretory luminal cells (De March et al., 1998; Isaacs and Coffey, 1989; Schaklen and Van Leenders, 2003). These cell types represent progressive stages of differentiation, being distinguished not only by morphology and localization, but also by the pattern of gene expression, by the responses to physiological stimuli and by cellular plasticity. The secretory luminal cells are androgen dependent and also are the most abundant epithelial cells in both normal and hyperplastic prostates (Liu et al., 1997). Thus, prostatic involution caused by androgen ablation is largely due to the loss of these cells by apoptosis (Banerjee et al., 2000, Isaacs and Coffey 1989, Kerr and Seale 1973, Kyprianou et al. et al., 2003). In addition, other cells from the acinar epithelium are the neuroendocrine cells, which are not subject to androgenic action (Schaklen and Van Leenders, 2003), and the basal cell population, which represents a pool of stem cells essential for epithelial renewal (Garcia-Florez and Carvalho, 2005).

Surrounding the acinar ducts, there is a vascularized stroma containing collagen and elastic fibers, fibroblasts, smooth muscle cells (smc), immune system cells as well as a wide spectrum of regulatory molecules, growth factors and remodeling enzymes (Carvalho and Line, 1996; McNeal, 1997). Collagen is secreted by the fibroblasts and smc and has a structural role, promoting cell adhesion, resistance and tissue integrity. However, elastic fibers are related to prostatic extensibility and deformation (Carvalho et al., 1997; Montes, 1992). Together, these fibrillar elements play an important role, providing the cellular plasticity required during the contraction of smc, involving the acinar ducts, and elastic restoration during the contraction return (Carvalho et al., 1997).

In all species, the prostate develops from the endodermal urogenital sinus (UGS). The prostate, as other organs, is developed by the association of an epithelial parenchyma with a fibromuscular stroma, and it is dependent on reciprocal interactions between the differentiating epithelium and the underlying mesenchyme (Marker et al., 2003). The androgenic action is fundamental for the development and maintenance of the prostate homeostasis. The processes of proliferation and cell death are directly or indirectly regulated by androgens. The testosterone enters into the prostate cells and is converted to dihydrotestosterone (DHT) by the enzyme 5-α-reductase (Jena and Ramanão, 2010). The effect of the androgens is mediated by their receptor (AR), whose activation leads to the expression of genes related to the secretory activity of growth factors (Doncajour and Cunha, 1973).

It is well known that, during development, mesenchymal cells are the first to express AR receptors (Cooke et al., 1991). The paracrine stimulation of the smc on the epithelium not only induces the differentiation and acquisition of ductal morphogenesis, but also is essential for its maintenance. Several studies pointed out that paracrine signs of epithelial origin, together with androgens, also influence the pattern of smc differentiation (Marker et al., 2003). It is noteworthy that the imbalance in these paracrine interactions is decisive in the establishment and progression of prostate diseases such as benign hyperplasia (BPH) and prostate carcinoma (PCa) (Cunha et al., 1996; Hayward et al., 1996; Tuxhorn et al., 2002). In this context, situations that influence the cell kinetic or stroma-epithelium signaling such as diet, diabetes and aging may contribute to disrupt prostate homeostasis.

METABOLIC DISEASES, AGING AND THE ROLE OF MELATONIN IN THE PROSTATE PHYSIOLOGY

Diabetes mellitus (DM) is a set of complex metabolic disorders, of genetic or environmental origin, characterized by the failure in the secretion of insulin and/or action and consequent hyperglycemia. DM affects 8.3% of the world population (around 387 million people) and approximately 5.1 million people between the ages of 20 and 80 died due to diabetes or related complications in 2014 (International Diabetes Federation, Diabetes Atlas, 2014). DM is usually associated with systemic complications such as infections, neuropathy and angiopathy, which can also affect the genital tract. Concerning the male reproductive function, some adverse effects of DM are well known, such as sexual impotence, reduced libido and impairment of the reproductive potential due to urinary infections and a decrease in the number and quality of sperm (Arrellano-Valdez et al., 2014; Kolodny et al. al., 1874; Stege and Rabe, 1997). Recently, in a cohort study, it was observed that poor glycemic control of type 2 diabetes was associated with lower serum PSA levels and smaller prostate volumes (Atalay et al., 2017).

Some clinical studies point to a lower risk of developing PCa in diabetic patients, especially regarding Type 1 DM (Fall et al., 2013, Moreu et al., 2011, Xu et al., 2013, Yu et al., 2013), due to insulin deficiency and reduced testosterone levels (Bhasin et al., 2007, Cap, 2012). However, other studies point to a greater susceptibility to prostate tumors in diabetic individuals (Hammarsten and Hogstedt, 2002; Park et al., 2014; Shiozawa and Horie et al., 2014). In addition, research developed with individuals affected by PCa undergoing androgen deprivation therapy has shown that there is a risk of these individuals developing insulin resistance and hyperglycemia (Basaria et al., 2006; Tsai et al., 2015; Tzortzis et al., 2017).

The effects of this metabolic disorder have been extensively investigated with the use of experimental models of induced or spontaneous diabetes. Experimental studies have shown that diabetic animals present early damage in their prostate histophysiology (Barbosa-Desongles et al., 2013; Cagnon et al., 2000; Yu et al., 2008). Both epithelial and stromal compartments are affected by this disease resulting in gland atrophy, impairment of secretory activity, increase of apoptosis and decrease in the proliferation of epithelial cells. The remodeling of extracellular matrix components, and phenotypic changes in stromal cells such as dilation of organelles and approximation of cytoplasmic dense bodies (observed in eletronic microscopy) were also observed in the prostate of diabetic rats (Arcolino et al., 2010; Cagnon et al., 2000; Carvalho et al. 2003; Fávaro et al., 2009; Fávaro e Cagnon, 2010; Gobbo et al., 2012a; Ribeiro et al., 2006; 2008; 2009; Suthagar et al., 2009).

Our research group found in previous studies a higher incidence of malignant and premalignant lesions in the prostate of diabetic animals after 3 months of aloxan-induced diabetes (Ribeiro et al., 2008). Although the effects of diabetogenic drugs cannot be neglected in this case, it is worth mentioning that experimental diabetes causes the remodeling of the stromal collagen fibers, inflammatory processes and the alteration of adhesion proteins, factors that favor the establishment of intraepithelial neoplasia and adenocarcinoma (Cagnon et al., 2000; Cagnon e Fávaro, 2009; Ribeiro et al., 2006). Moreover, this metabolic disorder negatively affects the gland's maturation and alters the expression of multiple genes in the prostate (Soudamani et al., 2005; Ye et al., 2011).

The occurrence of DM modulates prostate changes, reducing levels of androgens and promoting low intracellular glucose levels (Chandrashekar et al., 1991; Yono et al., 2005). The testosterone withdrawal happens due to the imbalance of the pituitary-gonadal axis and the lower secretion of luteinizing hormone LH, the hormone that stimulates androgen synthesis by Leydig cells (Bebakar et al., 1990; Bhasin et al. 2007; Wang et al. 2000). Due to this androgenic scarcity, the expression of the androgen receptor (AR) in the prostate is altered (Arcolino et al., 2010; Gobbo et al., 2012b; Tesone et al., 1980). Even combined treatment with testosterone and insulin was unable to normalize the levels of cell proliferation in diabetic mice (Favaro et al., 2009).

Many diabetes-related consequences are caused by the high glycemic levels that it promotes, so it is important to investigate the consequences of prolonged hyperglycemia. High glycemic levels favor the production of free radicals, consequently increasing the oxidative stress. The oxidative status in DM is due to impairment of the mitochondrial electron transfer, the activation of polyol pathways, the catalysis of cyclooxygenase intermediate products and enhanced non-enzymatic glycation products (AGE) (Jang et al., 2010; Montillla et al., 1998; Sivitz et al., 2010). In turn, AGEs produced by non-enzymatic glycation lead to the generation of reactive oxygen species (ROS), the activation of Bax, and the expression of pro-apoptotic and pro-inflammatory genes, such as c-Jun N-terminal kinase (Buccellato et al., 2004; Du et al., 1998; Wautier et al., 2001). Thus, as confirmed by cDNA microarray analysis, DM can alter the expression of multiple genes, particularly those related to cell proliferation and differentiation, oxidative stress biomarkers, DNA damage repair and apoptosis (Ye et al., 2011). In addition, as glucose oxidation increases in the cytoplasm, the respiratory chain reaches its maximum capacity and the electrons are directed back to the III complex, increasing the production of ROS due to higher flux of electrons. The oxidative stress and resulting cellular damages are of prominent importance in the etiology of cancer in several organs (Alpay et al., 2016; Gago-Domingo et al., 2005; Miar et al., In press; Spector et al. 2000), including the prostate adenocarcinoma (Kotrikadze et al., 2008; Paschos et al., 2013;Yu et al., 2013).

Considering this scenario, the effectiveness of various antioxidants against oxidative stress caused by DM has been evaluated (Cameron and Cotter, 1999; Fernandes et al., 2011; Gobbo et al., 2012b; Muralidhara, 2007a, b; Rahimi et al., 2005). In our laboratory, we recently evaluated the protective effect of ascorbic acid on the prostatic antioxidant system, as well as its impact on cell death and proliferation in rat prostate gland (Gobbo et al., 2012b). It was found that one month of diabetes did not alter the biomarkers of oxidative stress in the plasma, increasing only the levels of activity of the enzymes catalase and GST. However, this study showed that one month of diabetes already compromises some biomarkers of the prostate antioxidant system and increases the immunoexpression of AGEs by the gland. The treatment with vitamin C normalized the increased activity of GST and decreased epithelial apoptosis in the prostate after one month of streptozotocin-induced diabetes. These data indicate that even weak antioxidants can prevent and even reverse the initial damage of DM.

In the same line of research, focusing on the interference of antioxidant substances in damages caused in the prostate by experimental diabetes, we propose to evaluate the effects of melatonin (MLT). The synthesis of MLT by the pineal gland is affected by hyperglycemia (Amaral et al., 2014). Diabetic individuals do not have a MLT-controlled circadian rhythm (O'Brien et al., 1986). MLT causes the inhibition of cyclic AMP (cAMP) pathway, which decreases insulin release (Pechke et al., 2002) and seems to promote the improvement of insulin sensitivity by phosphorylation of components of the insulin signaling pathway such as insulin receptor substrates 1 and 2 (IRS-1 and IRS-2) (Zanuto et al., 2013). Peschke and colleagues reported that MLT influences the synthesis of insulin (Peschke et al., 2002), and that insulin secretion obeys a circadian rhythm that is inversely proportional to the MLT release peak (Bazwinky-Wutschke et al., 2012).

At first, we examined the effects of low doses of MLT on the prostate response to diabetes (Gobbo et al., 2015a). For this, MLT was administered to Wistar rats through drinking water (10 µg/kg b.w), since weaning to adult age. DM was induced at 13 weeks of age by streptozotocin and short-term (one week) and long-term (four weeks) diabetes were evaluated. The prostate exhibited a 3-fold increase in gluthatione peroxidase (GPx) activity at short-term DM and a 2-fold increase in catalase (CAT) at long-term DM as well as an increase in GST in both stages of the disease. Comparing the biomarker of oxidative stress in the testis, epididymis and prostate, we have found that the rat prostate antioxidant system is more vulnerable to oxidative stress due to hyperglycemia than the testis and epididymis (Gobbo et al., 2015a). The administration of MLT normalized the GST activity in both ages and mitigated the increase GPx in the short-term diabetic animals and attenuated the increase of CAT in the prostate of long-term diabetic animals, even at low dosages. It is worth mentioning that that GST is an important component in the defense against oxidative damage in the rat prostate (Gobbo et al., 2012b; 2015a).

These findings agree with previous data that report a pivotal role of GST isoforms in the healthy prostate and in the disease progression (De Marzo, et al., 2007; Parson et al., 2001).

We used the same protocol mentioned above (Gobbo et al., 2015a) to discriminate the prostate tissue response to MLT treatment under diabetes and its possible influence on the prostate maturation. Thus, it was observed that low doses of MLT sustained the proliferative activity of the prostate after two months of experimental DM. This maintenance was associated with an improvement of the androgen synthesis that is known to be impaired in chronic DM, as quoted above (Arcolino et al., 2010; Gobbo et al., 2012b; 2015b). The non-diabetic animals, which received only MLT, displayed a 10% reduction of positive-cells for androgen receptor (AR), and their testosterone synthesis decreased. However, these alterations did not interfere with the proliferative and apoptotic response in the prostate of the younger rats. This suppressive action of MLT on androgen levels of healthy rats was expected and it may be due to the modulation of the gonadotropin-inhibitory hormone (GnIH) and its respective receptor (McGuire et al., 2011), the decrease in LH levels (Yilmaz et al., 2000) or the direct action of MLT in interstitial testicular cells through MLT receptors (Costa et al., 2016; Dubocovich et al., 2005).

In addition, it was observed that the prostate's response to MLT treatment differs depending on the progression of the disease, following alterations of testosterone levels, because modifications in proliferation and apoptosis rates and improvement of circulating testosterone were not found in the animals part of the short-term experiment. The immunohistochemical data showed a significant reduction of the MLT receptor type 1B (MTR1B) in the long-term diabetic group in relation to the untreated diabetic group, an alteration which can support the increased proliferation index found in these animals. The MTR1 is related to the antiproliferative response of prostate cancer cells to MLT, which leads to the down-regulation of activated AR signaling and upregulation of $p27^{kip1}$ in 22Rv1 cells (Shiu et al., 2013). Also, regarding the tissue response of diabetic prostate to MLT administration, preliminary results showed that treatment of diabetic rats with MLT counteracted the majority of diabetic histological changes in the prostate, in particular the epithelial and smooth muscle atrophy, and reduced the incidence of inflammatory infiltrates. Nevertheless, it did not prevent the development of microinvasive carcinoma (Gobbo et al., submitted).

It is difficult to discriminate the direct influence of this hormone in androgen-dependent organs, such as the prostate. Hevia et al. (2015) investigated the effects of MLT on culture media with different glucose concentrations and found that hyperglycemia favored the uptake of MLT by LNCaP and PC3 cells. The incubation of LNCaP cells in a hyperglycemic medium caused a reduction of AR expression through the activation of the nuclear factor kappa beta (NFκβ) by phosphorylation (Barbosa Desongles et al., 2013). To better clarify the implications of MLT from its secondary effect on androgen levels and the interferences of hyperglycemia and AR expression, *in vitro* experiments were performed to examine the MLT influence (5 mM and 10 mM) on the cell proliferation in human prostate epithelium cells (PNT1A) and prostate cancer cell lineages (androgen-dependent; 22Rv1 and androgen-independent, PC3 cancer cells) (Gobbo et al., 2015b) under hyperglycemic condition. The glucose concentration used in the medium (450 mg/dL) was equal to the glycemic index of the diabetic animals in the present investigation. The prolonged incubation in the hyperglycemic medium impaired the mitosis of 22Rv1 and PNT1A cells but improved this parameter for PC3 cells. This result was similar to that observed in the *in vivo* experiments, because an increase in mitosis was elicited at longer pre-incubation times with hyperglycemic medium and MLT. Since PC3 cells are androgen-independent prostate cancer cells, our findings indicate that this improvement of PC3 cell viability probably is not related to AR expression (Gobbo et al., 2015b).

Like diabetes, obesity commonly triggers insulin resistance due to changes in the pattern of adipocytokines released by the adipose tissue (Zeyda e Stulnig, 2007). Several studies regarding diets rich in fatty acids reported detrimental effect on the prostate homeostasis and have been implicated in PCa development at a worse prognosis (Chan et al., 2005; Giovannucci et al., 2003; Romero et al., 2003). In this context, nature of the lipid seems to be a relevant factor, since diets rich in saturated fatty acids lead to an increase in circulating levels of androgens, cell proliferation, as well as androgen and peroxisome receptor proliferation which are not observed in the case of the intaking of unsaturated fatty acids (Escobar et al., 2009). Although an inverse correlation was reported between increased insulin and circulating androgen levels (Pasquali et al., 1995), histological and cellular alterations caused by excess of lipids in the diet as well as the mechanisms that trigger them are still poorly understood.

Regarding the several effects of obesity and diet in the health, our laboratory research has extensively investigated their consequences on the prostate histophysiology in rodents. The treatment of adult rats fed for 17 weeks with a high-fat diet (20% saturated fat) increased the cell proliferation index and the incidence of premalignant prostatic lesions which may be related to the estrogen receptor (ER) and PI3K pathway (Ribeiro et al., 2012b). The application of the same protocol for obesity induction also revealed the occurrence of smooth muscle and prostatic epithelial cell hyperplasia as well as increased expression of Fibroblast Growth Factor 2 (FGF-2). These findings (Ribeiro et al., 2012a) may explain the high frequency of BPH in men with insulin resistance.

Another study from our laboratory showed that the high intake of lipids during pubertal maturation leads to increased cell proliferation and a decrease in the apoptotic rate causing epithelial hyperplasia in the ventral prostate (Pytlowanciv et al., submitted). These data are in line with previous studies that reported an increase in the prostate ventral lobe weight after high lipid intake during prostate maturation in rodents (Cai et al., 2001; Escobar et al., 2009). Prolonged consumption of a high-fat diet also provoked greater deposition of fibrillar and glycosaminoglycan components, increased the expression of metalloproteinases as well as glutathione S-transferase activity and lipid peroxidation in the ventral prostate of rats (Silva et al., 2015). Altogether, our data supports that obesity or even an increase in the intake of lipids is prejudicial to the prostate functions due to alterations that may encourage the onset of pathological lesions in the gland and increase PCa risk. In this line, how to attenuate such effects of obesity remains a challenge. The supplementation with MLT may be a good strategy to mitigate prostate changes due to the indole pleotropic functions, mainly antioxidant, antiproliferative and insulin signaling modulation.

The role of MLT in obesity has been extensively reviewed (Cardinali and Vigo, 2017; Cipolla-Neto et al., 2014). The anti-obesity effect of MLT has been suggested since its supplementation therapy in young rodents with intact pineal production decreased body weight and visceral fatness without reducing food intake (Cipolla-Neto et al., 2014; Prunet-Marcassus et al., 2003). Furthermore, indole administration exerted hyperinsulinemia attenuation, decreased triglyceride and leptin levels, normalized glucose (Prunet-Marcassus et al., 2003; Rasmussen et al., 1999; Sartori et al., 2009) and in cases after a pinealectomy, it caused overweight, but the supplementation completely reversed this effect (Cipolla-Neto et al., 2014). Obesity is also known to stimulate pro-inflammatory response due to the release of adipocytokines such as interleukin-1 (IL-1), interleukin-6 (IL-6) and Tumor Necrose Factor alpha (TNF-α) (Cardinali et al., 2017). In this context, MLT has been reported to down-regulate these cytokines and up-regulate anti-inflammatories such as interleukin-10 (IL-10) (Kireev et al., 2008). Regarding insulin signaling, indole played a key role by upregulating IRS-1 which seems to be crucial to alleviate insulin resistance (Du e Wei, 2014). Due to these multiple functions, the MLT acts as a "buffer" to the immune system (Carrillo-Vico et al., 2013) and has a great effect on the prostate gland since even low-grade inflammations may favor PCa development.

According to the literature, there is a prevalence of BPH cases in obese and hyperglycemic elderly men (Bourke and Griffin, 1966; Dahle et al., 2002). The prostate is among the organs most susceptible to senescence and is submitted to benign or malignant tissue alterations with the advancement of age (Banerjee et al., 2001; Cunha et al., 2004; Pegorin De Campos et al., 2006). BPH is detected in 70% of men over 60 years old (Ying et al., 2008) and prostate adenocarcinoma is the most common solid tumor and the second leading cause of cancer death in many countries (Jemal et al., 2009). The prostate gland, as mentioned above, is directly regulated by sex steroids and imbalances in testosterone/estrogen ratio due to aging can lead to onset and progression of lesions (Cohen and Rokhlin, 2009; Cunha, 1996; Cunha et al., 2004; Cunha et al., 1987; Hsing et al., 2002). Previous studies of our laboratory with gerbils (*Meriones Unguiculatus*) at 18 months of age revealed that alterations occur mainly in the epithelium, provoking the formation of intraepithelial neoplasia (46.6%), carcinomas and adenocarcinomas

(26.67%), stromal cell hyperplasia (20%) and increased lipofuscin granules in the cytoplasm (Campos et al., 2008). Moreover, was observed an imbalance in the proliferation and cell death indexes and an increase in the aggressiveness of the prostatic lesions in the aging rats (Arcolino et al., 2010; Bostwick et al., 1993; Tang e Porter, 1997), as well as a decrease of AR expression and 5-α reductase activity (Motta et al., 1986; Shain and Nitchuk, 1979).

Is worth mentioning that MLT secretion decays substantially with the advancement of age and this decrease is closely related to the antioxidant capacity of human plasma (Benot et al., 1999). There are two major declines in MLT levels: during puberty and the transition from adulthood to elderhood, there being a reduction of around 50% in the last case (Hardeland, 2012; Touitou, 2001). The lower secretory capacity of the pineal gland may be due to several causes, such as a decrease in the amount and noradrenergic sensitivity of receptors and pinealocytes (Greenberg and Weiss, 1978), the reduction of N-acetyltransferase activity, the enzyme of MLT synthesis pathway (Selmaoui and Touitou, 1999), the progressive degeneration of central nervous system or the nerve of pineal gland transmission axis (Srinivasan et al., 2005; Wu et al., 2006) as well as calcification of the gland itself (Kunz et al., 1999). Although the amplitude of MLT secretion in senile individuals oscillates significanty, nocturnal levels of this hormone are very close to daytime levels and there are no noticeable differences between day and night indole levels. Therefore, these low levels of MLT in the elderly lead to disturbances in their circadian rhythm that culminate in temporal desynchronization inducing different chronopathologies.

The occurrence of obesity during aging may lead to the worsening of some features common to both situations, such as insulin resistance and glucose intolerance (Barzilai et al., 1998; Nishimura et al., 1988; Watve and Yajnik, 2007). However, increase in the production of free radicals, inherent with aging, as well as the decrease of the antioxidant defense capacity (Stadtman, 2006; Starke-Reed and Oliver, 1989; Weinert and Timiras, 2003) may also favor the induction of insulin resistance because oxidative stress may activate kinases such as p38, mitogen-activated protein kinase (MAPK), and JNK, which influence the signal transduction of some proteins of insulin signaling pathway, including IRS. This association between oxidative stress and insulin resistance has been demonstrated in several tissues such as muscle, fat, liver and the endocrine pancreas (Blair et al., 1999; Tirosh et al., 1999; Tirosh et al., 2000). A study performed in our laboratory research with senile rodents fed for a long time (38 weeks) on a high-fat diet for obesity induction, exhibited several prostate alterations such as acinar atrophy due to the reduction in the epithelial height, collagen fibers remodeling and a strong fall in GPx activity, tightly related to a predisposition for prostatic lesions (Tamarindo et al., submitted). In addition, MLT, offered in drinking water at a dose of 100 μg/body weight/day, avoided prostatic atrophy in obese senile rats by maintaining the epithelial thickness affected by the diet, as well as stimulating the GST activity in the prostate

(Tamarindo et al., submitted). It is worth to emphasize that as obesity effects already discussed, the indole was able to attenuate the effects inherent to aging such as formation of prostatic calculi, as well as reduce the frequency of AR-positive cells and stimulate cell proliferation (Tamarindo et al., submitted). Aging also increases the synthesis of inflammatory mediators even in the absence of stress or acute infection, an event called by many authors "inflamm-aging" (Boren and Gershwin, 2004; Cevenini et al., 2013; Franceschi et al., 2000). In this context, outside the scope of our investigations, the MLT supplementation may also operate as a buffer for the immune system, as mentioned previously, and contribute to prostate hemostasis in aged subjects whose PCa risk is increased.

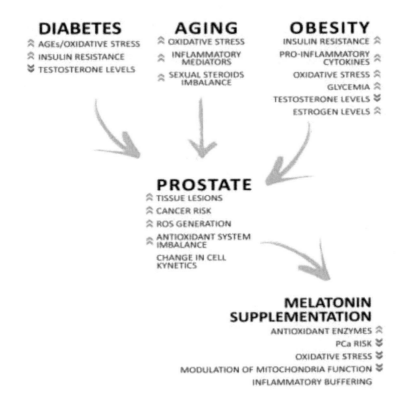

Figure 1. The interrelation between the metabolic disturbances, their effects on the prostate gland and the effective role of MLT.

In this line, under experimental conditions concerning metabolic syndromes, MLT has positive effects on the recovery of prostate damage avoiding the formation of environments favorable to the development of lesions and even decreasing the risk of PCa, summarized in Figure 1. Furthermore, it seems that the oxidative stress plays an important role in this property of the hormone, as well as influences the androgen regulation.

CANCER AND MELATONIN CHEMOPREVENTIVE EFFECT IN THE PROSTATE PHYSIOLOGY UNDER EXPERIMENTAL CONDITIONS

MLT supplementation has been suggested in the last decades as a potential strategy for tumor growth inhibition (Shiu, 2007; Stehle et al., 2011). Regarding the prostate, this hypothesis is corroborated by clinical and experimental studies that have reported a correlation between low serum levels of the indole and the development of PCa. It is well known that patients diagnosed with PCa have lower indole plasma concentrations, as well as that in situations where there is a progression of BPH to adenocarcinoma the decrease in hormone levels is remarkable (Bartsch et al., 1992; Shiu, 2007). Interestingly, 75% of PCa incidence occurs in men over 65 years old (Shiu, 2007) and, as mentioned above, with aging there is ablation in the MLT synthesis due to the pineal gland's impairment (Iguchi et al., 1982; Srinivasan et al., 2005; Waldhauser et al., 1988) as well as imbalance in the mitochondrial functions due to higher ROS generation (Scialò et al., 2017). Moreover, studies with pinealectomized animals have reported an association between the absence of pineal MLT with the development and progression of different tumors, affecting, among other areas, the prostate (Blask et al., 2011; Cutando et al., 2012). In this context, indole administration was able to suppress tumor growth after inoculation of LNCaP cells, even after castration (Siu et al., 2002). Therefore, this evidence along with the literature, indicates that, due to its pleiotropic properties, supplementation with MLT in at-risk age groups (30-40 years) with higher probability of Prostatic Intraepithelial Neoplasia (PIN) formation, may be a strategy for reducing the risk of developing PCa (Sakr et al., 1996, 1993). MLT role in preventing early premalignant changes has been the subject of study in our research.

MLT action can be exerted in a manner dependent or not of membrane receptors coupled to G protein sensitization, expressed in the prostate (Gilad et al., 1998). It is known that at low concentrations, similar to levels found in plasma, MLT acts mainly by the activation of these receptors (Reiter et al., 2005). According to reviews found in literature (Reiter et al., 2017; Tam et al., 2007), the mechanisms triggered by the hormone that result in its anticlonogenic effect in PCa cells are those of interference in androgen receptor signaling pathway (AR), cell cycle arrest and decrease of angiogenesis due to VEGF (Vascular Endothelial Growth Factor) downregulation in xenograft models (Paroni et al., 2014; Sohn et al., 2015). The antiproliferative potential of MLT has been associated mainly with the sensitization of MTR1, responsible for co-activating PKC (Protein Kinase C) and PKA (Protein Kinase A) in parallel, leading to upregulation of $p27^{kip1}$, and reducing the mitotic rate (Tam et al., 2008, 2007). Furthermore, the increased activation of PKC via MTR1 is related to AR nuclear exclusion and reduction of its transcriptional activity (Lupowitz et al., 2001; Rimler et al., 2002). However, the androgen independent cells *in vitro* models of high aggressiveness of PCa, exhibit undetectable levels of MTR1 and AR and, when exposed to MLT, suggest other pathways of mitotic inhibition that are not yet fully understood (Marelli et al., 2000; Reiter et al., 2017). Although MLT exerts its antitumor effect mainly through these pathways, some studies have also revealed its ability to sensitize neoplastic cells due to the suppression of survival pathways, such as AKT, enhancing the suppressive effect of other compounds (Gao et al., 2017; Lu et al., 2016).

MELATONIN: CORRELATIONS BETWEEN OXIDATIVE STRESS, MITOCHONDRIAL FUNCTIONS AND PROSTATE PHYSIOLOGY

At high concentrations, close to those found in most tissues, the MLT can cross the plasma membrane and exert action independently of receptors, attenuating the oxidative stress, for example (Reiter et al., 2005). As an antioxidant, it neutralizes ROS directly or through its metabolites N1-acetyl-N2-formyl-5-methoxykynuramine (AFMK) and N1-acetyl-5-methoxykynuramine (AMK) (Galano et al., 2013; Matuszak et al., 1997). In addition, MLT is able to promote increased activity and expression of antioxidant enzymes, such as CuZnSOD (Superoxide dismutase 1), MnSOD (Superoxide dismutase 2), GST and GPx in a tissue-dependent manner whereas it downregulates pro-oxidant enzymes in different cell types (Acuña-Castroviejo et al., 2005; Iñarrea et al., 2011). One of the mechanisms for this action is that the indole modulates the antioxidant defense system by inhibiting calmodulin, decreasing activation of kinases, inactivating RORα (Orphan Receptor Alpha) in the nucleus increasing expression of antioxidant enzymes (Tomás-Zapico and Coto-Montes, 2005). However, besides these well-known routes, its role as an oxidative stress attenuator may also be due to the modulation of the mitochondrial function. MLT capability to influence the mitochondrial membrane potential has been reported as tightly related to ROS generation (López et al., 2009). Furthermore, the indole can affect mitochondrial function by elevating ATP synthesis due to an increase in the activity of Electron Transport Chain (ETC) complexes I, III and IV (Acuña-Castroviejo et al., 2001; Martín et al., 2000; 2002). Therefore, the literature offers strong evidence that mitochondrial function modulation by MLT may be tightly related to oxidative stress, but this is not completely understood regarding the prostate gland, including its correlation with cell death and proliferation rates.

Our recent studies with prostate cells indicated a correlation between the increase in oxidative stress and the desacceleration of cell proliferation. We have sought to understand how the modulation of the redox status by MLT may influence the survival of prostate cells with high proliferative potential (PNT1A). Results showed that the indole was able to reduce oxidative stress in the PNT1A line due to the modulation of the mitochondrial physiology, mainly through the improvement of the oxidative phosphorylation (Tamarindo et al., submitted). Decreased generation of ROS after exposure to MLT may be related to changes in the inner mitochondrial membrane fluidity, as previously described by López and colleagues (2009) at the same concentration used in our studies (100µM). Moreover, the hormone may still reduce oxidative stress by attenuating the proton leak in the mitochondria, as described in adipocytes (Jimenéz-Aranda et al., 2014).

It is worth mentioning that with the improvement of the mitochondrial coupling, the morphology of the organelle changed. Prostatic cells exposed to the indole exhibited a smaller area and perimeter, which may elucidate a quality system responsible for total or partial removal of mitochondrial fractions that detain higher oxidative stress and are harmful to the cell or even related to proliferation stimulation (Tamarindo et al., submitted). On the other hand, it is not clear yet if the effects on the mitochondrial physiology may be a consequence of the early changes in organelle dynamics or modulation of antioxidant defense system. Our results also suggest that the antimitogenic effect was associated with cytosolic ATP increases due to respiratory control improvement. According to several authors, MLT is capable of raising the ADP/O ratio, increasing the ATP production and promoting increased expression of ATP synthase β subunit (Castillo et al., 2005; Chuffa et al., 2016; López et al., 2009; Martín et al., 2002). In this line, studies with prostatic cells have reported that the administration of extracellular ATP inhibits the growth of tumor cells (Zhang et al., 2013) and acts suppressing AKT pathway activation (Yang et al., 2017), which also showed decreased activation after exposure to MLT in our *in vitro* results. Some authors associate ATP increased intracellular production to apoptosis rather than necrosis (Tatsumi et al., 2003; Zamaraeva et al., 2005), a phenotype which was not observed in microscopic analysis after our incubation with indole. Therefore, decreases in cellular proliferation may be associated to the reduction of oxidative stress which influences mitotic stimulation and cell survival pathways. Together, our recent data points to the role of mitochondrial bioenergetics in cellular arrest and suggests that MLT acts as an effective antiproliferative agent.

It is worth mentioning that PNT1A cells exposed to MLT exhibited down-regulation of the AKT / mTOR pathway simultaneously to the modulation of mitochondrial physiology. It is known that AKT downregulation, as well as proliferation decrease, can also be associated with the upregulation of glycogen synthase kinase 3 beta (GSK3β) (Benelli et al., 2010). Moreover, increased GSK3β activation, an enzyme whose agonists have been investigated as potential PCa therapeutic agents, is related to cell arrest, mainly due to cyclin D1 degradation (Diehl et al., 1998), which would be correlated with decreased proliferation due to delay in G0 to G1, even in the prostate (Moretti et al., 2000; Siu et al., 2002).

Finally, our studies have shown that the improvement in oxidative phosphorylation, the mitochondrial morphology modulation as well as the suppression of survival pathways were not dependent on MTR1 and MTR2 receptors regarding PNT1A. The MLT, due to uptake by prostate cells (Hevia et al., 2010), revealed pathways other than the sensitization of membrane receptors and plays a key role in the modulation of mitochondrial activity and oxidative stress.

Conclusion

Metabolic disorders have a negative effect on the prostate functions due to changes they promote in the morphology, to the inflammatory environment, to oxidative stress and to proliferation-related pathways that may favor the increase of tissue lesions or even prostate cancer risk. Likewise, aging causes alterations in sexual steroid levels, generates antioxidant system impairment and increases insulin resistance and pro-inflammatory mediators that may disrupt prostate homeostasis. Our recent contributions together with the literature, have demonstrated that MLT supplementation attenuates effects of diabetes mellitus type 1 and 2 due to its anti-inflammatory, antioxidant and energy metabolism regulation properties. In addition, the indole may also exhibit chemopreventive potential through the modulation of mitochondria bioenergetics associated to proliferation rates of cells with malignant or premalignant profile. More studies are needed to better comprehend the role of the MLT in the prostate functions, but indole therapy may be a good strategy in the prevention of prostate disorders caused mainly due to aging and metabolic dysfunction. Moreover, there is scarce information on the consequences of MLT administration on the prostate and this should be carefully analyzed, especially when MLT is intended to be applied as a coadjutant in the treatment of prostate cancer for patients with metabolic disturbances, such as diabetes.

References

Acuña-Castroviejo, D., Escames, G., López, L. C., Hitos, A. B., León, J. (2005). Melatonin and Nitric Oxide: Two Required Antagonists for Mitochondrial Homeostasis. *Endocrine,* v. 27, n. 2, p. 159–168.

Acuña-Castroviejo, D., Martín, M., Macías, M., Escames, G., León, J., Khaldy, H., Reiter, R. J. (2001). Melatonin, mitochondria, and cellular bioenergetics. *Journal of pineal research,* v. 30, n. 2, p. 65–74.

Alpay, M., Backman, L. R., Cheng. X., Dukel, M., Kim, W. J, Ai., L., Brown, K. D. Breast (2015). Oxidative stress shapes breast cancer phenotype through chronic activation of ATM-dependent signaling. *Cancer Res Treat.* v.151, n. 1, p. 75-87.

Arcolino, F. O., Ribeiro, D. L., Gobbo, M. G., Taboga, S. R., Góes, R. M. (2010). Proliferation and apoptotic rates and increased frequency of p63-positive cells in the prostate acinar epithelium of alloxan-induced diabetic rats. *Int J Exp Path.* v. 91, n. 2, p.144–154.

Arellano-Valdez, F., Urrutia-Osorio, M., Arroyo, C., Soto-Veja, E. (2014). A comprehensive review of urologic complications in patients with diabetes. *Springerplus.* v. 23, n. 3, p. 549.

Atalay, H. A., Akarsu, M., Canat, L., Ülker, V, Alkan, İ., Ozkuvancı, U4. (2017). Impact of poor glycemic control of type 2 diabetes mellitus on serum prostate-specific antigen concentrations in men. *Prostate* v. 5, 3, p. 104-109.

Banerjee, P. P., Banerjee, S., Brown, T. R. (2001). Increased androgen receptor expression correlates with development of age-dependent, lobe-specific spontaneous hyperplasia of the brown Norway rat prostate. *Endocrinology,* v. 142, n. 9, p. 4066–75.

Banerjee, P. P., Banerjee, S., Lai, J. M., Strandberg, J. D., Zirkin, B. R., Brown, T. R. (1998). Age dependent and lobe-specific spontaneous hyperplasia in the Brown Norway rat prostate. *Biol Reprod.* v. 59, p. 1163-1170.

Barbosa-Desongles, A., Hernández, C., De Torres, I., Munell, F., Poupon, M. F., Simó, R., Selva, D. M. (2013). Diabetes protects from prostate cancer by downregulating androgen receptor: new insights from LNCaP cells and PAC120 mouse model. *PloS One.* v.8, n. 9, e74179.

Bartsch, C., Bartsch, H., Schmidt, A., Ilg, S., Bichler, K. H., Flüchter, S. H. (1992). Melatonin and 6-sulfatoxymelatonin circadian rhythms in serum and urine of primary prostate cancer patients: Evidence for reduced pineal activity and relevance of urinary determinations. *Clinica Chimica Acta,* v. 209, n. 3, p. 153–167.

Barzilai, N., Banerjee, S., Hawkins, M., Chen, W., Rossetti, L. (1998). Caloric restriction reverses hepatic insulin resistance in aging rats by decreasing visceral fat. *The Journal of clinical investigation,* v. 101, n. 7, p. 1353–1361.

Basaria, S., Muller, D. C., Carducci, M. A., Egan, J., Dobs, A. S. (2006). Hyperglycemia and insulin resistance in men with prostate carcinoma who receive androgen-deprivation therapy. *Cancer.* v. 106, p.,581–588.

Bazwinsky-wutschke, I., Wolgast, S., Muhlbauer E. et al. (2012). Phosphorylation of cyclic AMP-response element–binding protein (CREB) is influenced by melatonin treatment in pancreatic rat insulinoma b-cells (INS-1). *J Pineal Res.* v. 53, p. 344–357.

Bebakar, W. M. W., Honour, J. W., Foster, D., Liu, Y. L., Jacobs, H. S. (1990). Regulation of testicular function by insulin and transforming growth factor-β. *Steroids.* v. 55, n.6, p. 266-270.

Benelli, R., Monteghirfo, S., Venè, R., Tosetti, F., Ferrari, N. (2010). The chemopreventive retinoid 4HPR impairs prostate cancer cell migration and invasion by interfering with FAK/AKT/GSK3beta pathway and beta-catenin stability. *Molecular cancer,* v. 9, p. 142.

Benot, S., Goberna, R., Reiter, R. J., Garcia-Mauriño, S., Osuna, C., Guerrero, J. M. (1999). Physiological levels of melatonin contribute to the antioxidant capacity of human serum. *Journal of pineal research*, v. 27, n. 1, p. 59–64.

Bhasin, S., Enzlin, P., Coviello, A., Basson, R. (2007). Sexual dysfunction in men and women with endocrine disorders. *Lancet.* n. 369, p. 597–611.

Blair, A. S., Hajduch, E., Litherland, G. J., Hundal, H. S. (1999). Regulation of glucose transport and glycogen synthesis in L6 muscle cells during oxidative stress. Evidence for cross-talk between the insulin and SAPK2/p38 mitogen-activated protein kinase signaling pathways. *The Journal of biological chemistry*, v. 274, n. 51, p. 36293–9, 17.

Blask, D. E., Hill, S. M., Dauchy, R. T., Xiang, S., Yuan, L., Duplessis, T., Mao, L., Dauchy, E., Sauer, L. A. (2011). Circadian regulation of molecular, dietary, and metabolic signaling mechanisms of human breast cancer growth by the nocturnal melatonin signal and the consequences of its disruption by light at night. *Journal of pineal research*, v. 51, n. 3, p. 259–69.

Boren, E., Gershwin, M. E. (2004). Inflamm-aging: autoimmunity, and the immune-risk phenotype. *Autoimmunity reviews*, v. 3, n. 5, p. 401–6.

Bostwick, D. G., Amin, M. B., Dundore, P., Marsh, W., Schultz, D. S. (1993). Architectural patterns of high-grade prostatic intraepithelial neoplasia. *Human pathology*, v. 24, n. 3, p. 298–310.

Bourke, J. B., Griffin, J. P. (1966). Hypertension, diabetes mellitus, and blood groups in benign prostatic hypertrophy. *British journal of urology*, v. 38, n. 1, p. 18–23.

Buccellato, L. J., Tso, M., Akinci, O. I., et al. (2004). Reactive oxygen species are required for hyperoxia-induced Bax activation and cell death in alveolar epithelial cells. *J Biol Chem*, v. 279, n. 8, p. 6753-6760.

Cagnon, V. H., Camargo, A. M., Rosa, R. M., Fabiani, R., Padovani, C. R., Martinez, F. E. (2000). Ultrastructural study of the ventral lobe of the prostate of mice with streptozotocin induced diabetes (C57BL/6J). *Tissue & Cell*, v. 32, n. 4, p.275–283.

Cai, X., Haleem, R., Oram, S., Cyriac, J., Jiang, F., Grayhack, J. T., Kozlowski, J. M., Wang, Z. 92001). High fat diet increases the weight of rat ventral prostate. *The Prostate*, v. 49, n. 1, p. 1–8, 15.

Cameron, N. E., Cotter, M. A. (1999). Effects of antioxidants on nerve and vascular dysfunction in experimental diabetes. *Diabetes Res. Clin. Pract,* 45,137–146.

Campos, S. G. P., Zanetoni, C., Scarano, W. R., Vilamaior, P. S. L., Taboga, S. R. (2008). Age-related histopathological lesions in the Mongolian gerbil ventral prostate as a good model for studies of spontaneous hormone-related disorders. *International Journal of Experimental Pathology*, v. 89, n. 1, p. 13–24.

Cáp, J. (2012). Androgen deficit and diabetes. *Vnitr Lek*, v. 58, n.3, p. 228-231.

Cardinali, D. P., Vigo, D. E. (2017). Melatonin, mitochondria, and the metabolic syndrome. *Cellular and Molecular Life Sciences*, n. 123456789, p. 1–14.

Carrillo-Vico, A., Lardone, P., Álvarez-Sánchez, N., Rodríguez-Rodríguez, A., Guerrero, J. (2103). Melatonin: Buffering the Immune System. *International Journal of Molecular Sciences,* v. 14, n. 4, p. 8638–8683.

Carvalho, H. F., Camargo, M., Cagnon, V. H., Padovani, C. R. (2003). Effects of experimental diabetes on the structure and ultrastructure of the coagulating gland of C57BL/6J and NOD mice. *Anat Rec. Discoveries in Molecular, Cellular, and Evolutionary Biology.* v. 270, n. 2, p. 129–136.

Carvalho, H. F., Line, S. R. P. (1996). Basement membrane associated changes in the rat ventral prostate following castration. *Cell Biol Int.* v.20, p. 809-819.

Carvalho, H. F., Vilamaior, P. S. L., Taboga, S. R. (1997). Elastic system of the rat ventral prostate and its modifications following orchiectomy. *Prostate.* v. 321, p. 27-34.

Castillo, C., Salazar, V., Ariznavarreta, C., Vara, E., Tresguerres, J. A. F. (2005). Effect of melatonin administration on parameters related to oxidative damage in hepatocytes isolated from old Wistar rats. *Journal of Pineal Research,* v. 38, n. 4, p. 240–246.

Cevenini, E., Monti, D., Franceschi, C. (2013). Inflamm-ageing. *Current Opinion in Clinical Nutrition and Metabolic Care,* v. 16, n. 1, p. 14–20.

Chan, J. M., Gann, P. H., Giovannucci, E. L. (2005). Role of diet in prostate cancer development and progression. *Journal of Clinical Oncology,* v. 23, n. 32, p. 8152–8160.

Chuffa, L. G. A., Lupi Júnior, L. A., Seiva, F. R. F., Martinez, M., Domeniconi, R. F., Pinheiro, P. F. F., Dos Santos, L. D., Martinez, F. E. (2016). Quantitative Proteomic Profiling Reveals That Diverse Metabolic Pathways Are Influenced by Melatonin in an in Vivo Model of Ovarian Carcinoma. *Journal of Proteome Research,* v. 15, n. 10, p. 3872–3882.

Cipolla-Neto, J., Amaral, F. G., Afeche, S. C., Tan, D. X., Reiter, R. J. (2014). Melatonin, energy metabolism, and obesity: A review. *Journal of Pineal Research,* v. 56, n. 4, p. 371–381.

Cohen, M. B., Rokhlin, O. W. (2009). Mechanisms of prostate cancer cell survival after inhibition of AR expression. *Journal of Cellular Biochemistry,* v. 106, n. 3, p. 363–371.

Cocke, P. S., Young, P., Cunha, G. R. (1991). Androgen receptor expression in developing male reproductive organs. *Endocrinology.* v. 128, p. 2867-2873.

Cunha, G. R. (1996). Growth factors as mediators of androgen action during male urogenital development. *The Prostate. Supplement,* v. 6, p. 22–5.

Cunha, G. R., Cooke, P. S., Kurita, T. (2004). Role of stromal-epithelial interactions in hormonal responses. *Archives of histology and cytology,* v. 67, n. 5, p. 417–434.

Cunha, G. R., Donjacour, A. A., Cooke, P. S., Mee, H., Bigsby, R. M., Higgins, S. J., Sugimura, y. (1987). The Endocrinology and Developmental Biology of the Prostate*. *Endocrine Reviews,* v. 8, n. 3, p. 338–362.

Cutando, A., López-Valverde, A., Arias-Santiago, S., DE Vicente, J., DE Diego, R. G. (2012). Role of melatonin in cancer treatment. *Anticancer research,* v. 32, n. 7, p. 2747–.

Da Costa, C. F., Gobbo, M. G., Taboga, S. R., Pinto-Fochi, M. E., Góes, R. M. (2016). Melatonin intake since weaning ameliorates steroidogenic function and sperm motility of streptozotocin-induced diabetic rats. *Andrology.* v. 4, n. 3, 526-541.

Dahle, S. E., Chokkalingam, A. P., Gao, Y. T., Deng, J., Stanczyk, F. Z., Hsing, A. W. (2002). Body size and serum levels of insulin and leptin in relation to the risk of benign prostatic hyperplasia. *The Journal of urology,* v. 168, n. 2, p. 599–604.

De Marzo, A. M., Nakai, Y., Nelson, W. G. Inflammation, atrophy, and prostate carcinogenesis. *Urol Oncol.* V. 2, n.5, pp. 398-400, 2007.

Di Francesco, and Tenaglia, R. L. (2017). Metabolic Syndrome and Aggressive Prostate Cancer at Initial Diagnosis. *Horm Metab Res.* n. 49, n. 7, p. 507-509.

Diehl, J. A., Cheng, M. G., Roussel, M. F., Sherr, C. J. (1998). Glycogen synthase kinase 3 beta regulates cyclin D1 proteolysis and subcellular localization. *Genes & Development,* v. 12, n. 22, p. 3499–3511.

Doncajour, A. A., Cunha G. R. (1983). Assessment of prostatic protein secretion in tissue recombinants made of urogenital sinus mesenchyme and urothelium from normal or androgen-insensitive mice. *Endocrinology,* v. 131, p. 2342-2350.

Du, X., Stotcklauser-Farber, K., Rosen, P., et al. 1998. Generation of reactive oxygen intermediates, activation of NF-kappaB, and induction of apoptosis in human endothelial cells by glucose: role of nitric oxide synthase? *Free Radic Biol Med.* v. 27, n. 7-8, pp. 752-763.

Du, Y., Wei, T. (2014). Inputs and outputs of insulin receptor. *Protein & Cell,* v. 5, n. 3, p. 203–213.

Dubocovich, M. L., Markowska, M. (2005). Functional MT1 and MT2 melatonin receptors in mammals. *Endocrine.* v. 27, n. 2, p. 101-110.

Escobar, E. L. O., Gomes-Marcondes, M. C. C., Carvalho, H. F. (2009). Dietary fatty acid quality affects AR and PPAR?? levels and prostate growth. *Prostate,* v. 69, n. 5, p. 548–558.

Fall, K., Garmo, H., Gudbjornsdottir, S., Stattin, P., Zethelius, B. (2013). Diabetes Mellitus and Prostate Cancer Risk; A Nationwide Case-Control Study within PCBaSe Sweden. *Cancer Epidemiol Biomarkers Prev.* v. 22, 6, p. 1102-1109.

Fávaro, W. J., Cagnon, V. H. A. (2010). Effect of combined hormonal and insulin therapy on the steroid hormone receptors and growth factors signalling in diabetic mice prostate. *Int J Exp Path.* n.91, n.6, p. 537–545.

Fávaro, W. J., Padovani, C. R., Cagnon, V. H. (2009). Ultrastructural and proliferative features of the ventral lobe of the prostate in non-obese diabetic mice (NOD). following androgen and estrogen replacement associated to insulin therapy. *Tissue Cell.* v. 41, n. 2., p. 119–132.

Franceschi, C., Bonafè, M., Valensin, S., Olivieri, F., De Luca, M., Ottaviani, E., De Benedictis, G. Inflamm-aging. (2000). An evolutionary perspective on immunosenescence. *Annals of the New York Academy of Sciences,* v. 908, p. 244–54.

Galano, A., Tan, D. X., Reiter, R. J. (2013). On the free radical scavenging activities of melatonin's metabolites, AFMK and AMK. *Journal of pineal research,* v. 54, n. 3, p. 245–57.

Gao, Y., Xiao, X., Zhang, C., Yu, W., Guo, W., Zhang, Z., Li, Z., Feng, X., Hao, J., Zhang, K., Xiao, B., Chen, M., Huang, W., Xiong, S., Wu, X., Deng, W. (2017). Melatonin synergizes the chemotherapeutic effect of fluorouracil in colon cancer by suppressing PI3K/AKT and κB/iNOS signaling pathways. *J. Pineal Res.* 2017; 62:e12380.

Garcia-Flórez, M., Carvalho, H. F. (2005). Prostatic epithelial cell. In: *Cells - a multidisciplinary approach.* Manole, Barueri. p.335-345.

Gilad, E., Pick, E., Matzkin, H., Zisapel, N. (1998). Melatonin receptors in benign prostate epithelial cells: evidence for the involvement of cholera and pertussis toxins-sensitive G proteins in their signal transduction pathways. *The Prostate,* v. 35, n. 1, p. 27–34.

Giovannucci, E., Rimm, E. B., Liu, Y., Leitzmann, M., Wu, K., Stampfer, M. J., Willett, W. C. (2003). Body mass index and risk of prostate cancer in U.S. health professionals. *Journal of the National Cancer Institute,* v. 95, n. 16, p. 1240–4.

Gobbo, M. G., Dizeyi, N., Abrahamsson, P. A., Bertilsson, P. A., Masitéli, V. S., Taboga, S. R., Góes, R. M. (2015). Influence of melatonin on proliferative and apoptotic responses of prostate under normal and hyperglycemic condition. *J Diabetes Res.* 538529.

Gobbo, M. G., Ribeiro, D. L., Taboga, S. R., de Almeida, E. A., Góes, R. M. (2012b). Oxidative stress markers and apoptosis in the prostate of diabetic rats and the influence of vitamin C treatment. *J Cell Biochem.* v. 113, n. 7, p. 2223–2233.

Gobbo, M. G., Taboga, S. R., Ribeiro, D. L., Góes, R. M. (2012a). Short-term stromal alterations in the rat ventral prostate following alloxan-induced diabetes and the influence of insulin replacement. *Micron.* v. 43, n. 2-3, p. 326–333.

Gobbo, M. G.; Tamarindo, G. H.; Ribeiro, D. L; De Campos, S. G. P.; Taboga, S. R.; Góes, R. M. *Pathological lesions and global DNA methylation in rat prostate under streptozotocin-induced diabetes and melatonin treatment,* submitted.

Greenberg, L. H., Weiss, B. (1978). beta-Adrenergic receptors in aged rat brain: reduced number and capacity of pineal gland to develop supersensitivity. *Science* (New York, N.Y.), v. 201, n. 4350, p. 61–3.

Hammarsten, J., Hogstedt, B. (2002). Hyperinsulinaemia as a risk factor for developing benign prostatic hyperplasia. *Eur. Urol.* v. 39, p. 151–158.

Hardeland, R. (2012). Melatonin in aging and disease -multiple consequences of reduced secretion, options and limits of treatment. *Aging and disease,* v. 3, n. 2, p. 194–225.

Hayward, S. W., Baskin, L. S., Haughney, P. C., Foster, B. A., Cunha, A. R., Dahiya, R., Prins G. S., Cunha, G. R. (1996). Stromal development in the rat ventral prostate, anterior prostate and seminal vesicle. *Acta Anat.* v. 155, p. 94-103.

Hevia, D., Mayo, J. C., Quiros, I., Gomez-Cordoves, C., Sainz, R. M. (2010). Monitoring intracellular melatonin levels in human prostate normal and cancer cells by HPLC. *Analytical and Bioanalytical Chemistry,* v. 397, n. 3, p. 1235–1244.

Hsing, A. W., Reichardt, J. K. V., Stanczyk, F. Z. (2002). Hormones and prostate cancer: Current perspectives and future directions. *The Prostate,* v. 52, n. 3, p. 213–235.

Iguchi, H., K. H., K., H., I. (1982). Age-Dependent Reduction in Serum Melatoniin Concentrations in Healthy Human Subjects. *Journal of Clinical Endocrinology and Metabolism,* v. 55, n. 1, p. 27–29.

Iñarrea, P., Casanova, A., Alava, M. A., Iturralde, M., Cadenas, E. (2011). Melatonin and steroid hormones activate intermembrane Cu,Zn-superoxide dismutase by means of mitochondrial cytochrome P450. *Free Radical Biology and Medicine,* v. 50, n. 11, p. 1575–1581.

International Diabetes Federation. Diabetes: facts and figures. IDF. *Diabetes Atlas* (2014). 7th edn. International Diabetes Federation, Brussels. Accessed on August, 2017.

Isaacs, J. T., Coffey, D. S. (1989). Etiology and disease process of benign prostatic hyperplasia. *Prostate Suppl,* v. 2, p. 33-50.

Jemal, A., Siegel, R., Ward, E., Hao, Y., Xu, J., Thun, M. J. (2009). Cancer Statistics, both sexes female both sexes estimated deaths. *CA Cancer J Clin,* v. 59, n. 4, p. 1–25.

Jena, A. V. G., Ramarão. P. (2010). Insulin-resistance and benign prostatic hyperplasia: The connection. *Eur. J. Pharmacol.* 651, p. 75-81.

Jiménez-Aranda, A., Fernández-Vázquez, G., Mohammad A. Serrano, M., Reiter, R. J., Agil, A. (2014). Melatonin improves mitochondrial function in inguinal white adipose tissue of Zücker diabetic fatty rats. *Journal of Pineal Research,* v. 57, n. 1, p. 103–109.

Kerr, J. F. R., Searle, J. (1973). Deletion of cells by apoptosis during castration–induced involution of the rat prostate. *Virchows Arch* (B). v. 1, p. 87-102.

Kireev, R. A., Tresguerres, A. C. F., Garcia, C., Ariznavarreta, C., Vara, E., Tresguerres, J. A. F. (2008). Melatonin is able to prevent the liver of old castrated female rats from oxidative and pro-inflammatory damage. *Journal of pineal research,* v. 45, n. 4, p. 394–402.

Kolodny, R. C., Kahn, C. B., Goldstein, H. H., Barnett, D. M. (1974). Sexual dysfunction in diabetic men. *Diabetes,* n. 23, p. 306-309.

Kotrikadze, N., Alibegashvili, M., Zibzibadze, M., Abashidze, N., Chigogidze, T., Managadze, L., Artsivadze, K. (2008). Activity and content of antioxidant enzymes in prostate tumors. *Exp. Onc.* V. 30, n. 3, p. 244-247.

Kunz, D., Schmitz, S., Mahlberg, R., Mohr, A., Stöter, C., Wolf, K. J., Herrmann, W. M. (1999). A new concept for melatonin deficit: On pineal calcification and melatonin excretion. *Neuropsycho-pharmacology,* v. 21, n. 6, p. 765–772.

Kyprianou, N., Isaacs, J. T. (1988). Activation of programmed cell death in the rate ventral prostate after castration. *Endocrinol*, 22,552-562.

Liu, A. Y., True, L. D., Latray, L. (1997). Cell-cell interaction in prostate gene regulation and cytodifferentiation. *Proc Natl Acad Sci* USA. v. 94,10705-10710.

López, A., García, J. A., Escames, G., Venegas, C., Ortiz, F., López, L. C., Acuña-Castroviejo, D. (2009). Melatonin protects the mitochondria from oxidative damage reducing oxygen consumption, membrane potential, and superoxide anion production. *Journal of Pineal Research,* v. 46, n. 2, p. 188–198.

Lu, Y., Chen, D., Wang, D., Chen, L., Mo, H., Sheng, H., Bai, L., Wu, Q., Yu, H., Xie, D., Yun, J., Zeng, Z., Wang, F., Ju, H., Xu, R. (2016). Melatonin enhances sensitivity to fluorouracil in oesophageal squamous cell carcinoma through inhibition of Erk and Akt pathway. *Nature Publishing Group,* v. 7, n. 10, p. e2432-12.

Lupowitz, Z., Rimler, A, Zisapel, N. (2001). Evaluation of signal transduction pathways mediating the nuclear exclusion of the androgen receptor by melatonin. *Cellular and molecular life sciences: CMLS*, v. 58, n. 14, p. 2129–35.

Marelli, M. M., Limonta, P., Maggi, R., Motta, M., Moretti, R. M. (2000). Growth-inhibitory activity of melatonin on human androgen-independent DU 145 prostate cancer cells. *The Prostate,* v. 45, n. 3, p. 238–44.

Marker, P. C., Donjacour, A. A., Dahiya, R., Cunha, G. R. (2003). Hormonal cellular and molecular control of prostatic development. *Develop Biol.* v. 253, p.165-174.

Martín, M., Macías, M., Escames, G., Reiter, R. J., Agapito, M. T., Ortiz, G. G., Acuña-Castroviejo, D. (2000). Melatonin-induced increased activity of the respiratory chain complexes I and IV can prevent mitochondrial damage induced by ruthenium red in vivo. *Journal of pineal research,* v. 28, n. 4, p. 242–8.

Martín, M., Macías, M., León, J., Escames, G., Khaldy, H., Acuña-Castroviejo, D. (2002). Melatonin increases the activity of the oxidative phosphorylation enzymes and the production of ATP in rat brain and liver mitochondria. *International Journal of Biochemistry and Cell Biology*, v. 34, n. 4, p. 348–357.

Matuszak, Z., Reszka, K., Chignell, C. F. (1997). Reaction of melatonin and related indoles with hydroxyl radicals: EPR and spin trapping investigations. *Free radical biology & medicine,* v. 23, n. 3, p. 367–72.

McGuire, N. L., Kangas, K., & Bentley, G. E. (2011). Effects of melatonin on peripheral reproductive function: regulation of testicular GnIH and testosterone. *Endocrinology,* v. 152, n. 9, p. 3461–3470.

McNeal J. E. (1997). Prostate. In: *Histology for Pathologists.* Lippincott-Raven Publishers, Philadelfia. Cap. 42.

Montilla, P. L., Vargas J. F., Túnez, I. F., et al. (1998). Oxidative stress in diabetic rats induced by streptozotocin: protective effects of melatonin. *J Pineal Res.* v. 25, n.2, pp 94-100.

Moretti, R. M., Marelli, M. M., Maggi, R., Dondi, D., Motta, M., Limonta, P. (2000). Antiproliferative action of melatonin on human prostate cancer LNCaP cells. *Oncology reports,* v. 7, n. 2, p. 347–51.

Motta, M., Zoppi, S., Martini, L. (1986). In vitro metabolism of testosterone in the rat prostate: influence of aging. *Journal of steroid biochemistry,* v. 25, n. 5B, p. 897–903.

Muralidhara, S. B. (2007a). Occurrence of oxidative impairments, response of antioxidant defences and associated biochemical perturbations in male reproductive milieu in the Streptozotocin-diabetic rat. *Int J Androl.* v.30, p. 508-518.

Muralidhara, S. B. (2007b). Early oxidative stress in testis and epididymal sperm in streptozotocin-induced diabetic mice: Its progression and genotoxic consequences. Re*prod Toxicol,* v. 23, p. 578-587.

Nishimura, H., Kuzuya, H., Okamoto, M., Yoshimasa, Y., Yamada, K., Ida, T., Kakehi, T., Imura, H. (1988). Change of insulin action with aging in conscious rats determined by euglycemic clamp. *The American journal of physiology,* v. 254, n. 1 Pt 1, p. E92-8.

O'brien, I. A., Lewin, I. G., O'hare, J. P., Arendt, J., Corral, R. J. (1986). Abnormal circadian rhythm of melatonin in diabetic autonomic neuropathy. *Clin Endocrinol* (Oxf), v. 24, p. 359-364.

Park, J., Cho, S. Y., Lee, Y. J., Lee, S. B., Son, H., Jeong, H. (2014). Poor glycemic control of diabetes mellitus is associated with higher risk of prostate cancer detection in a biopsy population. *PLoS One.* v. 8, n. 9,

Paroni, R., Terraneo, L., Bonomini, F., Finati, E., Virgili, E., Bianciardi, P., Favero, G., Fraschini, F., Reiter, R. J., Rezzani, R., Samaja, M. (2014). Antitumour activity of melatonin in a mouse model of human prostate cancer: Relationship with hypoxia signalling. *Journal of Pineal Research,* v. 57, n. 1, p. 43–52.

Parsons, J. K., Nelson, C. P, Gage, W. R., et al. (2001). GSTA1 Expression in normal, preneoplastic, and neoplastic. *Human prostate Tissue. Prostate.* V. 49, n.1, p 30-37.

Paschos, A., Pandya, R., Duivenvoorden, W. C., Pinthus, J. H. (2013). Oxidative stress in prostate cancer: changing research concepts towards a novel paradigm for prevention and therapeutics. *Prostate Cancer Prostatic Dis.* v. 16, n. 3, p 217-225.

Pasquali, R., Casimirri, F., De Iasio, R., Mesini, P., Boschi, S., Chierici, R., Flamia, R., Biscotti, M., Vicennati, V. (1995). Insulin regulates testosterone and sex hormone-binding globulin concentrations in adult normal weight and obese men. *The Journal of clinical endocrinology and metabolism,* v. 80, n. 2, p. 654–8.

Pegorin De Campos, S. G., Zanetoni, C., Góes, R. M., Taboga, S. R. (2006). Biological behavior of the gerbil ventral prostate in three phases of postnatal development.

Anatomical Record - Part A Discoveries in Molecular, Cellular, and Evolutionary Biology, v. 288, n. 7, p. 723–733.

Prunet-Marcassus, B., Desbazeille, M., Bros, A., Louche, K., Delagrange, P., Renard, P., Casteilla, L., Pénicaud, L. (2003). Melatonin reduces body weight gain in Sprague Dawley rats with diet-induced obesity. *Endocrinology,* v. 144, n. 12, p. 5347–52.

Rahimi, R., Nikfar, S., Larijani, B., Abdollahi, M. (2005). A review on the role of antioxidants in the management of diabetes and itscomplications. *Biomed Pharmacother,* v. 59, n. 7, p. 365-373.

Rasmussen, D. D., Boldt, B. M., Wilkinson, C., Yellon, S. M., Matsumoto, A. M. (1999). Daily Melatonin Administration at Middle Age Suppresses Male Rate Visceral Fat, Plasma Leptin, and Plasma Insulin to Youthful Levels. *Endocrinology,* v. 140, n. 2, p. 1009–1012.

Reiter, R. J., Tan, D. X., Maldonado, M. D. (2005). Melatonin as an antioxidant: Physiology versus pharmacology. *Journal of Pineal Research,* v. 39, n. 2, p. 215–216.

Reiter, R., Rosales-Corral, S., Tan, D. X., Acuna-Castroviejo, D., Qin, L., Yang, S. F., Xu, K. (2017). *Melatonin, a Full Service Anti-Cancer Agent: Inhibition of Initiation, Progression and Metastasis.* doi.org, v. 18, n. 4, p. 843.

Ribeiro, D. L., Pinto, M. E., Maeda, S. Y., Taboga, S. R., Góes, R. M. (2012a). High fat-induced obesity associated with insulin-resistance increases FGF-2 content and causes stromal hyperplasia in rat ventral prostate. *Cell and Tissue Research,* v. 349, n. 2, p. 577–588.

Ribeiro, D. L., Pinto, M. E., Rafacho, A., Bosqueiro, R., Maeda, S. Y., Anselmo-franci, J. A., Taboga, O. R., Go, R. M. (2012b). *High-Fat Diet Obesity Associated with Insulin Resistance Increases Cell Proliferation, Estrogen Receptor, and PI3K Proteins in Rat Ventral Prostate.* v. 33, n. 5, p. 854–865.

Ribeiro, D. L., Candido, E. M., Caldeira, E., Manzato, A. J., Taboga, S. R., Cagnon, V. H. A. (2006). Prostatic stromal microenvironment and experimental diabetes. *Eur J Histochem,* v. 5, p. 51-60.

Ribeiro, D. L., Marques, S. F., Alberti, S., Spadella, C. T., Manzato, A. J., Taboga, S. R., Dizeyi, N., Abrahamsson, P. A., Góes, R. M. (2008). Malignant lesions in the ventral prostate of alloxan-induced diabetic rats. *Int J Exp Pathol.* v. 89, n. 4, p. 276-283.

Ribeiro, D. L., Taboga, S. R., Góes, R. M. (2009). Diabetes induces stromal remodelling and increase in chondroitin sulfate proteoglycans of the rat ventral prostate. *Int J Exp Pathol.* v. 90, n. 4, p. 400-411.

Rimler, A., Culig, Z., Lupowitz, Z., Zisapel, N. (2002). Nuclear exclusion of the androgen receptor by melatonin. *Journal of Steroid Biochemistry and Molecular Biology,* v. 81, n. 1, p. 77–84.

Romero Cagigal, I., Ferruelo Alonso, A., Berenguer Sánchez, A. (2003). Diet and prostate cancer. *Actas urologicas espanolas,* v. 27, n. 6, p. 399–409.

Sakr, W. A., Grignon, D. J., Haas, G. P., Heilbrun, L. K., Pontes, J. E., Crissman, J. D. (1996). Age and racial distribution of prostatic intraepithelial neoplasia. *European urology*, v. 30, n. 2, p. 138–44.

Sakr, W. A., Haas, G. P., Cassin, B. F., Pontes, J. E., Crissman, J. D. (1993). The frequency of carcinoma and intraepithelial neoplasia of the prostate in young male patients. *The Journal of urology*, v. 150, n. 2 Pt 1, p. 379–85.

Sartori, C., Dessen, P., Mathieu, C., Monney, A., Bloch, J., Nicod, P., Scherrer, U., Duplain, H. (2009). Melatonin improves glucose homeostasis and endothelial vascular function in high-fat diet-fed insulin-resistant mice. *Endocrinology*, v. 150, n. 12, p. 5311–7.

Schalken, J. A., Van leenders, G. (2003). Cellular and molecular biology of the prostate: stem cell biology. *Urology*. v. 62, p. 11-20.

Scialò, F., Fernández-Ayala, D. J., Sanz, A. (2017). Role of Mitochondrial Reverse Electron Transport in ROS Signaling: Potential Roles in Health and Disease. *Frontiers in Physiology*, v. 8, n. June, p. 1–7.

Selmaoui, B., Touitou, Y. (1999). Age-related differences in serum melatonin and pineal NAT activity and in the response of rat pineal to a 50-Hz magnetic field. *Life sciences*, v. 64, n. 24, p. 2291–7.

Shain, S. A., Nitchuk, W. M. (1979). Testosterone metabolism by the prostate of the aging AXC rat. *Mechanisms of ageing and development*, v. 11, n. 1, p. 9–22.

Shiozaw S., Horie, S. (2014). Prostate cancer and metabolic syndrome *Nihon Rinsho*, v.72, n. 12, p. 2234-2240.

Shiu, S. Y. W. (2007). Towards rational and evidence-based use of melatonin in prostate cancer prevention and treatment. *Journal of Pineal Research*, v. 43, n. 1, p. 1–9.

Shiu, S. Y. W., Leung, W. Y., Tam, C. W., Liu, V. W. S., Yao, K. M. (2013). Melatonin MT1 receptor-induced transcriptional up-regulation of p27(Kip1) in prostate cancer antiproliferation is mediated via inhibition of constitutively active nuclear factor kappa B (NF-κB): potential implications on prostate cancer chemoprevention an. *J Pineal Res* v. 54, n. 1, p. 69–79.

Silva, S. A., Gobbo, M. G., Pinto-Fochi, M. E., Rafacho, A., Taboga, S. R., Almeida, E. A., Góes, R. M., Ribeiro, D. L. (2015). Prostate hyperplasia caused by long-term obesity is characterized by high deposition of extracellular matrix and increased content of MMP-9 and VEGF. *International Journal of Experimental Pathology*, v. 96, n. 1, p. 21–30.

Siu, S. W. F., Lau, K. W., Tam, P. C., Shiu, S. Y. W. (2002). Melatonin and prostate cancer cell proliferation: interplay with castration, epidermal growth factor, and androgen sensitivity. *The Prostate*, v. 52, n. 2, p. 106–22.

Sivitz, W. I., Yorek, M. A. Mitochondrial dysfunction in diabetes: from molecular mechanisms to functional significance and therapeutic opportunities. *Antioxid Redox Signal*, v. 12, n. 4, p. 537-577, 2010.

Sohn, E. J., Won, G., Lee, J., Lee, S., Kim, S. H. (2015). Upregulation of miRNA3195 and miRNA374b mediates the anti-angiogenic properties of melatonin in hypoxic PC-3 prostate cancer cells. *Journal of Cancer,* v. 6, n. 1, p. 19–28.

Soudamanim, S., Yuvaraj, S., Malinim, T., Balasubramanianm, K. (2005). Experimental diabetes has adverse effects on the differentiation of ventral prostate during sexual maturation of rats. *Anat. Rec.* 287, 1281–1289.

Srinivasan, V., Pandi-Perumal, S. R., Maestroni, G. J., Esquifino, A. I., Hardeland, R., Cardinali, D. P. (2005). Role of melatonin in neurodegenerative diseases. *Neurotoxicity research,* v. 7, n. 4, p. 293–318.

Stadtman, E. R. (2006). Protein oxidation and aging. *Free radical research,* v. 40, n. 12, p. 1250–8.

Starke-Reed, P. E., Oliver, C. N. (1989). Protein oxidation and proteolysis during aging and oxidative stress. *Archives of biochemistry and biophysics,* v. 275, n. 2, p. 559–67.

Stege, R. W., Rabe, M. B. (1997). The effect of diabetes mellitus on endocrine and reproductive function. *Proc Soc Exp Biol Med.* v. 214, p. 1-11.

Stehle, J. H., Saade, A., Rawashdeh, O., Ackermann, K., Jilg, A., Sebestény, T., Maronde, E. (2011). A survey of molecular details in the human pineal gland in the light of phylogeny, structure, function and chronobiological diseases. *Journal of Pineal Research*, v. 51, n. 1, p. 17–43.

Sugimura, Y., Cunha, G. R., Donjacour, A. A. (1986). Morphological and histological study of castration-induced degeneration and androgen-induced regeneration in the mouse prostate. *Bio. Reprod,* V. 34, P.973-983.

Suthagar, E., Soudamani, S., Yuvaraj, S., Ismail Khan, A., Aruldhas, M. M., Balasubramanian, K. (2009). Effects of streptozotocin (STZ)-induced diabetes and insulin replacement on rat ventral prostate. *Biomedicine Pharmacotherapy,* v. 63,. N. 1, p. 43–50.

Tan, C. W., Chan, K. W., Liu, V. W. S., Pang, B., Yao, K. M., Shiu, S. Y. W. (2008). Melatonin as a negative mitogenic hormonal regulator of human prostate epithelial cell growth: Potential mechanisms and clinical significance. *Journal of Pineal Research*, v. 45, n. 4, p. 403–412.

Tam, C. W., Mo, C. W., Yao, K. M., Shiu, S. Y. W. (2007). Signaling mechanisms of melatonin in antiproliferation of hormone-refractory 22Rv1 human prostate cancer cells: implications for prostate cancer chemoprevention. *Journal of Pineal Research,* v. 42, n. 2, p. 191–202.

Tamarindo, G. H., Gobbo, M. G., Pytlowanciv, E. Z, Taboga, S. R, Almeida, E. A., Goes, R. M. (2017). Melatonin increases Glutathione-S-transferase (GST) and reduces corpora amylacea in prostate of old rats regardless long-term high-fat diet. *Prostate*, submitted.

Tamarindo, G. H., Ribeiro, D. L., Gobbo, M. G., Guerra; L. H. A., Rahal, P., Taboga, S. R., Carvalho, H. F., Gadelha, F. R., Góes, R. M. (2017). Melatonin alone or combined to Docosahexaenoic acid exert antiproliferative effect by modulating oxidative stress and mitochondria bioenergetics of PNT1A cells. *Journal of Pineal Research,* submitted.

Tang, D. G., Porter, A. T. (1997). Target to apoptosis: a hopeful weapon for prostate cancer. *The Prostate,* v. 32, n. 4, p. 284–93.

Tatsumi, T., Shiraishi, J., Keira, N., Akashi, K., Mano, A., Yamanaka, S., Matoba, S., Fushiki, S., Fliss, H., Nakagawa, M. (2003). Intracellular ATP is required for mitochondrial apoptotic pathways in isolated hypoxic rat cardiac myocytes. *Cardiovascular Research,* v. 59, n. 2, p. 428–440.

Tirosh, A., Potashnik, R., Bashan, N., Rudich, A. (1999). Oxidative stress disrupts insulin-induced cellular redistribution of insulin receptor substrate-1 and phosphatidylinositol 3-kinase in 3T3-L1 adipocytes. A putative cellular mechanism for impaired protein kinase B activation and GLUT4 translocation. *The Journal of biological chemistry,* v. 274, n. 15, p. 10595–602.

Tirosh, A., Rudich, A., Bashan, N. (2000). Regulation of glucose transporters-- implications for insulin resistance states. *Journal of pediatric endocrinology & metabolism: JPEM,* v. 13, n. 2, p. 115–33.

Tomás-Zapico, C., Coto-Montes, A. (2005). A proposed mechanism to explain the stimulatory effect of melatonin on antioxidative enzymes. *Journal of Pineal Research,* v. 39, n. 2, p. 99–104.

Touitou, Y. (2001). Human aging and melatonin. Clinical relevance. *Experimental Gerontology,* v. 36, n. 7, p. 1083–1100.

Tuxhorn, J. A., Ayala, G. E., Smith, M. J., Dang, T. D., Rowley, D. R. (2002). Reactive stroma in human prostate cancer: induction of myofibroblast phenotype and extracelular matrix remodeling. *Clin Cancer Res,* v. 8, p. 2912-2923.

Tzortzis, V, Samarinas, M., Zachos. I., Oeconomou, A., Pisters, L. L., Bargiota, (2017). A. dverse effects of androgen deprivation therapy in patients with prostate cancer: focus on metabolic complications. *Hormones.* v.16, n.2, p. 115-123.

Untergasser, G., Madersbacher, S., Berger, P. (2005). Benign prostatic hyperplasia: age-related tissue-remodeling. *Exp Gerontol.* v. 40, p. 121-128.

Waldhauser, F., Weiszenbacher, G., Tatzer, E., Gisinger, B., Waldhauser, M., Schemper, M., Frisch, H. (1988). Alternations in nocturnal serum melatonin levels in humans with growth and aging. *Journal of Clinical Endocrinology and Metabolism,* v. 66, n. 3, p. 648–652.

Wang. Z., Ikeda, K., Wada, Y., Foster, H. E., Weiss, R. M., Latifpour, J. (2000). Expression and localization of basic fibroblast growth factor in diabetic rat prostate. *JU Int.* 85(7), p. 945-952.

Watve, M. G., Yajnik, C. S. (2007). Evolutionary origins of insulin resistance: a behavioral switch hypothesis. *BMC evolutionary biology,* v. 7, n. 1, p. 61.

Wautier, M. P., Chappey, O., Corda, S., Stern, D. M., Schmidt, A. M., Wautier, J. L. (2001). Activation of NADPH oxidase by AGE links oxidant stress to altered gene expression via RAGE. *Am J Physiol Endocrinol Metab,* v. 280, p. 685–694.

Weinert, B. T., Timiras, P. S. (2003) Physiology of Aging. Invited review: Theories of aging. *Journal of applied physiology,* v. 95, p. 1706–1716.

Wu, Y. H., Fischer, D. F., Kalsbeek, A., Garidou-Boof, M. L., Van der Vliet, J., Van Heijningen, C., Liu, R. Y., Zhou, J. N., Swaab, D. F. (2006). Pineal clock gene oscillation is disturbed in Alzheimer's disease, due to functional disconnection from the "master clock". *FASEB journal: official publication of the Federation of American Societies for Experimental Biology,* v. 20, n. 11, p. 1874–6.

Yang, Z., Liu, Y., Shi, C., Zhang, Y., Lv, R., Zhang, R., Wang, Q., Wang, Y. (2017). Suppression of PTEN/AKT signaling decreases the expression of TUBB3 and TOP2A with subsequent inhibition of cell growth and induction of apoptosis in human breast cancer MCF-7 cells via ATP and caspase-3 signaling pathways. *Oncology Reports,* p. 1011–1019.

Ye, C., Li, X., Wang, Y., Zhang, Y., Cai, M., Zhu, B., Wen, X. (2011). Diabetes causes multiple genetic alterations and downregulates expression of DNA repair genes in the prostate. Laboratory Investigation; *Journal of Technical Methods and Pathology.* v. 91. n. 9, p. 1363–1374.

Yilmaz, B., Kutlu, S., Mogulkoç, R., Canpolat, S., Sandal, S., Tarakçi, B., & Kelestimur, H. (2000). Melatonin inhibits testosterone secretion by acting at hypothalamo-pituitary-gonadal axis in the rat. *Neuro En*docrinology *Letters,* v. 21, n. 4, p. 301–306.

Ying, W. L., Tam, N. N. C., Evans, J. E., Green, K. M., Zhang, X., Ho, S. M. (2008). Differential proteomics in the aging Noble rat ventral prostate. *Proteomics,* v. 8, n. 13, p. 2750–2763.

Yono, M., Pouresmail, M., Takahashi, W., Flanagan, J. F., Weiss, R. M., Latifpour, J. (2005). Effect of insulin treatment on tissue size of the genitourinary tract in BB rats with spontaneously developed and streptozotocin-induced diabetes. *Naunyn-Schmiedeberg's Archives of Pharmacology,* v. 372, n. 3, p. 251–255.

Yu, O. H., Foulkes, W. D., Dastani, Z., Martin, R. M., Eeles, R., Richards, J. B. (2013). An assessment of the shared allelic architecture between type 2 diabetes and prostate cancer. *Cancer Epidel Biomarkers Prev.* v. 8, p. 1473-1475.

Zamaraeva, M. V, Sabirov, R. Z., Maeno, E., Ando-Akatsuka, Y., Bessonova, S. V, Okada, Y. (2005). Cells die with increased cytosolic ATP during apoptosis: a bioluminescence study with intracellular luciferase. *Cell death and differentiation,* v. 12, n. 11, p. 1390–7.

Zanuto, R., Siqueira-Filho, M. A., Caperuto, L. C., Bacurau, R. F., Hirata, E., Peliciari-Garcia, R. A., do Amaral, F. G., Marçal, A. C., Ribeiro, L. M, Camporez, J. P., Carpinelli, A. R., Bordin, S., Cipolla-Neto, J., Carvalho,. (2013). Melatonin improves insulin sensitivity independently of weight loss in old obese rats. *J Pineal Res,* v. 2, p. 156-165.

Zeyda, M., Stulnig, T. M. (2007). Adipose tissue macrophages. *Immunology Letters,* v. 112, n. 2, p. 61–67.

Zhang, Y., Chin-Quee, K., Riddle, R. C., Li, Z., Zhou, Z., Donahue, H. J. (2013). BRMS1 Sensitizes Breast Cancer Cells to ATP-Induced Growth Suppression. *BioResearch open access,* v. 2, n. 2, p. 77–83.

In: Melatonin: Medical Uses and Role in Health and Disease ISBN: 978-1-53612-987-8
Editors: Lore Correia and Germaine Mayers © 2018 Nova Science Publishers, Inc.

Chapter 4

MELATONIN RECEPTORS, BEHAVIOUR AND BRAIN FUNCTION

Olakunle J. Onaolapo[1,*], *MBBS, PhD*
and *Adejoke Y. Onaolapo*[2,†], *MBBS, PhD*

[1]Department of Pharmacology, Faculty of Basic Medical Sciences, Ladoke Akintola University of Technology, Osogbo, Osun State, Nigeria
[2]Department of Anatomy, Faculty of Basic Medical Sciences,
Ladoke Akintola University of Technology, Ogbomoso, Oyo State, Nigeria

ABSTRACT

Melatonin is a tryptophan-derived molecule that is critical to the transduction of circadian and seasonal information. It is also known to play crucial roles in several physiological processes, including the regulation of behavioural and cognitive processes in humans and rodents. There are evidences that a number of physiological and behavioural effects of melatonin in mammals are mediated by specific G-protein coupled receptors (GPCRs); melatonin (MT) $_1$ and $_2$, which are expressed in several locations in the mammalian central nervous system. In this chapter, we review the roles of melatonin receptors in health and disease, with specific references to their involvement in mood, anxiety-related and neurodegenerative disorders; and the possibilities of melatonin receptors as mediators of melatonergic therapeutics.

ABBREVIATIONS

IUPHAR   International Union of Basic and Clinical Pharmacology
cAMP   Cyclic Adenosine Monophosphate

---

[*] Email: olakunleonaolapo@yahoo.co.uk.
[†] Email: adegbayibiy@yahoo.com.

C-Terminal	Carboxyl Terminal
4P-PDOT	4-Phenyl-2-propionamidotetralin
mRNA	messenger RiboNucleic Acid
HT	Hydroxy Tryptamine
D	Dopamine
6-OHDA	6-Hydroxy Dopamine Hydrochloride
Aβ	Amyloid Beta

1. INTRODUCTION

The neurohormone melatonin (*N*-acetyl-5-methoxytryptamine) is mainly synthesised by the pineal gland, with lesser contributions by a number of other tissues, such as the retina [1]. The secretion of melatonin is regulated by circadian and seasonal variations in daylight length; and in both nocturnal and diurnal species, melatonin production by the pineal gland occurs during the dark phase at night with its production acutely suppressed by light. Melatonin is important in the regulation of the mammalian biologic rhythms [2, 3] (sleep-wake cycle, seasonal adaptation and pubertal development). The functions of melatonin have been shown to span different aspects of biologic processes in humans or rodents; with influences on modulation of memory formation, [4-6] control of body posture and balance [7], locomotion, [5, 6] nociception, [8] depression, [9, 10] anxiety [6, 11, 12] and disorders like schizophrenia [12-15]. Melatonin has also been associated with neuroprotection, [16] neurogenesis, [17, 18] maintenance of oxidant/antioxidant balance, [12, 15, 19] modulation of cardiovascular and/or immune system, [20] bone formation, tumour suppression and diabetes modulation [21]. Melatonin exerts its effects on biologic systems via its ability to bind to melatonin receptors, intracellular proteins like calmodulin [16] and orphan nuclear receptors; [22] or by exerting an antioxidant effect in tissues or organs [22]. Melatonin's anti-proliferative effect in cancer management has been associated with its ability to directly antagonise the binding of calcium to intracellular proteins like calmodulin, [16, 22] while its immune-modulatory effects have been linked to its interaction with retinoid-related orphan nuclear hormone receptors.

1.1. Melatonin Receptors

1.1.1. Historical Perspective

The first melatonin receptor (Mel$_{1C}$), which was found to be expressed only in non-mammalian species like birds and fishes was cloned from *Xenopus laevis* immortalised melanophores in 1994 [23]. Subsequently, there was the pharmacological characterisation of the first mammalian melatonin receptor, [24] the cloning of the first human melatonin receptor (Mel$_{1A}$), [25], and then the second melatonin receptor (Mel$_{1B}$) [26]. Historically, melatonin's ability to aggregate melanosomes of amphibian dermal melanophores was one of the observations that led to the postulation of theories supporting the presence of melatonin receptors, and the foundation for possible structure-activity relationships of melatonin analogues [27]. It also helped to show that melatonin receptors were coupled to a pertussis toxin-sensitive G-protein [28]. [^{3}H] melatonin was employed as a radioligand to label membranes' binding sites in bovine cerebral cortex, hypothalamus, and cerebellum [29] and the discovery of the first functional melatonin receptor followed [24, 30, 31]. Melatonin's inhibition of retinal dopamine release (via its activity at presynaptic heteroreceptors) was used to establish potency of several agonists and putative antagonists; and with this came the discovery of luzindole, the first competitive melatonin antagonist [31]. In 1984, another radioligand (2-[^{125}I] iodomelatonin) was introduced as a tracer in melatonin radioimmunoassay; [32] and it led to receptor localisation [33] and receptor characterization in native tissues [34-36].

1.1.2. Classification of Melatonin Receptors

Melatonin receptors are currently named and classified according to the operational and structural criteria developed by the IUPHAR Committee on Receptor Nomenclature and Drug Classification [37, 38]. However, initially, they were classified using results obtained from in vitro radioligand binding to native tissues and bioassays; [31, 36] using classic pharmacological criteria. In the first classification, two putative receptors (ML$_1$ and ML$_2$) were distinguished based on pharmacological and kinetic differences observed at 2-[^{125}I] iodomelatonin binding sites [39, 40]. Pharmacological profile of the ML$_1$ receptors (2-iodomelatonin > melatonin ≫ *N*-acetylserotonin) showed that the binding affinity of iodomelatonin to mammalian retina, pars tuberalis and rabbit retina (presynaptic receptor) was greater than melatonin and the binding affinity of iodomelatonin or melatonin was significantly greater than N-acetylserotonin (another endogenous ligand) binding [31, 38, 40, 41]. For ML$_2$ receptors, the pharmacological profile showed that iodomelatonin binding to hamster brain membranes could be distinguished by the binding affinity of *N*-acetylserotonin, which showed equal affinity with melatonin [42, 43]. Subsequently, two mammalian G protein-coupled melatonin receptors Mel1a (MT1) and Mel1b (MT2) which exhibited distinct molecular structure and chromosomal localization [44, 45] were cloned [25, 26, 46]. Radioligand binding profile of 2-^{125}I-Iodomelatonin to both melatonin receptors showed similarities to the general pharmacology of the ML$_1$ type [44]. These two melatonin receptors were later distinguished based on their binding affinity for specific ligands [47, 48].

To date, classification of melatonin receptors takes into consideration information taken from the pharmacological, structural and functional profile of these receptors; and IUPHAR guidelines [37, 38]. Melatonin receptors are therefore named after their endogenous ligand (melatonin), abbreviated as "MT" with each receptor subtype denoted by a numerical subscript (MT$_1$, MT$_2$). Species orthologues are denoted by a recommended lower-case prefix (h, human; o, ovine; r, rat; m, mouse; e.g., rMT$_1$) [38, 49].

1.1.2.1. MT$_1$, MT$_2$ and MT$_3$ Receptors

Melatonin receptors are classified as either MT$_1$ (Mel1$_a$,) or MT$_2$ (Mel$_{1b}$) [38]. These two receptors belong to the G-protein-coupled receptor (GPCR) superfamily of seven transmembrane domains and show high homology at the amino-acid level [50]. (Activation of either G receptors results in intracellular signaling via the modification of the activities of adenylate cyclase, phospholipase C/A$_2$, and guanylyl cyclise [16].

MT$_1$ refers to the first mammalian melatonin receptor which was first cloned in frogs [23, 25] and later in other mammals [25]. The MT$_1$ receptor which comprises of 350 amino acids is a G$_i$ protein-coupled receptor (specifically G$_{i\alpha2}$, G$_{i\alpha3}$, and G$_{q/11}$) and is particularly linked to the pertussis-toxin sensitive G proteins. This receptor lowers cAMP levels [51] by the inhibition of adenylate cyclase in recombinant expression systems and native tissues [38]. The MT$_1$ receptors when coupled to G$_{q/11}$ activates phospholipase-C, increasing calcium levels which then goes on to activate signaling by intracellular proteins like calmodulin (CaM), calreticulin and CaM kinases [16, 22, 52]. The MT$_2$ receptor, which is the second mammalian melatonin receptor, was cloned by Reppert et al. [26] from brain, retina and the human pituitary gland. MT$_2$ consists of 363 amino acids and shows 60% homology to the MT$_1$ receptor [44]. The MT$_2$ receptor is also a G$_{i/o}$ protein-coupled receptor that inhibits adenylate cyclase or guanylyl cyclase production in recombinant systems, and stimulates phospholipase C activity in native tissue and the suprachiasmatic nucleus (SCN) [38].

The existence of a third membrane-bound melatonin receptor (*MT$_3$* receptor) was accepted at a time [38]. It was referred to in uppercase italics (*MT$_3$*) because details of its structural characterisation were lacking at the time. The *MT$_3$* binding site has distinct pharmacology with similar affinity for two endogenous indoles, melatonin and its precursor, *N*-acetylserotonin as observed with the ML$_2$ receptor type. The characterisation of *MT$_3$* as a melatonin binding site on cytosolic enzyme, quinone reductase [53, 54] led to its removal from the IUPHAR nomenclature. The *MT$_3$* receptor is expressed in the retina of the rabbit [55] as well as hamster tissues, with high levels found in the liver, kidneys, heart, adipose tissue, and the brain [53]. Melatonin, resveratrol, and compound S29434 are three compounds that show affinity for the *MT$_3$* receptor [56].

1.1.2.2. Other Melatonin-Related Receptors

A melatonin-related receptor now known as the GPR50 was cloned in 1996. It was initially classified as a member of the melatonin receptor subfamily, as a result of its high homology (shares 45% amino acid sequence identity) and the presence of shared signature characteristics with the other two melatonin receptors (MT$_1$ and MT$_2$); although at the time, its function was poorly understood. The GPR50 receptor is an X-linked inherited G-protein binding receptor located on the X chromosome (Xq28), having 618 amino acids [57] and believed to be the mammalian orthologue of the non-mammalian MEL1c receptor [58]. Although little is known about the function of the GPR50 receptor, studies have shown that a deletion mutation in GPR50 may be associated with bipolar disorder and major depression [59]. This receptor subtype which is now classified as an orphan GPCR, has now been discovered not to bind melatonin, [22] although it increases the effectiveness of melatonin binding to the MT$_1$ receptor [60, 61]. Studies have also revealed that GPR50 forms heterodimers with MT$_1$ or MT$_2$, but not β$_2$-adrenoceptors or CCR5 chemokine receptors. It also has the capacity to inhibit agonist binding of MT$_1$ but not MT$_2$, preventing G proteins coupling or B-arrestin binding to the heterodimers [60, 62]. Research into the structural and functional relationships that exist between GPR50 and the MT$_1$ are ongoing; however, to date, we are aware that the formation of heterodimers between MT$_1$/GPR50 is constitutive; and believed to depend on levels of expression of both proteins in tissues [50]. Secondly, although MT$_1$ activity was undetectable in cells found to be expressing GPR50, MT$_1$ receptors became fully functional once the GPR50 receptors were silenced, suggesting that endogenous GPR50 expression levels may indeed regulate MT$_1$ activity; and thirdly, there are speculations that the function of MT$_1$ in the heterodimer might also be regulated post-translationally by GPR50 C-tail proteolysis, since studies using heterologous expression revealed that MT$_1$ receptor remains fully functional when engaged in heterodimers with a GPR50 with a truncated C-terminal [60].

The orphan nuclear receptors of the retinoic acid receptor family, {retinoid Z receptor (RZR) and retinoid acid receptor-related orphan receptor (ROR)} are other receptors that have been linked to melatonin. While speculations exist regarding melatonin's direct interaction with nuclear receptors, studies by Becker-Andre and his co-workers support these theories [63, 64]. Studies by Carlberg et al. [64] and Wiesenberg et al. [65] reported that melatonin binds to RZRα and RORα1 subtypes of the retinoic acid receptor family, although a few researchers have disputed these findings, [66, 67] while some question the possibility of melatonin being a native ligand at these receptors [68]. However, activation of these receptors has been associated with immune system modulation with suggestions that they regulate cytokine production by immune cells upon melatonin binding [69].

1.1.2.3. Distribution of Melatonin Receptors in Peripheral Tissues

There are considerable interspecies variations in the density, location and expression of melatonin receptors [40]. In both humans and rodents, melatonin receptors have been found in the epithelial cells of the prostate gland, breast, myometrium, granulosa cells of the ovary, placenta, and fetal kidney. In the gastrointestinal tract, melatonin receptors are also found in the liver, gallbladder, jejunal and colonic mucosa [16].

1.2. Melatonin Receptor and Behaviour

Melatonin {acting via two G protein-coupled receptors (MT$_1$ and MT$_2$) receptors} modulates and controls several behaviours and brain functions [11, 70]. Initially, the lack of selective melatonin receptor ligands and/or the dearth of studies assessing behavioural changes in melatonin receptor knockout (MT$_1^{-/-}$, MT$_2^{-/-}$) mice limited our knowledge of the impact of melatonin/melatonin receptors interactions in behaviour and brain function. However, with the advent of selective MT$_1$ [71] and MT$_2$ [70] receptor ligands, researchers are now able to explore (independent of each other) the roles of both melatonin receptors in brain functions [72].

1.2.1. Melatonin Receptor and Sleep

Studies examining the specific roles of melatonin (MT$_1$ and MT$_2$) receptors on sleep (using selective receptor agonist or genetic manipulations) revealed that contrary to previously-held beliefs supporting the notion that sleep control was entirely under the purview of MT$_1$ receptor, whilst MT$_2$ receptors controlled circadian phase-shifts [73]; MT$_2$ receptors also modulate sleep, particularly non rapid eye movement sleep (NREMS) without affecting rapid eye movement sleep (REMS), and also decrease the amount of NREMS without affecting REMS [11, 74].

1.2.2. Melatonin Receptors, Depression and Anxiety

The antidepressant [75, 76] and/or anxiolytic [6, 10, 12, 76, 77] properties of melatonin or agomelatine (a non-selective melatonin receptor agonist), or their ability to potentiate the anxiolytic effects of diazepam [78] have been demonstrated using commonly validated animal models, in stressed and non stressed animals. However, the role of the melatonin receptors in modulating these behaviours is still being studied.

In evaluating the possible roles of MT$_1$ or MT$_2$ receptors in anxiety, using selective MT$_2$ receptor antagonist (4P-PDOT) and luzindole (non selective MT$_1$/MT$_2$ antagonist); Ochoa-Sanchez et al. [74] demonstrated that of the two, MT$_2$ receptors were responsible for melatonin's anxiolytic effects. However, using melatonin receptor (MT$_2^{-/-}$) knockout mice, Comai and Gobbi [11] revealed that the number of open-arm entries and time spent in the open arm was not altered, although the MT$_2^{-/-}$ mice spent more time in the centre compared to controls (wild-type mice), an observation that was suggestive of a learning and memory deficit more than an anxiety-related effect; since time spent in the central platform of the elevated plus maze has been associated with the process of decision-making [79]. Nevertheless, these studies helped to reach the conclusion that the genetic manipulation of MT$_2$ receptors showed complex paradigm-specific interference in the control of anxiety-related behaviours [11].

Studies evaluating the role of melatonin receptors in the antidepressant effects of melatonin, using melatonin receptor agonist (S20394) revealed that repeated but not acute administration induced antidepressant-like effects in Flinder Sensitive Line rats; a rodent model of enhanced behavioural despair [80]. There have been suggestions that the antidepressant-like effects of melatonin are due to its activities at dopamine or serotonin receptors. However, deductions from recent studies have revealed that melatonin receptors were involved in the antidepressant-like effect of not only melatonin [81], but also of standard antidepressants, like fluoxetine or desipramine [82]. During a study in mice, Sumaya et al. [81] documented the MT_2 receptor-mediated antidepressant effect of luzindole; following the inability of luzindole to decrease duration of immobility in the forced swim test in $MT_2^{-/-}$ mice, despite displaying antidepressant activity via its inhibition of endogenous melatonin in wild-type mice. In 2006, Imbesi and co-workers [82] reported brain-region specific alteration in melatonin receptor (MT_1 and MT_2) mRNA following prolonged treatment with antidepressants, which was absent following single drug injection. This further affirms the role of melatonin receptors in the control of depression. Studies investigating the antidepressant effects of agomelatine [75, 82-85] have reported that its antidepressant and/or psychotropic effects are as a result of synergism between its melatonergic and 5-HT_{2C} receptor properties. This opinion has been further buttressed by studies that have demonstrated the formation of heterodimers between melatonin receptors and serotonin (5HT_{2C}) receptors [86]. Studies have also shown that the formation of MT_2/5-HT_{2C} heterodimers is more efficient than MT_1/5-HT_{2C} heterodimers or 5-HT_{2C} homodimers. Kamal et al. [86] demonstrated that administration of melatonin or agomelatine activated G1 signaling via its action at the MT_2/5-HT_{2C}, causing the inhibition of forskolin-stimulated cAMP production, while melatonin increases inositol phosphate production by transactivating the Gq pathway via MT2/5-HT_{2C} heterodimers signalling that may be important in the antidepressant function of melatonin or its receptor agonists. In a recent study using MT_1 receptor knockout mice, Comai and co-workers [87] revealed evidence that strongly defines a role for the MT_1 receptors in the aetiology of depression, especially melancholic depression. This buttressed results of two earlier studies; the first being a post-mortem study by Wu et al. [88] that reported an increase in MT_1 receptors density in the SCN of depressive patients; and a second where de Bodinat and co-workers [84] demonstrated that agomelatine showed higher affinity for MT_1 receptors compared to MT_2 or 5-HT_{2c} receptors; although the consensus remains that the antidepressant activity of agomelatine is related to its activity at multiple (MT_1, MT_2, 5-HT_{2c}, and 5-HT_{2b}) receptors sites [11, 83, 89, 90].

1.2.3. Melatonin Receptors, Learning and Memory-Formation

The regulation of learning and memory-formation has been linked to endogenous melatonin secretion, either via direct or indirect influences of the circadian or light/dark cycles [91]. The inhibition of endogenous melatonin synthesis via circadian rhythms directly facilitates long-term potentiation and hence, memory formation; while the night-time peak in melatonin levels is inhibitory on memory consolidation [92]. However, the nootropic effects of exogenous melatonin has been demonstrated repeatedly, using validated learning and memory paradigms like the Y-maze/radial arm maze, [5, 6, 12] verbal association task, [93] novel object recognition test, [4] olfactory social memory test in rats, [94] and also the enhancement of performance and improvement of memory acquisition in conditions of stress [95].

Studies have demonstrated that melatonin regulates memory formation, acting on hippocampal neurons by increasing the firing state of neurons in the cornus ammonis (CA1) region, promoting synaptic transmission [96, 97] and inhibition of long-term potentiation (LTP) in the CA1 dendritic in mouse hippocampal brain slices [98]. MT_1 and MT_2 receptors are expressed in the hippocampus [74]; and the role of melatonin receptors in memory was demonstrated from post-mortem studies carried out in subjects with Alzheimer disease, which revealed increased MT_1 and decreased MT_2 receptor immunoreactivity [99, 100] in the hippocampi of these subjects. Using knockout mice, MT_2 receptor's role in memory formation was demonstrated with the inhibition of LTP in MT_2 knockout but not in MT_1 knockout mice [98]. Larson et al. [101] also demonstrated a decrease in hippocampal LTP in MT_2 knockout C3H/HeN mice compared to wild-type mice; and learning deficits in the elevated plus maze in MT_2 knockout mice. More recently, O'Neal-Moffitt et al. [102] demonstrated improved spatial and reference memory performance, and basic memory function in MT_1/MT_2 double knockout mice. In zebrafish, Rawashdeh et al. [103] demonstrated melatonin receptor involvement in memory, with reversal of melatonin's ability to suppress long-term memory formation through the administration of the melatonin receptor antagonist luzindole. Accumulating evidence on melatonin's effects on memory strongly suggests the importance of MT_2 receptor [11].

1.2.4. Melatonin Receptors, Locomotion and Dopamine Receptors

There is increasing evidence to suggest the role of melatonin in the modulation of locomotor behaviour. In animals, exogenous administration of melatonin has been shown to decrease locomotor activity [5, 6, 12, 35, 104]. The mechanism by which melatonin exerts an inhibitory effect on locomotor behaviour is still being investigated. However, the initial speculation was that for melatonin to exert any influence on locomotor response, it would have to interact with the extrapyramidal system. Several neurotransmitters (dopamine, glutamate, γ-amino butyric acid and acetylcholine) are involved in the initiation and execution of motor behaviour [105]. The first inkling of possible melatonin dopamine-receptor relationships was had from studies that showed that iontophoretic treatment with melatonin decreased neuronal firing in rat striatal neurons; an effect reversed following treatment with a D_2 antagonist [106]. In another study, administration of melatonin in drinking water resulted in an increase in the affinity of D_2 striatal receptor binding to a D_2 antagonist [107]. In 2005, Uz and co-workers [108] demonstrated the possible role melatonin receptors might play in movement and/or reward system following the observation of MT_1 receptor expression in the caudate nucleus, nucleus accumbens, putamen, prefrontal cortex, ventral tegmental area, substantia nigra and amygdala; areas linked to the central dopaminergic (nigro-striatal and mesolimbic) pathways, which are strongly linked to movement and reward systems [108]. The presence of MT_1 receptors in the nigrostriatal and mesolimbic pathways suggested that melatonin has a role in regulating dopamine-related behavioural modifications. Also in an earlier study, Sumaya and co-workers [105] had observed melatonin ameliorated apomorphine (a D_2 agonist)-induced stereotypy and fluphenazine (a D_2 antagonist)-induced hypokinesia in rats. Another study demonstrated the involvement of the MT_1 and MT_2 receptors in modulating the development and expression of Methamphetamine-induced locomotor sensitisation in melatonin proficient C3H/HeN mice, via their interactions with dopamine receptors [109]. Numerous studies have shown that presynaptic melatonin heteroreceptors modulate calcium-dependent release of dopamine in the retina and brain, while also playing important roles in the modulation of dopaminergic transmission [24, 110, 111]. Activation of the somadendritic MT_1 receptors inhibit neuronal firing, while presynaptic MT_1/MT_2 heteroreceptors inhibit dopamine release in dopaminergic neurons originating in the ventral tegmental area (VTA) and/or substantia nigra pars compacta [108]. At different times, Dubocovich [24] and Zisapel [110] also reported that removal of presynaptic inhibitory MT_1 and MT_2 hetero-receptors on dopaminergic neuron receptors could increase the apoptotic effects of drugs like methamphetamine by increasing dopamine release.

1.3. Melatonin Receptors and the Mammalian Brain

The distribution of melatonin receptors within the brain and spinal cord is still being studied, owing mainly to the lack of selective antibodies for the melatonin receptors. Receptor autoradiography with the nonselective 2-[^{125}I]iodomelatonin, or real-time quantitative reverse transcription–polymerase chain reaction (RT-PCR) are the two methods currently used for localising melatonin receptor mRNA in tissues [11]. Studies in rodents have identified melatonin receptors mRNA in the prefrontal cortex, SCN, cerebellar cortex, retina, preoptic area, hippocampus, olfactory bulb, substantia nigra, nucleus accumbens and ventral tegmental area [96, 112-114]; while in humans, melatonin receptors mRNA have been identified in the retina, SCN, hypothalamus, cerebellar cortex (Bergmann glia and other astrocytes), hippocampus, cerebral cortex, the nucleus basalis of Meynert, paraventricular nucleus, tuberomammillary nucleus, periventricular nucleus, ventromedial and dorsomedial nuclei, the diagonal band of Broca, amygdala, supraoptic nucleus, infundibular nucleus, sexually dimorphic nucleus, mammillary bodies, substantia nigra, cortical areas, paraventricular thalamic nucleus and the retina [57, 115, 116].

More specifically, MT_2 receptors in the hippocampus were detected on the cornus ammonis (CA3/4) pyramidal neurons, which receive excitatory glutamatergic inputs from the entorhinal cortex; while MT_1 receptors were found predominantly in the CA1 [22]. Polyclonal antibodies have also been used to demonstrate the presence of MT_2 receptors in the pars reticulata of the substantia nigra, reticular thalamus, hippocampal CA_2 and CA_3 regions, supraoptic nucleus and red nucleus [74]. In humans, polyclonal antibodies were used to demonstrate the presence of MT_2 receptors on the neurons and nerve fibres (but not on the glial cells) in the SCN, the supraoptic nucleus and the para-ventricular nucleus [88].

1.3.1. Melatonin Receptor and Brain Function in Health

The identification and cloning of melatonin receptors jump-started our understanding of melatonin's role in sleep and a number of other biologic processes. The binding of melatonin as well as its receptor mRNA levels, follow a circadian rhythm, with levels of expression modulated by light and the concentration of melatonin in the plasma. The expression of MT_1mRNA and ^{125}I-melatonin binding in the SCN and pars tuberalis of rats and the Siberian hamster varies daily; with elevated levels occurring during daytime; exposure to light at night also increases ^{125}I-melatonin binding, coinciding with the suppression of melatonin synthesis [117]. This suggests that a melatonin-mediated down-regulation of the melatonin receptors occur, [16] a theory that has being the conclusion of a number of other studies [117, 118]. MT_1 and MT_2 receptors also exhibit diurnal rhythm with low levels observed during the day and high levels at night [16].

MT_1 and MT_2 receptors are involved in a number of physiological processes such as the modulation of locomotor activity, anxiolysis, anti-nociception and anti-neophobic effects of melatonin [108]. The detection of MT_1 and MT_2 receptor mRNA in the SCN, led to studies which revealed that melatonin, via its activities at the MT_1 receptor, mediated the acute inhibition of SCN, suppressing its neuronal firing rate [119]; while at the MT_2 receptor, it mediates circadian rhythm phase-shifts [119-121]. MT_1 and MT_2 receptors are expressed in a number of brain regions where they subserve diverse functions; in the pituitary gland, activation of MT_1 receptors in the pars tuberalis inhibits prolactin secretion, [122] in the anterior pituitary, melatonin inhibits the expression of the clock genes (*Per1* and *Per2*), [119, 123] it also phase-delays the circadian rhythm of plasma corticosterone, increasing the amplitude of thyrotropin rhythms, [123] and in cerebral arteries it mediates vasoconstriction [119]. Activation of the MT_2 receptors in the cerebral arteries induces vasodilation. The melatonin receptors are also involved in the regulation of blood pressure [124] and control of seasonal and non-seasonal reproduction [125].

In the eyes, immunocytochemical studies have revealed the distribution of MT_1 and MT_2 receptors in the cornea, choroid, sclera, photoreceptors, retinal pigment epithelium, retinal ganglion cells and blood vessels [126, 127].

Evidence suggesting that melatonin may act directly on photoreceptors, especially in the phototransduction pathways of the rod were brought to light with the observation that MT_1 receptors were expressed in photoreceptor cells in the human eye [127]. MT_1 receptors are also expressed in horizontal, amacrine and ganglion cells of the human eye, [127] while MT_2 are distributed in the sclera, lens and retinal pigmented epithelium [128].

The expression of MT_1 receptors in a number of dopamine-containing amacrine cells in the human retina reveals the possible role that may be played by melatonin in the modulation of dopamine release in the retina [129]. In separate studies, Dubocovich, [24] Schiller et al. [130] and Jaliffa et al. [131] demonstrated the inhibitory effect of melatonin on dopamine release; while a few other studies reported that retinal dopamine in turn inhibits retinal melatonin synthesis, [131] suggesting that dopamine and melatonin may be mutually-inhibitory retinal signals for day and night [16].

1.3.2. Melatonin Receptor and Brain Function in Disease

The expression of melatonin receptors in the human brain has been reported to be associated with some pathologies like Alzheimer disease, [99] depression [61] and sleep disorders [132]; while the deletion mutant of GPR50 has been genetically linked with psychiatric disorders like bipolar disorder and major depression [59].

Expression of melatonin receptors in the brain has been associated with Alzheimer disease; where studies have demonstrated alterations in the distribution of MT_1 and MT_2 receptor immunoreactivity in post-mortem brain from subjects with the disease, when compared against age-matched controls. In 2002 [99] and 2005 [100] respectively, Savaskan and co-workers reported an increase in MT_1 immunoreactivity, and a decrease in MT_2 immunoreactivity in post-mortem hippocampal tissue of Alzheimer disease patients. In 2006, Brunner and co-workers [133] reported a decrease in MT_1 and MT_2 receptor immunoreactivity in the pineal gland and occipital cortex of Alzheimer disease patients, compared to control [133]; and in 2007, Wu et al. [134] also reported a decrease in the expression of MT_1 receptors in the SCN.

In depressive patients, Imbesi et al. [82] reported significant brain region-specific alteration in MT_1 and MT_2 mRNA content following prolonged antidepressants therapy. In the hippocampus, chronic administration of desipramine, and clomipramine increased MT_1 mRNA content, while decreasing MT_2 mRNA content. In the striatum, prolonged administration of fluoxetine, desipramine, and clomipramine decreased MT_1 mRNA content; without altering striatal MT_2 mRNA, possibly demonstrating melatonin receptor involvement in depressive illness. In Parkinson disease (PD), Adi et al. [135] found evidence of a significant decrease in the expression of MT_1 and MT_2 receptors in amygdala and substantia nigra from post-mortem whole brain tissue of well-characterised PD patients, compared to age-matched controls.

1.4. MT₁ and MT₂ Receptors: Mediators of Melatonergic Therapeutics

A growing body of evidence has continued to demonstrate the complex roles of melatonin in the modulation of a diverse number of physiological and biological processes. Melatonin exerts its influence all over the body via its activities at melatonin receptors (MT₁ and MT₂), quinone reductase enzyme family (former MT₃ melatonin receptor subtype), the GPCR50 orphan receptor subunit, and a few orphan nuclear hormone receptors. The ability of melatonin receptors to couple to multiple and distinct signal transduction cascades and interact with other receptor subunits leading to cellular responses, makes them unique at the molecular level [52]. These receptors exhibit daily (24 hour light/dark) fluctuations in their sensitivity to specific cues which could then be modulated by either melatonin itself or by other cues (photoperiod) or ligands (e.g., oestrogen). Impairments in the interactions between melatonin and its receptors had been linked to a number of diseases/disorders. However, with advances in knowledge, targeted manipulations of these interactions are also yielding therapeutic benefits. Melatonin's therapeutic potential is limited by its high first-pass metabolism, short half-life, poor oral bioavailability [136] and its non-specific activities at multiple receptors. Hence, researches aimed at discovering and developing new classes of melatoninergic ligands with improved pharmacological properties such as receptor subtype selectivity and increased binding affinity are underway [3]. Over the years, such researches have yielded a number of agents; some of which had been granted approval for clinical use, while others are still in the pipeline.

1.4.1. Insomnia/Circadian Rhythm Disorders

Endogenous or exogenous melatonin/melatonin receptor interactions are strongly linked to both the promotion of sleep, or aetiopathogenesis of sleep disorders [137-140]. Melatonin's ability to synchronize or entrain the circadian clock by its direct effects on the SCN has led to researches aiming to correctly define melatonin/melatonin receptor interaction roles in the management of circadian rhythms disorders that occur in certain types of insomnia, jet lag and shift work. It is known that levels of endogenous melatonin decreases with age [132] and this age-associated decline may not be unrelated to the complaint of poor sleep quality in the elderly [141]. Melatonin controlled release tablets, which attempts to solve the problem of the rather short half-life of melatonin, was approved for insomnia treatment in patients over 55 years of age by the European Medicines Agency (EMA) [19]. Compared to other drugs, such as benzodiazepines, melatonin therapy is believed to be associated with less potential for hangover and withdrawal effects; and with much lesser tendency towards development of addiction.

Studies aimed at developing more melatonergic medications are ongoing; however, Ramelteon, a tricyclic, synthetic analogue of melatonin with higher receptor affinity and non-selective agonist activity at MT₁ and MT₂ receptors had been approved by the US

Food and Drug Administration (FDA) since 2005 for the management of insomnia [142]. In comparison to melatonin, ramelteon shows higher lipophilicity; also it does not bind with substantial affinity to quinone reductase 2, and seems to have no influence on calmodulin-dependent actions [143]. As the first melatonergic agonist approved for human use, it had been successfully used in treating elderly insomniacs, and was found to be promising for treating patients with primary insomnia and those suffering from circadian rhythm sleep disorders [144].

In individuals with visual impairment, loss of synchronisation of the endogenous circadian clock with the environmental light-dark cycle leads to a shifted or unstable timing of melatonin secretion; this has been linked sleep disorders, such as delayed sleep disorder or non-24 h sleep-wake disorder [110]. TIK-301 (LY 156735) is a melatonergic agonist/ serotonergic antagonist which was originally developed by Eli Lilly and Co (Indianapolis, IN, USA). In comparison to agomelatine (another melatonergic agonist and serotonergic antagonist), it demonstrates a powerful blockade of serotonergic 5-HT$_{2c}$ and 5-HT$_{2b}$ receptors [145]; hence, it is believed to have antidepressant effects. It was approved by the FDA for use in sleep disorders in visually-impaired individuals [19].

Tasimelteon (VEC162) is an orally-available melatonergic drug that underwent clinical tests to phase III, by Vanda Pharmaceuticals Inc (Washington, DC, USA), under license from Bristol-Myers Squibb Co (New York, USA) [146]. Trials showed that it improved sleep latency, sleep efficiency, and wake after sleep onset [147] It was eventually approved in both Europe and America for the treatment of non-24-hour sleep-wake rhythm disorder in totally-blind adults. Its sleep-initiating and antidepressant effects were also on trial [19]; however, its development for the treatment of major depressive disorders was discontinued in January 2013 by Vanda Pharmaceuticals, due to a failure of top-line results from a phase III study to meet the primary endpoint [143].

UCM765 is a moderately MT$_2$-selective partial agonist which had been reported to promote non-rapid eye movement sleep (NREMS) in rodents [148]. Its effect is nullified by the pharmacological blockage or genetic deletion of MT$_2$ receptors. UCM765's effects on sleep are different from those of non-selective MT$_1$/MT$_2$ agonists such as the structurally-related MT$_1$/MT$_2$ agonist UCM793 which is known to slightly decrease sleep onset, without affecting maintenance of NREMS [143]. Another selective MT$_2$ receptor partial agonist isUCM924 which has the same binding profile at the MT$_2$ receptor, but is more metabolically stable than UCM765; hence having a longer half-life [143]. Generally, selective MT$_2$ receptor agonists have high potentials for being developed as novel hypnotic agents.

1.4.2. Depression/Seasonal Affective Disorder/Anxiety

Agomelatine is a non-selective (MT$_1$/MT$_2$) melatonin receptor agonist, and a 5-HT$_{2c}$ antagonist. It is the first reported melatonergic drug having anxiolytic and antidepressant effects [16]. In 2008; it was approved by the EMA for the treatment of major depression in adults [11]. Lower doses are recommended for seasonal affective disorders. Unlike several conventional antidepressants, agomelatine has beneficial effects on sleep.

Luzindole is melatonin receptor antagonist with potential therapeutic applications. It has a much greater affinity for the MT_2 over the MT_1 receptor. Pre-clinical studies had demonstrated its antidepressant-like effect in behavioural paradigms such as the forced swim test in mice [81]; and its disruption of the circadian rhythm [149].

M3C is a new melatonin analogue which has been proposed for the treatment of anxiety. It was demonstrated to be effective as an anxiolytic-like agent in pinealectomised rats, at doses lower than any other melatonin analogues previously reported; also available data suggest that the anxiolytic properties of M3C may be mediated by MT_1 and MT_2 receptors [150].

1.4.3. Parkinson and Alzheimer Disease

Sleep-related symptoms, such as difficulty in initiating and maintaining sleep, excessive daytime somnolence, and parasomnias are common in patients with Parkinson Disease (PD) [151]. Melatonin is currently used in the treatment of sleep problems in PD patients; however, the finding of a reduced expression of melatonin MT_1 and MT_2 receptors in the amygdala and substantia nigra of patients with PD [135] indicates that there is a possibility that the melatonergic system may be involved in the overall pathophysiology of PD. Therefore, the potential use of melatonergic agents in PD is likely to go beyond management of sleep-related symptoms. In a preclinical study, administration of carefully-selected doses of melatonin analogues (ML-23 and S-20928) in a chronic, bilateral 6-OHDA model of PD was associated with improvements in behavioural parameters; however, in the same study, higher doses of S-20928 appeared to worsen symptoms [152].

Neu-P11 is an MT_1/MT_2 receptor agonist and a serotonin $5\text{-}HT_{1A/1D}$ receptor agonist, developed for the treatment of insomnia [153]. In rodents, Neu-P11 has been shown to promote sleep, improve insulin sensitivity, and exert antidepressant and anxiolytic activities [154] Neu-P11 was demonstrated to enhance rat's memory in the novel object recognition test, and improve neuronal and cognitive impairments in a rat model of AD (induced by intrahippocampal Aβ(1–42) injection); hence, It holds promise as a potential novel agent for the management of AD [4].

CONCLUSION

As more resources are devoted to understanding melatonin and melatonin/melatonin receptor interactions; the knowledge gained will be increasingly harnessed to better our understanding of the roles of melatonin and its receptors in physiological processes, and pathophysiology of central nervous system disorders. Hopefully, this will continue to culminate in the development of newer drugs with diverse applications.

REFERENCES

[1] Weston P, DiJon DH, Kwoon YW. Melatonin modulates M4-type ganglion-cell photoreceptors. *Neuroscience*. 2015;303: 178–188.

[2] Pandi-Perumal S, Srinivasan V, Maestroni G, Cardinali D, Poeggeler B, Hardeland R. Melatonin: Nature's most versatile biological signal? *FEBS J.* 2006;273: 2813–2838.

[3] Chan KH, Wong YH. A molecular and chemical perspective in defining melatonin receptor subtype selectivity. *Int J Mol Sci.* 2013;14:18385–18406.

[4] He P, Ouyang X, Zhou S, Yin W, Tang C, Laudon M, Tian S. A novel melatonin agonist Neu-P11 facilitates memory performance and improves cognitive impairment in a rat model of Alzheimer's disease. *Horm Behav.* 2013;64:1–7.

[5] Onaolapo OJ, Onaolapo AY, Akanni AA, Eniafe AL. Central depressant and nootropic effects of daytime melatonin in mice *Annal Neurosci* 2014;21:90-96.

[6] Onaolapo AY, Adebayo AN, Onaolapo OJ. Exogenous daytime melatonin modulates response of adolescent mice in a repeated unpredictable stress paradigm *Naunyn-Schmiedeberg's Arch. Pharmacol.* 2017a;390:149–61.

[7] Fraschini F, Cesarani A, Alpini D, Esposti D, Stankov BM. Melatonin influences human balance. *Biol. Signals Recept.* 1999;8: 111–119.

[8] Yoon MH, Park HC, Kim WM, Lee HG, Kim YO, Huang LJ Evaluation for the interaction between intrathecal melatonin and clonidine or neostigmine on formalin-induced nociception. *Life Sci.* 2008; 83(25-26):845-50.

[9] Detanico BC, Piato AL, Freitas JJ, et al. Antidepressant-like effects of melatonin in the mouse chronic mild stress model. *Eur J Pharmacol.* 2009;607:121–125.

[10] El Mrabet FZ, Ouakki S, Mes_oui A, El Hessni A, Ouichou A. Pinealectomy and exogenous melatonin regulate anxiety-like and depressive-like behaviors in male and female wistar rats. *Neurosci Med.* 2012;3:394–403.

[11] Comai S, Gobbi G. Unveiling the role of melatonin MT2 receptors in sleep, anxiety and other neuropsychiatric diseases: a novel target in psychopharmacology. *J Psychiatry Neurosci.* 2014;39:6–21.

[12] Onaolapo AY, Aina OA, Onaolapo OJ. Melatonin attenuates behavioural deficits and reduces brain oxidative stress in a rodent model of schizophrenia. *Biomed. Pharmacother.* 2017b;92: 373-83.

[13] Park HJ, Park JK, Kim SK, Cho AR, Kim JW, Yim SV, Chung JH. Association of polymorphism in the promoter of the melatonin receptor 1A gene with

schizophrenia and with insomnia symptoms in schizophrenia patients. *J Mol Neurosci.* 2011;45:304–08.

[14] Romo-Nava F, Alvarez-Icaza GD, Fresán-Orellana A, Saracco Alvarez R, Becerra-Palars C, Moreno J, Ontiveros Uribe MP, Berlanga C, Heinze G, Buijs RM. Melatonin attenuates antipsychotic metabolic effects: an eight-week randomized, double-blind, parallel-group, placebo-controlled clinical trial. *Bipolar Disord*, 2014;16:410-21.

[15] da Silva Araújo T, Maia Chaves Filho AJ, Monte AS, Isabelle de Góis Queiroz A, Cordeiro RC, de Jesus Souza Machado M, de Freitas Lima R, Freitas de Lucena D, Maes M, Macêdo D. Reversal of schizophrenia-like symptoms and immune alterations in mice by immunomodulatory drugs. *J Psychiatr Res.* 2017;84:49-58.

[16] Pandi-Perumal SR, Trakht I, Srinivasan V, Spence DW, Maestroni GJM, Zisapel N, Cardinali DP. Physiological effects of melatonin: Role of melatonin receptors and signal transduction pathways. *Prog. Neurobiol.* 2008;85:335–353.

[17] Liu J, Somera-Molina KC, Hudson RL, Dubocovich L. Melatonin potentiates running wheel-induced neurogenesis in the dentate gyrus of adult C3H/HeN mice hippocampus. *J Pineal Res.* 2013;54(2): 222–231.

[18] Tocharus C, Puriboriboon Y, Junmanee T, Tocharus J, Ekthuwapranee K, Govitrapong P. Melatonin enhances adult rat hippocampal progenitor cell proliferation via ERK signaling pathway through melatonin receptor. *Neuroscience.* 2014;275:314–321.

[19] Hardeland R, Madrid JA, Tan DX, Reiter RJ. Melatonin, the circadian multioscillator system and health: the need for detailed analyses of peripheral melatonin signaling. *J Pineal Res.* 2012;52:139–166.

[20] Sánchez-Barceló EJ, Mediavilla MD, Tan DX, Reiter RJ. Clinical uses of melatonin: Evaluation of human trials. *Curr. Med. Chem.* 2010;17:2070–2095.

[21] Lyssenko V, Nagorny CLF, Erdos MR, Wierup N, Jonsson A, Spégel P, Bugliani M, Saxena R, Fex M, Pulizzi N, et al. Common variant in MTNR1B associated with increased risk of type 2 diabetes and impaired early insulin secretion. *Nat. Genet.* 2009;41:82–88.

[22] Ekmekcioglu C. Melatonin receptors in humans: biological role and clinical relevance. *Biomed Pharmacother.* 2006;60:97–108.

[23] Ebisawa T, Karne S, Lerner MR, Reppert SM. Expression cloning of a high-affinity melatonin receptor from *Xenopus* dermal melanophores. *Proc Natl Acad Sci US.* 1994;91:6133–6137.

[24] Dubocovich ML. Melatonin is a potent modulator of dopamine release in the retina. *Nature.* 1983;306:782–784.

[25] Reppert SM, Weaver DR, Ebisawa T. Cloning and characterization of a mammalian melatonin receptor that mediates reproductive and circadian responses. *Neuron.* 1994;13:1177–1185.

[26] Reppert SM, Godson C, Mahle CD, Weaver DR, Slaugenhaupt SA, Gusella JF. Molecular characterization of a second melatonin receptor expressed in human retina and brain: the Mel1b melatonin receptor. *Proc Natl Acad Sci US.* 1995;92(19):8734-8.

[27] Heward CB, Hadley ME. Structure-activity relationships of melatonin and related indoleamines. *Life Sci* 1975;17:1167–1177.

[28] White BH, Sekura RD, Rollag MD. Pertussis toxin blocks melatonin-induced pigment aggregation in Xenopus dermal melanophores. *J Comp Physiol B* 1987;157:153–159.

[29] Cardinali DP, Vacas MI, Boyer EE. Specific binding of melatonin in bovine brain. *Endocrinology* 1979;105:437–441.

[30] Dubocovich ML. (Characterization of a retinal melatonin receptor. *J Pharmacol Exp Ther* 1985;234:395–401.

[31] Dubocovich ML. Luzindole (N-0774): a novel melatonin receptor antagonist. *J Pharmacol Exp Ther* 1988;246:902–910.

[32] Vakkuri O, Lämsä E, Rahkamaa E, Ruotsalainen H, Leppäluoto J. Iodinated melatonin: preparation and characterization of the molecular structure by mass and 1H NMR spectroscopy. *Anal Biochem* 1984;142:284–289.

[33] Vaněcek J, Pavlík A, Illnerová H. Hypothalamic melatonin receptor sites revealed by autoradiography. *Brain Res* 1987;435:359–362.

[34] Dubocovich ML, Takahashi JS. Use of 2-[^{125}I]iodomelatonin to characterize melatonin binding sites in chicken retina. *Proc Natl Acad Sci US* 1987;84:3916–3920.

[35] Sugden D. Melatonin: binding site characteristics and biochemical and cellular responses. *Neurochem Int* 1994;24:147–157.

[36] Dubocovich ML. Melatonin receptors: are there multiple subtypes? *Trends Pharmacol Sci* 1995;16:50–56.

[37] Ruffolo R, Humphrey P, Watson S, Spedding M. Revised NC-IUPHAR recommendations for nomenclature of receptors, in *The IUPHAR Compendium of Receptor Characterization and Classification* (Ruffolo R, editor. ed) pp 7–8, IUPHAR Media, London, UK, 2000.

[38] Dubocovich ML, Delagrange P, Krause DN, Sugden D, Cardinali DP, Olcese. International Union of Basic and Clinical Pharmacology. LXXV. Nomenclature, Classification, and Pharmacology of G Protein-Coupled Melatonin Receptors. *Pharmacol Rev.* 2010; 62(3): 343–380.

[39] Cardinali DP. Melatonin. A mammalian pineal hormone. *Endocr Rev* 1981;2:327–346.

[40] Morgan P, Barrett P, Howell H, Helliwell R. Melatonin receptors: Localization, molecular pharmacology and physiological significance. *Neurochem. Int.* 1994;24:101–146.

[41] Hagan RM, Oakley NR. Melatonin comes of age? *Trends Pharmacol Sci* 1995;16:81–83.

[42] Pickering DS, Niles LP. Pharmacological characterization of melatonin binding sites in Syrian hamster hypothalamus. *Eur J Pharmacol* 1990;175:71–77.

[43] Molinari EJ, North PC, Dubocovich ML. 2-[^{125}I]iodo-5-methoxycarbonylamino-N-acetyltryptamine: a selective radioligand for the characterization of melatonin ML2 binding sites. *Eur J Pharmacol* 1996;301:159–168.

[44] Reppert SM, Weaver DR, Ebisawa T, Mahle CD, Kolakowski LJ. Cloning of a melatonin-related receptor from human pituitary. *FEBS Lett.* 1996;386:219–224.

[45] Barrett P., Conway S., Jockers R., Strosberg A., Guardiola-Lemaitre B., Delagrange P., Morgan P. Cloning and functional analysis of a polymorphic variant of the ovine Mel 1a melatonin receptor. *Biochim. Biophys. Acta.* 1997;1356:299–307.

[46] Audinot V, Bonnaud A, Grandcolas L, Rodriguez M, Nagel N, Galizzi JP, Balik A, Messager S, Hazlerigg DG, Barrett P, et al. Molecular cloning and pharmacological characterization of rat melatonin MT1 and MT2 receptors. *Biochem Pharmacol* 2008;75:2007–2019.

[47] Browning C, Beresford I, Fraser N, Giles H. Pharmacological characterization of human recombinant melatonin mt(1) and MT(2) receptors. *Br J Pharmacol* 2000;129:877–886.

[48] Audinot V, Mailliet F, Lahaye-Brasseur C, Bonnaud A, Le Gall A, Amossé C, Dromaint S, Rodriguez M, Nagel N, Galizzi JP, et al. New selective ligands of human cloned melatonin MT1 and MT2 receptors. *Naunyn Schmiedebergs Arch Pharmacol* 2003;367:553–561.

[49] Dubocovich ML, Delagrange P, Olcese J. Melatonin receptors, in *IUPHAR database* (IUPHAR-DB). 2009, www.iuphar-db.org/DATABASE/.

[50] Jockers R, Maurice P, Boutin JA, Delagrange P. Melatonin receptors, heterodimerization, signal transduction and binding sites: what's new? *Br J Pharmacol.* 2008; 154(6): 1182–1195. doi: 10.1038/ bjp.2008.184.

[51] Capsoni S, Viswanathan M, De Oliveira AM, Saavedra JM. Characterization of melatonin receptors and signal transduction system in rat arteries forming the circle of Willis. *Endocrinology.* 1994;135:373–378.

[52] Witt-Enderby PA, Bennett J, Jarzynka MJ, Firestine S, Melan MA. Melatonin receptors and their regulation: biochemical and structural mechanisms. *Life Sci.* 2003;72:2183–2198.

[53] Nosjean O, Ferro M, Coge F, Beauverger P, Henlin JM, Lefoulon F, Fauchere JL, Delagrange P, Canet E, Boutin JA. Identification of the melatonin-binding site MT3 as the quinone reductase 2. *J Biol Chem.* 2000;275:31311–31317.

[54] Vincent L, Cohen W, Delagrange P, Boutin JA, Nosjean O. Molecular and cellular pharmacological properties of 5-methoxycarbonylamino-*N*-acetyltryptamine (MCA-NAT): a nonspecific MT3 ligand. *J Pineal Res*. 2010;48:222–229.

[55] Pintor J, Martin L, Pelaez T, Hoyle CH, Peral A. Involvement of melatonin MT(3) receptors in the regulation of intraocular pressure in rabbits. *Eur J Pharmacol*. 2001;416:251–254.

[56] Ferry G, Hecht S, Berger S, Moulharat N, Coge F, Guillaumet G, Leclerc V, Yous S, Delagrange P, Boutin JA. Old and new inhibitors of quinone reductase 2. *Chem Biol Interact*. 2010;186:103–109.

[57] Li DY, Smith DG, Hardeland R, Yang MY, Xu HL, Zhang L, Yin HD, Zhu Q Melatonin Receptor Genes in Vertebrates. *Int J Mol Sci*. 2013;14(6): 11208–11223

[58] Dufourny L, Levasseur A, Migaud M, et al. GPR50 is the mammalian ortholog of Mel1c: evidence of rapid evolution in mammals. *BMC Evol Biol*. 2008;8:105.

[59] Thomson PA, Wray NR, Thomson AM, et al. Sex-specific association between bipolar affective disorder in women and GPR50, an X-linked orphan G protein-coupled receptor. *Mol Psychiatry*. 2005;10:470–8.

[60] Levoye A. Dam J, Ayoub M, Guillaume J, Couturier C, Delagrange P, Jockers R. The orphan GPR50 receptor specifically inhibits MT1 melatonin receptor function through heterodimerization. *EMBO J*. 2006;25, 3012–3023.

[61] Hirsch-Rodriguez E, Imbesi M, Manev R, Uz T, Manev H. The pattern of melatonin receptor expression in the brain may influence antidepressant treatment. *Med Hypotheses*. 2007;69:120–4.

[62] Chaste P, Clement N, Mercati O, et al. Identification of pathway-biased and deleterious melatonin receptor mutants in autism spectrum disorders and in the general population. *PLoS One*. 2010;5:e11495. .doi.10.1371/journal.pone.0011495.

[63] Becker-André M, Wiesenberg I, Schaeren-Wiemers N, André E, Missbach M, Saurat J, Carlberg C Pineal gland hormone melatonin binds and activates an orphan of the nuclear receptor superfamily. *J. Biol. Chem*. 1994;269:28531–28534.

[64] Carlberg C, Hooft van Huijsduijnen R, Staple JK, DeLamarter JF, Becker-Andre M. RZRs, a new family of retinoid-related orphan receptors that function as both monomers and homodimers. *Mol Endocrinol*. 1994;8:757–770.

[65] Wiesenberg I, Missbach M, Kahlen JP, Schrader M, Carlberg C. Transcriptional activation of the nuclear receptor RZR alpha by the pineal gland hormone melatonin and identification of CGP 52608 as a synthetic ligand. *Nucleic Acids Res*. 1995;23:327–333.

[66] Dai J, Ram PT, Yuan L, Spriggs LL, Hill SM. Transcriptional repression of RORalpha activity in human breast cancer cells by melatonin. *Mol Cell Endocrinol.* 2001;176:111–120.

[67] Bitsch F, Aichholz R, Kallen J, Geisse S, Fournier B, Schlaeppi JM. Identification of natural ligands of retinoic acid receptor-related orphan receptor alpha ligand-binding domain expressed in Sf9 cells – a mass spectrometry approach. *Anal Biochem.* 2003;323:139–149.

[68] Jetten AM. Retinoid-related orphan receptors (RORs): critical roles in development, immunity, circadian rhythm, and cellular metabolism. *Nucl Recept Signal.* 2009;7:e003.

[69] Skwarlo-Sonta K, Majewski P, Markowska M, Oblap R., Olszanska B Bidirectional communication between the pineal gland and the immune system. *Can. J. Physiol. Pharmacol.* 2003, 81, 342–349.

[70] Mor M, Rivara S, Pala D, Bedini A, Spadoni G, Tarzia G. Recent advances in the development of melatonin MT(1) and MT(2) receptor agonists. *Expert Opin Ther Pat.* 2010;20:1059–77.

[71] Rivara S, Pala D, Lodola A, Mor M, Lucini V, Dugnani S, Scaglione F, Bedini A, Lucarini S, Tarzia G, Spadoni G. (2012) MT1-selective melatonin receptor ligands: synthesis, pharmacological evaluation, and molecular dynamics investigation of N-{[(3-O-substituted)anilino]alkyl}amides. *Chem Med Chem.* (11):1954-64.

[72] Zlotos DP (2012) Recent progress in the development of agonists and antagonists for melatonin receptors. *Curr Med Chem.* 2012; 19(21):3532-49.

[73] Dubocovich ML. Melatonin receptors: Role on sleep and circadian rhythm regulation. *Sleep Med. 2007*;8:34–42.

[74] Ochoa-Sanchez R, Comai S, Lacoste B, et al. Promotion of non-rapid eye movement sleep and activation of reticular thalamic neurons by a novel MT2 melatonin receptor ligand. *J Neurosci.* 2011;31:18439–52.

[75] Papp M, Gruca P, Boyer P-A, Mocaër E. (2003) Effect of Agomelatine in the Chronic Mild Stress Model of Depression in the Rat. *Neuropsychopharmacol.* 2003;28: 694–703.

[76] El Mrabet FZ, Lagbouri I, Mesfioui A, El Hessn A and Ouichou A. "The Influence of Gonadectomy on Anxiolytic and Antidepressant Effects of Melatonin in Male and Female Wistar Rats: A Possible Implication of Sex Hormones," *Neurosci. Med.* 2012; 2:162-173.

[77] Papp M, Litwa E, Gruca P, Mocaër E. Anxiolytic-like activity of agomelatine and melatonin in three animal models of anxiety. *Behav Pharmacol.* 2006;17:9–18.

[78] Loiseau F, Le Bihan C, Hamon M, Theibot M-H Effects of melatonin and agomelatine in anxiety-related procedures in rats: Interaction with diazepam *Eur. Neuropsychopharmacol.* 2006;16: 417–42.

[79] Rodgers RJ, Johnson NJ. Factor analysis of spatiotemporal and ethological measures in the murine elevated plus-maze test of anxiety. *Pharmacol Biochem Behav.* 1995;52:297–303.

[80] Overstreet DH, Pucilowski O, Retton MC, et al. Effects of melatonin receptor ligands on swim test immobility. *Neuroreport.* 1998;9:249–53.

[81] Sumaya IC, Masana MI, Dubocovich ML. The antidepressant-like effect of the melatonin receptor ligand luzindole in mice during forced swimming requires expression of MT2 but not MT1 melatonin receptors. *J Pineal Res.* 2005;39:170–7.

[82] Imbesi M, Uz T, Yildiz S, Arslan AD, Manev H Drug- and region-specific effects of protracted antidepressant and cocaine treatment on the content of melatonin MT(1) and MT(2) receptor mRNA in the mouse brain. *Int J Neuroprot Neuroregener.* 2006; 2:185-9.

[83] Bourin M, Mocaer E, Porsolt R. Antidepressant-like activity of S 20098 (agomelatine) in the forced swimming test in rodents: involvement of melatonin and serotonin receptors. *J Psychiatry Neurosci.* 2004;29:126–33.

[84] de Bodinat C, Guardiola-Lemaitre B, Mocaër E, et al. Agomelatine, the first melatonergic antidepressant: discovery, characterization and development. *Nat Rev Drug Discov.* 2010;9:628–42.

[85] Guardiola-Lemaitre B, De Bodinat C, Delagrange P, Millan M J, Munoz C, Mocaë E Agomelatine: mechanism of action and pharmacological profile in relation to antidepressant properties *Br J Pharmacol.* 2014;17: 3604–3619.

[86] Kamal M, Gbahou F, Guillaume JL, Daulat AM, Benleulmi-Chaachoua A, Luka M, Chen P, Kalbasi Anaraki D, Baroncini M, Mannoury la Cour C, Millan MJ, Prevot V, Delagrange P, Jockers R Convergence of melatonin and serotonin (5-HT) signaling at MT2/5-HT2C receptor heteromers. *J Biol Chem.* 2015; 290:11537-46.

[87] Comai S, Ochoa-Sanchez R, Dominguez-Lopez S, Bambico FR, Gobbi G (2015)Melancholic-Like Behaviors and Circadian Neurobiological Abnormalities in Melatonin MT$_1$ Receptor Knockout Mice *Int J Neuropsychopharmacol* 18: pyu075. doi: //doi.org/10.1093/ijnp/pyu075.

[88] Wu YH, Ursinus J, Zhou JN, Scheer FA, Ai-Min B, Jockers R, van Heerikhuize J, Swaab DF Alterations of melatonin receptors MT1 and MT2 in the hypothalamic suprachiasmatic nucleus during depression. *J Affect Disoders* 2013;148:357-67.

[89] Gałecka E, Szemraj J, Florkowski A, Gałecki P, Bieńkiewicz M, Karbownik-Lewińska M, Lewiński A Single nucleotide polymorphisms and mRNA expression for melatonin MT(2) receptor in depression. *Psychiatry Res.* 2011;189:472-4.

[90] Chenu F, Shim S, El Mansari M, Blier P Role of melatonin, serotonin 2B, and serotonin 2C receptors in modulating the firing activity of rat dopamine neurons. *J Psychopharmacol.* 2014;28:162-7.

[91] LeGates TA, Fernandez DC, Hattar S. Light as a central modulator of circadian rhythms, sleep and affect. *Nat. Rev. Neurosci.* 2014;15:443-454.

[92] Rawashdeh O, Maronde E. The hormonal Zeitgeber melatonin: role as a circadian modulator in memory processing. *Front. Mol. Neurosci.* 2012; 5: 27.

[93] Gorfine T, Zisapel N. Melatonin and the human hippocampus, a time dependent interplay. *J Pineal Res*, 2007;43:80–86.

[94] Argyriou A, Prast H, Philippu A. Melatonin facilitates short-term memory. *Eur J Pharmacol*, 1998;349:159–162.

[95] Rimmele U, Spillmann M, Bartschi C, Wolf OT, Weber CS, Ehlert U, Wirtz PH. Melatonin improves memory acquisition under stress independent of stress hormone release. *Psychopharmacol (Berl),* 2009;202:663–672..

[96] Musshoff U, Riewenherm D, Berger E, Fauteck JD, Speckmann EJ. Melatonin receptors in rat hippocampus: molecular and functional investigations. *Hippocampus.* 2002;12:165-73.

[97] El-Sherif Y, Tesoriero J, Hogan MV, Wieraszko A Melatonin regulates neuronal plasticity in the hippocampus. *J Neurosci Res.* 2003;72:454-60.

[98] Wang LM, Suthana NA, Chaudhury D, Weaver DR, Colwell CS. Melatonin inhibits hippocampal long-term potentiation. *Eur J Neurosci.* 2005;22:2231–37.

[99] Savaskan E, Ayoub MA, Ravid R, Angeloni D, Fraschini F, et al. Reduced hippocampal MT_2 melatonin receptor expression in Alzheimer's disease. *J Pineal Res.* 2005;38:10–1.

[100] Savaskan E, Olivieri G, Meier F, Brydon L, Jockers R, et al. Increased melatonin 1a-receptor immunoreactivity in the hippocampus of Alzheimer's disease patients. *J Pineal Res.* 2002; 32:59–62.

[101] Larson J, Jessen RE, Uz T, Arslan AD, Kurtuncu M, et al. Impaired hippocampal long-term potentiation in melatonin MT_2 receptor-deficient mice. *Neurosci Lett.* 2006;393:23–26.

[102] O'Neal-Moffitt G, Pilli J, Kumar SS, Olcese J. Genetic deletion of MT_1/MT_2 melatonin receptors enhances murine cognitive and motor performance. *Neuroscience.* 2014;277:506–21.

[103] Rawashdeh O, de Borsetti NH, Roman G, Cahill GM. Melatonin suppresses nighttime memory formation in zebrafish. *Science.* 2007;318:1144–46.

[104] Chuang J-I, Lin M-T. Pharmacological effects of melatonin treatment on both locomotor activity and brain serotonin release in rats. *J Pineal Res* 1994;17:11 – 6.

[105] Sumaya LC, Byers DM, Irwin LN, Del Val S, Moss DE. Circadian-dependent effect of melatonin on dopaminergic D2 antagonist-induced hypokinesia and

[105] agonist-induced stereotypies in rats. *Pharmacology, Biochemistry and Behavior* 2004;78:727–733.

[106] Escames G, Castroviejo DA, Vives F. Melatonin-dopamine interaction in the striatal projection area of sensorimotor cortex in the rat. *Neuro Report* 1996;7:597–600.

[107] Hamdi A. Melatonin administration increases the affinity of D2 dopamine receptors in the rat striatum. *Life Sci* 1998;63:2115–20.

[108] Uz T, Arslan AD, Kurtuncu M, et al. The regional and cellular expression profile of the melatonin receptor MT1 in the central dopaminergic system. *Brain Res Mol Brain Res*. 2005;136:45–53.

[109] Hutchinson AJ, Hudson RL, Dubocovich ML Genetic deletion of MT_1 and MT_2 melatonin receptors differentially abrogates the development and expression of methamphetamine-induced locomotor sensitization during the day and the night in C3H/HeN mice. *J Pineal Res*. 2012;53: 399–409.

[110] Zisapel N. Melatonin-dopamine interactions: from basic neurochemistry to a clinical setting. *Cellular Mol Neurobiol*. 2001;21:605–616.

[111] Mcclung CA. Circadian rhythms, the mesolimbic dopaminergic circuit, and drug addiction. *The Scientific World Journal*. 2007;7:194–202.

[112] Hardeland R, Cardinali DP, Srinivasan V., Spence DW, Brown G.M., Pandi-Perumal SR Melatonin—A pleiotropic, orchestrating regulator molecule. *Progr. Neurobiol*. 2011;93:350–384.

[113] Adi N, Mash DC, Ali Y, Singer C, Shehadeh L, Papapetropoulos S. Melatonin MT_1 and MT_2 receptor expression in Parkinson's disease. *Med Sci Monit* 2010;16: BR61–BR67.

[114] Drew J, Barrett P, Mercer J, Moar K, Canet E, Delagrange P, Morgan P. Localization of the melatonin-related receptor in the rodent brain and peripheral tissues. *J. Neuroendocrinol*. 2001;13:453–458.

[115] Mazzucchelli C, Pannacci M, Nonno R, et al. The melatonin receptor in the human brain: cloning experiments and distribution studies. *Brain Res Mol Brain Res*. 1996;39:117–26.

[116] Sallinen P, Saarela S, Ilves M, et al. The expression of MT1 and MT2 melatonin receptor mRNA in several rat tissues. *Life Sci*. 2005;76:1123–34.

[117] Masson-Pevet M, Gauer F, Schuster C, Guerrero HY. Photic regulation of MT1 melatonin receptors and 2-iodomelatonin binding in the rat and Siberian hamster. *Biol. Signals Recept*. 2000;9:188–196.

[118] Schuster C, Gauer F, Malan A, Recio J, Pevet, Masson-Pevet M. The circadian clock, light/dark cycle and melatonin are differentially involved in the expression of daily and photoperiodic variations in mt(1) melatonin receptors in the Siberian and Syrian hamsters. *Neuroendocrinol*. 2001;74,:55–68.

[119] Dubocovich ML, Rivera-Bermudez MA, Gerdin MJ, Masana MI. Molecular pharmacology, regulation and function of mammalian melatonin receptors. *Front Biosci*. 2003;8:d1093–108.

[120] von Gall C, Stehle JH, Weaver DR. Mammalian melatonin receptors: molecular biology and signal transduction. *Cell Tissue Res*. 2002;309:151–62.

[121] Jin X, von Gall C, Pieschl RL, Gribkoff VK, Stehle JH, Reppert SM, Weaver DR. Targeted disruption of the mouse Mel(1b) melatonin receptor. *Mol Cell Biol*. 2003;23:1054–1060.

[122] Morgan PJ. The pars tuberalis: the missing link in the photoperiodic regulation of prolactin secretion? *J Neuroendocrinol*. 2000;12:287–95.

[123] Jiménez-Ortega V, Barquilla PC, Pagano ES, Fernández-Mateos P, Esquifino AI, Cardinali DP. Melatonin supplementation decreases prolactin synthesis and release in rat adenohypophysis: correlation with anterior pituitary redox state and circadian clock mechanisms. *Chronobiol Int*. 2012;29:1021-35.

[124] Doolen S., Krause D., Dubocovich M., Duckles S. Melatonin mediates two distinct responses in vascular smooth muscle. *Eur. J. Pharmacol. 1998*;345:67–69.

[125] González-Arto M, Vicente-Carrillo A, Martínez-Pastor F, Fernández-Alegre E, Roca J, Miró J, Rigau T, Rodríguez-Gil JE, Pérez-Pé R, Muiño-Blanco T, Cebrián-Pérez JA, Casao A (2016) Melatonin receptors MT1 and MT2 are expressed in spermatozoa from several seasonal and non seasonal breeder species. *Theriogenol*. 2016;86:1958-68.

[126] Rada JA, Wiechmann AF Melatonin receptors in chick ocular tissues: implications for a role of melatonin in ocular growth regulation. *Invest. Ophthalmol. Vis. Sci*. 2006;47:25–33.

[127] Scher J, Wankiewicz E, Brown GM, Fujieda H. MT1 melatonin receptor in the human retina: expression and localization. *Investig. Ophthalmol. Vis. Sci*. 2002;43:889–897.

[128] Wiechmann AF, Udin SB, Summers Rada JA. Localization of Mel1b melatonin receptor-like immunoreactivity in ocular tissues of Xenopus laevis. *Exp. Eye Res*. 2004;79:585–594.

[129] Fujieda H, Scher J, Hamadanizadeh SA, Wankiewicz E, Pang SF, Brown GM. Dopaminergic and GABAergic amacrine cells are direct targets of melatonin: immunocytochemical study of MT1 melatonin receptor in guinea pig retina. *Vis. Neurosci*. 2000;17:63–70.

[130] Schiller ED, Champney TH, Reiter CK, Dohrman DP. Melatonin inhibition of nicotine-stimulated dopamine release in PC12 cells. *Brain Res*. 2003;966:95-102.

[131] Jaliffa CO, Lacoste FF, Llomovatte DW, Sarmiento MI, Rosenstein RE. Dopamine decreases melatonin content in golden hamster retina. *J. Pharmacol. Exp. Ther*. 2000;293:91–95.

[132] Galley HF, Lowes DA, Allen L, et al. Melatonin as a potential therapy for sepsis: a phase I dose escalation study and an ex vivo whole blood model under conditions of sepsis. *J Pineal Res.* 2014;56:427–438.

[133] Brunner P, Sözer-Topcular N, Jockers R, Ravid R, Angeloni D, Fraschini F, Eckert A, Müller-Spahn F, Savaskan E. Pineal and cortical melatonin receptors MT1 and MT2 are decreased in Alzheimer's disease. *European J. Histochem.* 2006;50:311-316.

[134] Wu YH, Zhou JN, Van Heerikhuize J, Jockers R, Swaab DF. Decreased MT1 melatonin receptor expression in the suprachiasmatic nucleus in aging and Alzheimer's disease. *Neurobiol Aging.* 2007; 28:1239-47.

[135] Adi N, Mash DC, Ali Y, Singer C, Shehadeh L, Papapetropoulos S (2010) Melatonin MT1 and MT2 receptor expression in Parkinson's disease. *Med Sci Monit.* 2010;16:BR61-7.

[136] Uchikawa O, Fukatsu K, Tokunoh R, Kawada M, Matsumoto K, Imai Y, Hinuma S, Kato K., Nishikawa H, Hirai K., *et al.* Synthesis of a novel series of tricyclic indan derivatives as melatonin receptor agonists. *J. Med. Chem.* 2002, *45*, 4222–4239.

[137] Bartlett DJ, Biggs SN, Armstrong SM. Circadian rhythm disorders among adolescents: assessment and treatment options. *Med J Aust.* 2013;199:S16–S20.

[138] Lahteenmaki R, Puustinen J, Vahlberg T, et al. Melatonin for sedative withdrawal in older patients with primary insomnia: a randomized double-blind placebo-controlled trial. *Br J Clin Pharmacol.* 2014;77:975–985.

[139] Amstrup AK, Sikjaer T, Mosekilde L, et al. The effect of melatonin treatment on postural stability, muscle strength, and quality of life and sleep in postmenopausal women: a randomized controlled trial. *Nutr J.* 2015;14:102.

[140] Chang YS, Lin MH, Lee JH, et al. Melatonin supplementation for children with atopic dermatitis and sleep disturbance: a randomized clinical trial. *JAMA Pediatr.* 2016;170:35–42.

[141] Leger D, Laudon M, Zisapel N. Nocturnal 6-sulfatoxymelatonin excretion in insomnia and its relation to the response to melatonin replacement therapy. *Am J Med.* 2004;116:91–95.

[142] McGechan A, Wellington K. Ramelteon. *CNS Drugs. 2005*;19:1057–1065.

[143] Carocci A, Catalano A, Sinicropi MS Melatonergic drugs in development *Clinical Pharmacology: Advances and Applications Volume* 2014;6:127—137.

[144] Hardeland R, Poeggeler B, Srinivasan V, Trakht I, Pandi-Perumal SR, Cardinali DP. Melatonergic drugs in clinical practice. *Arzneimittelforschung.* 2008;58:1-10.

[145] Landolt HP, Wehrle R. Antagonism of serotonergic 5-HT2A/2C receptors: mutual improvement of sleep, cognition and mood? *Eur J Neurosci.* 2009;29:1795–809.

[146] Rajaratnam SMW, Polymeropoulos MH, Fisher DM, et al. Melatonin agonist tasimelteon (VEC-162) for transient insomnia after sleep-time shift: two randomised controlled multicentre trials. *Lancet.* 2009;373:482–491.

[147] Ohta T, Murao K, Miyake K, Takemoto K. Melatonin receptor agonists for treating delirium in elderly patients with acute stroke. *J Stroke Cerebrovasc Dis.* 2013;2.

[148] Comai S, Gobbi G. Unveiling the role of melatonin MT2 receptors in sleep, anxiety and other neuropsychiatric diseases: a novel target in psychopharmacology. *J Psychiatry Neurosci.* 2014;39:6–21.

[149] Dubocovich ML. Selective MT2 melatonin receptor antagonists block melatonin-mediated phase advances of circadian rhythms. *FASEB J.* 1998;12:1211–1220.

[150] Bustamante-García R, Lira-Rocha AS, Espejo-González O, Gómez-Martínez AE, Picazo O. Anxiolytic-like effects of a new 1-N substituted analog of melatonin in pinealectomized rats. *Prog Neuropsychopharmacol Biol Psychiatry.* 2014;51:133–139.

[151] Ceravolo R., Rossi C., Kiferle L., Bonuccelli U. (2010) Nonmotor symptoms in Parkinson's disease: the dark side of the moon. *Future Neurology* 2010;5:851–871.

[152] Willis GL, Robertson AD Recovery of experimental Parkinson's disease with the melatonin analogues ML-23 and S-20928 in a chronic, bilateral 6-OHDA model: a new mechanism involving antagonism of the melatonin receptor. *Pharmacol Biochem Behav.* 2004;79:413-29.

[153] Paulis L, Simko F, Laudon M. Cardiovascular effects of melatonin receptor agonists. *Expert Opin Investig Drugs.* 2012;21:1661–1678.

[154] Tian SW, Laudon M. Antidepressant- and anxiolytic effects of the novel melatonin agonist Neu-P11 in rodent models. *Acta Pharmacol Sin.* 2010;31:775–783.

In: Melatonin: Medical Uses and Role in Health and Disease ISBN: 978-1-53612-987-8
Editors: Lore Correia and Germaine Mayers © 2018 Nova Science Publishers, Inc.

Chapter 5

THE IMMUNOMODULATORY PROPERTIES OF MELATONIN

Sylwia Mańka[*] and Ewa Majewska
Department of Pathophysiology and Clinical Immunology,
Chair of Basic Sciences, Medical University of Lodz, Lodz, Poland

ABSTRACT

According to available literature, melatonin plays an important role in many aspects of human physiology, however this review will focus just on its effect on the immune system response. Despite a large number of studies proving a relationship between melatonin and the immune system, the definitive effect of its influence on immunity and a mechanism of action still remains arguable. Melatonin mediates between neurohormonal and immune systems by connecting the development of the immune reaction with melatonin secretion induced by inflammatory mediators. Melatonin is a kind of 'immunological buffer', which stimulates the immune processes, particularly in patients with immune disorders but also suppresses immunity in the case of its excessive activation. Furthermore, it modulates the apoptosis of leukocytes and the course of inflammatory responses, acting both as its activator and inhibitor. This immunomodulatory function, as well as its antioxidant properties make the hormone a potentially interesting therapeutic agent. A wide range of data suggests that melatonin might find widespread clinical use as a supportive therapy in the treatment of inflammatory conditions, bacterial and viral infections, cancer and autoimmune diseases.

Throughout the present chapter, we discuss the current knowledge about the immunomodulatory properties of melatonin and its effect on non-specific, humoral and cellular immune responses with particular emphasis on the production of reactive oxygen species and cytokines as well as on apoptosis. Finally, we briefly outline the clinical use of melatonin in several immune conditions, such as inflammation and cancer.

Keywords: melatonin, immune system, immunomodulation

---

[*] Corresponding Author Email: sylwia.manka@umed.lodz.pl.

INTRODUCTION

The first reports about a biologically active substance located in the pineal gland were published in 1917 by McCord and Allen, who observed that bovine pineal extracts added to water caused the whitening of frog *Rana pipiens* skin [1]. Many years later, in 1958, a group of dermatologists under the direction of Lerner, isolated and determined the chemical structure of a substance from the bovine pineal gland and named it melatonin (MLT) [2]. The initial compound for MLT synthesis is the endogenous tryptophan which is converted into serotonin as a result of hydroxylation and decarboxylation. In further stages of the serotonin metabolism, serotonin N-acetyltransferase (NAT) and hydroxyindole-O-methyltransferase (HIOMT) are involved. As a result, serotonin is converted into 5-methoxy-N-acetyltryptamine, i.e., melatonin [3]. MLT is a hormone produced mainly by the pineal gland and its biosynthesis occurs with a circadian rhythm, regulated by the master clock localized in the suprachiasmatic nucleus (SCN) of the hypothalamus [4]. The secretion of pineal melatonin is closely dependent on the time of day or night. The level of MLT persists at low levels during daytime (10-20 pg/ml) and rises in the late evening, reaching the maximum values (80-120 pg/ml) between midnight and 3.00 a.m. [5]. The duration of nightly elevated melatonin secretion is shorter during the period of a long day in spring/summer months [6, 7, 8] and the acrophase of MLT release was achieved earlier in summer as compared to winter [9]. The concentration of melatonin varies with age. The peak of MLT secretion is between 4 and 7 years of age, followed by significant decreasing levels during puberty, which remains relatively stable up to the age of 45 years and gradually decreases thereafter, so in old age the differences between day and night concentrations are slight [5]. In addition to the pineal gland, MLT is also produced by other tissues like the gastrointestinal tract, retina, skin, and immune system cells (i.e., bone marrow, lymphocytes, mastocytes or platelets [3, 10, 11]). It is important that pineal melatonin synthesis and secretion are synchronized with the light-dark cycle while the extra-pineal melatonin synthesis does not show day/night fluctuations and does not affect melatonin levels measured in the blood, which implies that extra-pineal MLT has a local, autocrine or paracrine action in these cells [10, 11, 12].

The primary physiological role of melatonin (also called the 'hormone of darkness') is to provide the body with the necessary information about the daily and seasonal environmental conditions and to coordinate the biological rhythms, especially the sleep-wake cycle rhythm [4]. It plays a role as a "clock" and "calendar" for the organism. The increased concentration of MLT informs about darkness and the time of its duration. Respectively, the prolonging or shortening of an elevated MLT period are the signs of the coming autumn-winter and spring-summer months [13]. In addition to the chronobiological activity, melatonin participates in the regulation of blood pressure and renal function, thermoregulation, gastrointestinal function and modulation of inflammatory and immune processes [3].

Recent studies have shed new light on the role of melatonin in immunomodulation. The first suggestion of a possible relationship between MLT and the immune system was revealed in 1926 after the observation by Berman who noted a marked improvement in the resistance against infection in kittens fed with the extract of pineal glands from young bulls [14]. Later, in studies performed in 1986, Maestroni et al. demonstrated that inhibition of melatonin synthesis in mice results in the impairment of both cellular and humoral immunity [15]. Nowadays, the impact of MLT on the immune system has been supported by several findings. A correlation has been demonstrated between melatonin production and circadian and seasonal fluctuations in immune functions. Pinealectomy or administration of MLT in many experimental approaches, both *in vivo* and *in vitro,* have been shown to result in changes in immune functions and additionally, receptors of melatonin have been found in leukocytes (in T lymphocytes, monocytes and neutrophils) [16, 17, 18]. Despite the significant progress made in recent years in understanding the effects of melatonin on the immune system, many of its properties and details of those interactions still remain unexplained. The difficulty in the unambiguous determination of the effects of melatonin on the immune system is related to the fact that these effects are various and dependent on many factors, e.g., sex, age, time of day or season and the dose and manner of melatonin administration [14]. Some authors claim that MLT exhibits a stimulating effect on both the innate and adaptive immune system, whereas others describe its anti-inflammatory properties [19-23]. To explain these opposite effects, there are some hypotheses suggesting that melatonin acts as an 'immunological buffer', which stimulates the immune processes under physiologic or immunodeficiency conditions (induced by the immunosuppression, stress or advanced age) or inhibits them in the case of exacerbated activation of the immune system [24] (Figure 1).

Melatonin regulates the course of inflammatory response, acting both as its activator and inhibitor [25, 26]. Depending on the phase of inflammation, MLT differentially modulates the proinflammatory cytokine synthesis, phospholipase A2 and lipoxygenase activity and regulates the leukocyte survival. It is suggested that by activating the synthesis of inflammatory mediators such as IL-1, TNF-α and metabolites of arachidonic acid, MLT might promote the early phases of inflammation and contribute to its attenuation. The anti-inflammatory action of melatonin results from the suppression of inflammatory mediator synthesis as well as from its antioxidant and anti-apoptotic properties [25].

Certain disturbances in MLT synthesis and secretion have been involved in the pathogenesis of human disorders like cancer, and several autoimmune or neurodegenerative diseases [27-30]. One of the reasons for the increasing incidence of neoplastic and infectious diseases in elderly people is the proposed decline in immunity correlated with age-related impairment of MLT production [31, 32].

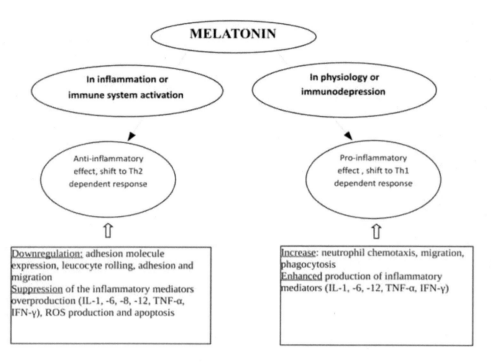

Figure 1. Melatonin as an 'immunological buffer'.

In turn, decreased melatonin production has been suggested as a cause of increased risk of cancer in night workers [33, 34]. On the other hand, there are promising results of studies on its potential application in pharmacotherapy in the above-mentioned diseases [30, 35, 36].

This chapter presents the current state of knowledge about melatonin's functional relationship with the immune system and its immunomodulatory effects and potential action mechanisms exerted on cells that belong to the innate immunity, such as neutrophils and monocytes-macrophages, as well as on cells involved in the adaptive immune responses, i.e., lymphocytes. In addition, it briefly summarizes and highlights the potential modulatory effects of melatonin in several pathological conditions including inflammation and cancer.

MECHANISMS OF MELATONIN ACTION

The main cause of MLT pleiotropic effects is the multitude of intracellular mechanisms of hormone action. Generally, there are four types of MLT action: binding to the membrane receptors, nuclear receptors, and intracellular proteins, and also by its antioxidant action [37].

The pharmacological classification by the IUPHAR (International Union of Basic and Clinical Pharmacology) committee distinguishes three types of receptors for melatonin; MT1, MT2 and MT3 [38]. The first two listed receptors belong to the superfamily of G-protein coupled receptors (GPCRs), whereas MT3 receptors are related to the family of quinone reductases [39]. Through the interaction with membrane receptors, MLT regulates several second messengers such as cAMP (cyclic adenosine monophosphate), cGMP (cyclic guanosine monophosphate), DAG (diacylglycerol), IP$_3$ (inositol triphosphate), arachidonic acid and intracellular Ca^{2+} concentration, mediating signal transduction from activated receptors [40]. Melatonin can also bind to and activate ROR/RZR receptors (*retinoid orphan receptors/retinoid Z receptors*), which are located in the nucleus [37].

The melatonin receptors are clearly widespread and multifunctional, allowing melatonin to act as a pleiotropic modulator. They are localized beside the central nervous system in the majority of cells of the digestive tract, gonads, cardiovascular and immune systems [41]. MT1/MT2 and ROR / RZR receptors are localized in bone marrow cells, neutrophils, monocytes, T cells and NK cells, which confirms the fact that melatonin can directly affect immune system cells [17, 18, 42].

Owing to its lipophilic nature, MLT can easily cross through the cell membranes and act on many proteins in the cytoplasm, among others on calmodulin modifying Ca^{2+}-dependent cellular processes. It is believed that this mechanism of MLT action could explain the anti-proliferative effects of the hormone on cancer cells [43].

Moreover, MLT modifies intracellular processes in receptor-independent ways relating to its antioxidant properties and ability to deactivate free radicals, such as singlet oxygen, nitric oxide (NO·), radical hydroxide (OH·), superoxide radical anion (O$_2$·$^-$), peroxynitrite (ONOO$^-$) and peroxyl radical (LOO·). Products of the reaction are; 2-, 3-, 6-, 7-hydroxymelatonin, cyclic 2-, 3-hydroxymelatonin, N(1)-acetyl-N(2)-formyl-5-methoxykynuramine (AFMK) and N(1)-acetyl-5-methoxykynuramine (AMK) [44] which can also neutralize ROS, enhancing the anti-oxidative potential of MLT [45].

THE EFFECT OF MELATONIN ON FUNCTIONING AND APOPTOSIS OF CHOSEN INNATE AND ADAPTIVE IMMUNE CELLS

Melatonin and Neutrophils

Neutrophils are the basis for an adequate response to pathogens and are the key cells that cause inflammation but so far, little attention has been devoted to the impact of melatonin on the cells' function. The results of studies carried out mainly on animal models are difficult to transfer onto human data, and furthermore, the results are often ambiguous.

Neutrophil migration to the inflammatory sites is the fundamental process of innate immunity against pathogen invasion [46]. Experiments in *in vivo* animal studies demonstrated that neutrophil migration exhibits rhythmicity, i.e., more neutrophils are recruited to the injury site at night than in the daytime [47, 48]. The authors observed that treatment with MLT could facilitate rhythmic neutrophil migration, likely by an increased secretion of TNF-α and IL-8 [47].

Melatonin in exacerbated immune responses can act in opposition. MLT inhibits the neutrophil migration to the injury site in a concentration- and time-dependent manner in fish, which could be responsible for the inflammatory response suppression [49]. Consistent results, i.e., the inhibition of neutrophil infiltration to the inflammatory site, were also observed in heatstroke rats, as well as in liver and lung injuries [50-52]. Decreasing the neutrophil infiltration is also one of many mechanisms responsible for the protective effect of MLT on acute pancreatitis [53].

Despite the fact that the leukocyte infiltration is a key feature of inflammatory and immune responses, there is little information available on the effect of melatonin on human leukocyte chemotaxis. On the one hand, the physiological concentration of MLT has *in vitro* and *in vivo* chemotactic effects on human neutrophils under physiological conditions [54], but on the other hand, MLT strongly inhibits the production of the chemokines such as IL-8 and TNF-α in human neutrophils, inducing chemotaxis [55].

Apart from chemotaxis, endothelial adherence and permeability are the other important processes involved in the recruitment of leucocytes to the injury site. MLT inhibits both leukocyte rolling and adhesion to endothelial cells via receptor-mediated mechanisms during acute inflammation in rats. Doses of MLT used in experiments were in the range of the nocturnal serum concentrations, suggesting that melatonin plays a physiological role in modulating acute inflammation [56]. These results explain why the weaker inflammatory acute response is observed when the inflammatory stimulus is applied at night rather than during the day [57]. It was demonstrated that MLT inhibits leukocyte-endothelial interactions and downregulates the expression of adhesion molecules, such as ICAM-1 and PECAM-1 [58]. Less neutrophils adhere to the endothelial cells obtained from rats killed during nighttime and the endothelial cells express a lower level of ICAM-1 and PECAM-1, participating in the leukocyte migration compared to the cells obtained during daytime [58, 59]. Moreover, MLT reduces vascular permeability induced by leukotriene B4 [60] and also inhibits L-selectin shedding from the surface of human neutrophils [61].

Suppression of the nocturnal melatonin secretion in rats, induced by a long-term exposure to light, downregulates the average circadian level of phagocytosis, whereas melatonin treatment significantly restores this process to the previous activity. The above observations suggest that the physiological role of MLT is to maintain the phagocytic activity of neutrophils [62]. Results of the *in vivo* and *in vitro* studies in birds demonstrated that melatonin enhances the phagocytosis activity of heterophils, cells performing similar functions as neutrophils [63, 64]. As regards the effect of MLT on phagocytosis of the human neutrophils, it has been shown that MLT does not exert any significant effect on the process of phagocytosis [65].

Melatonin regulates the respiratory burst of PMA-stimulated human neutrophils in a dual dose-dependent manner, i.e., low concentrations of MLT (10nM) intensify the intracellular production of reactive oxygen species (ROS), whereas a high concentration (2mM) limits this process [66]. The results suggest that MLT stimulates the intracellular oxygen potential through the protein kinase C activation, while the inhibition of ROS generation can be a result of melatonin's direct impact on the transmembrane potential of neutrophils [61, 66]. MLT decreases superoxide production ($O_2^{\cdot-}$) in human neutrophils activated by *Candida albicans* [67] and reduces the level of UV-induced ROS in a dose-dependent manner in IL-3-stimulated neutrophils [68]. Similarly, Espino et al. demonstrated that MLT reduces the intracellular ROS generation in human neutrophils induced by calcium signaling or TNF-α [69, 70]. On the contrary, MLT enhances the superoxide release from neutrophils and this effect is equivalent to those of PMA-stimulated neutrophils [71] or MLT has no effect on the production of $O_2^{\cdot-}$ [65].

One of the physiological functions of neutrophils is their microbicidal activity. The effect of MLT on this function is ambiguous. Some authors observed that MLT increased bactericidal properties of neutrophils towards enteropathogenic *Escherichia coli* (EPEC) inducing superoxide production [71]. On the contrary, MLT could inhibit *Staphylococcus aureus* killing by neutrophils through the inhibition of the hypochlorous acid (HOCl) formation, followed by inhibition of its highly reactive derivative formation, such as singlet oxygen [65].

When the functions of neutrophils against the pathogen are performed or when they are aged, the cells die by apoptosis. Extending the lifespan of neutrophils during inflammation is critical for the efficient destruction of pathogens, whereas apoptosis is a key mechanism for the elimination of neutrophils from the inflammatory site and the process attenuation [72, 73]. In leukocytes, MLT has an anti-apoptotic property, however, the mechanism has not been fully understood and has become the strong focus of attention for many researchers. MLT is able to reduce caspase-3 and -9 activities evoked by calcium signaling and hence, to reduce apoptosis in human neutrophils. The authors suggest that this inhibitory effect of MLT is associated with the blockade of the mitochondrial permeability transition pore (mPTP) opening and with the inhibition of Bax activation [74]. The same group of researchers later documented that the protective effect of MLT against the apoptosis of human neutrophils induced by the intracellular calcium overload or TNF-α/CHX is likely dependent on melatonin's anti-oxidant actions. More importantly, MLT alone does not affect the rates of apoptosis compared with control and only decreases apoptosis induced by pro-apoptotic agents [69, 70].

Additionally, it has been shown that melatonin *in vitro* decreases the apoptosis of neutrophils from patients with acute pancreatitis [75].

Melatonin and Monocytes/Macrophages

Melatonin stimulates hemopoiesis, affecting the production of progenitor cells for granulocytes and macrophages (CFU-GM) [76]. The nighttime peak of MLT secretion coincides with the peak of CFU-GM proliferation, and pinealectomy causes a decrease in the expression of the CFU-GM activity in rat bone marrow cell cultures [77, 78]. Supplementation of a diet with MLT increases the population of monocytes in both the bone marrow and spleen in mice. These stimulating effects of MLT may be either a result of its direct action on melatonin receptors or may result from the increased sensitivity of monocytes to stimulants, such as IL-3, IL-4, IL-6, and GM-CSF, released from the MLT-activated T cells [79].

Melatonin shows dual effects on pro-inflammatory cytokine synthesis, i.e., under physiological conditions MLT induces this process, whereas it acts contrary in the exacerbated immune response. Melatonin activates monocytes and induces their cytotoxic response by stimulating IL-1ß synthesis. This effect is noticeable only when LPS-stimulated cells are pretreated with melatonin, otherwise IL-1 is not secreted. The authors suggest that melatonin is necessary for priming the monocytes for a subsequent activation by LPS [21]. Melatonin also stimulates the synthesis of IL-6, -12 and TNF-α by receptor-mediated mechanisms [20, 42]. Both the membrane and nuclear melatonin receptors have been detected in monocytes and in the human monocytic cell line U937 [18, 80]. The receptor expression depends on the monocyte state of maturation and *in vitro* experiments showed that its expression negatively correlated with the monocyte differentiation and maturation [42].

An increasing number of publications demonstrate the anti-inflammatory properties of melatonin, which is primarily due to its decreasing impact on the exacerbated production of pro-inflammatory cytokines in inflammatory conditions. MLT inhibits TNF-α, IL-1β, -6, -8, -10 expression in LPS-stimulated RAW264.7 macrophages [81] and decreases TLR-3-mediated TNF-α expression in respiratory syncytial virus-infected RAW 264.7 macrophages [82]. Additionally, melatonin inhibits the production of IL-6 at both gene transcription and translation levels in the same macrophage cell line stimulated by LPS [83]. However, some authors, in contrast to previously described reports, established that MLT does not alter the LPS-stimulated TNF-α, IL-1β and IL-6 production in RAW 264.7 and in rodent macrophages [84, 85].

There are reports showing that MLT transiently stimulates the production of intracellular reactive oxygen species (ROS) in U937 monocytic cell lines [86, 87]. The authors suggest that the pro-radical action of MLT is not a result of a direct chemical pro-oxidant activity of melatonin (and/or its metabolites), but rather the effect related to its ability to elicit the activation of some cellular radical producing enzymes (e.g., lipoxygenases, cyclooxygenases, NO-synthase, NADPH-oxidase) [87]. Likewise, Radogna et al. postulated that MLT activates phospholipase A2 (PLA2) and 5-

lipoxygenase (5-LOX), which is a consequence of its binding to calmodulin. These enzymes are responsible for releasing high doses of arachidonic acid from phospholipids, which are metabolizing to 5-hydroxyeicosatetraenocic acids (5-HETEs), accompanied by the ROS generation [26]. The consequence of 5-lipoxygenase activation by MLT is the induction of the leukotriene B4 production, which promotes diapedesis by vessel permeabilization and acts as a chemoattractant [88]. These observations suggest that the pro-radical action of MLT associated with 5-LOX activation and promotion of inflammatory products could play an important role in promoting early phases of inflammation [25, 26]. In addition, MLT induces the production of ROS, such as O_2^-, H_2O_2 and NO_2 in human monocytes [21].

Similar to neutrophils, melatonin has the ability to reduce ROS in pathological states associated with high oxidant production. MLT and its metabolites such as N1-acetyl-N2-formyl-5-methoxykynuramine (AFMK) and N1-acetyl-5-methoxykynuramine (AMK), attenuate LPS-induced upregulation of cyclooxygenase-2 (COX-2) and inducible nitric oxide synthase (iNOS) in RAW264.7 macrophages [23, 81]. The authors revealed that the above effects of melatonin action occur through the mechanism involving the attenuation of NF-κB activation via modulation of toll-like receptor-4-mediated signaling pathways [81]. Likewise, in LPS-stimulated murine macrophages, MLT inhibits the expression of inducible isoform of nitric oxide synthase (iNOS) and, consequently, suppresses the nitric oxide production via the inhibition of nuclear factor-κB (NF-κB) activation [89]. In turn, Choi et al. support the opinion that MLT downregulates the production of NO by preventing the NF-κB translocation to the nucleus rather than through the suppression of its activation [83]. In addition, MLT downregulates the superoxide anion production in naphthalene-induced oxidative stress in cultured J774 macrophages [90]. However, MLT has no effect on the PMA-stimulated H_2O_2 and O_2^- production by macrophage cell lines [84], which is in contrast to the above-mentioned point of view.

Macrophages are considered professional antigen-presenting cells (APCs), which influence the process of adaptive immunity presenting antigens to, and activating T lymphocytes [91]. MLT enhances antigen presentation by splenic mouse macrophages to T cells via increasing the MHC II class molecule expression and production of IL-1 and TNF-α [92]. MLT activates human monocytes to produce IL-12 causing the Th1 lymphocyte activation [20].

Circulating monocytes have a short period of time before undergoing spontaneous apoptosis, however, during inflammation or their differentiation into macrophages, the life span of cells can be extended by preventing the activation of the apoptotic program [93]. MLT prevents the apoptosis induced by UVB irradiation in U937 cell line, which is linked to its anti-oxidant ability and also related to the prevention of mitochondrial alterations [94]. MLT inhibits apoptosis induced by pro-apoptotic agents via interaction with MT1/MT2 plasma membrane receptors in U937 monocytes [95, 96]. The authors suggest that the anti-apoptotic effects of melatonin occur by modulation of the balance between the pro-apoptotic Bax and the anti-apoptotic Bcl-2 protein expression. MLT promotes Bcl-2 re-localization to mitochondria, which induces Bax sequestration, thereby impairing its activation/dimerization. In addition or in part, this impairment could be a result of the radical scavenging ability of MLT [95]. In the latest studies, Radogna et al. demonstrate that the anti-apoptotic effect of MLT in U937 cell line and monocytes requires interaction with two different pathways triggered by high (MT1/MT2 receptors) and low affinity (calmodulin/LOX) targets. These pathways converge within mitochondria, where Bcl-2 interacts with Bax, sequesters and inactivates it [97].

Melatonin and Lymphocytes

The majority of published data on the immunomodulatory properties of melatonin refer to lymphocytes, probably because the lymphocytes are considered as the most important cells of the immune system.

Melatonin either increases the non-stimulated T-lymphocyte proliferation or does not exert any effect [92, 98, 99]. Physiological concentrations of MLT increase significantly the proliferation of human peripheral blood lymphocytes, possibly via the suppression of IL-10 production [99]. In contrast to the above-mentioned effect, an anti-proliferative effect of MLT on mitogen-stimulated peripheral lymphocytes or lymphocytes derived from tumors has also been described in *in vitro* studies. Markowska et al. demonstrate that melatonin inhibits the proliferative response of PHA-activated chicken lymphocytes, and importantly, MLT effect is hormone- and mitogen dose-dependent [98]. Similarly, melatonin inhibits proliferation in human and mouse lymphocytes and T-lymphoblastoid cell lines activated with PHA and ConA [100, 101] and suppresses the proliferation of Th1-hybridoma cells by inhibiting IL-2 secretion in these cells [102].

Melatonin affects the synthesis and secretion of many cytokines such as tumor necrosis factor (TNF)-α, interferon (IFN)-γ, IL- 1, -2, -6 and -12 by human blood mononuclear cells [19, 20].   García-Mauriño et al. reported that MLT can activate human Th1 lymphocytes by enhancing IL-2 production via the nuclear receptor-mediated mechanism. The effects described above were observed only when cells were either not stimulated or only slightly activated by low concentrations of PHA, whereas in the presence of high concentrations of mitogen, no effect was observed [19]. It has been shown that both resting and PHA-stimulated human lymphocytes are able to synthesize and release large amounts of melatonin, by which receptor-mediated action mechanisms may play a crucial role in modulating the IL-2/IL-2 receptor system. This was confirmed by the studies demonstrating that blocking endogenous MLT synthesis results in a decrease in both IL-2 production and IL-2 receptor expression and the addition of exogenous MLT restores these parameters. Carrillo-Vico et al. suggest that lymphocyte-synthesized melatonin might be involved in the regulation of the human immune system, where it probably acts through the intra-, auto- and/or paracrine mechanisms and interferes with the exogenous MLT effect on IL-2 production [11, 103]. The findings that melatonin can stimulate IL-2 production have led to the examination of the MLT impact on enhancing T-cell immunity in cancer patients [104-106].

In addition, the peak secretion of the pro-inflammatory cytokines such as IFN-γ, TNF-α and IL-1 occurs at night and coincides with the peak of MLT production [107]. In turn, clinical studies have shown that HIV-1-infected patients have significantly reduced melatonin levels correlating with diminished IL-12 concentration. These findings suggest that this reduction may be linked to the impairment of Th1 immune responses [108].

There are also observations in contrast to the previously mentioned data showing that melatonin promotes also the Th1 cell-mediated cellular response in physiological state. Physiological and pharmacological doses of MLT enhance IL-4 synthesis in bone marrow lymphocytes [109] and the deficiency of MLT in pinealectomized rats increases IFN-γ synthesis and decreases IL-10 levels, suggesting that MLT shifts the balance towards the Th-2 type immune response in certain clinical states [110].

Most data demonstrates that MLT has the anti-inflammatory effect via enhancement of Th2-dependent immunity in the case of exacerbated activation of the immune system. MLT administration *in vivo* acts on Th-2 cells, enhancing IL-4, -10 synthesis and inhibiting the secretion of IL-2, IFN-γ and TNF-α in antigen-primed mice [111, 112]. MLT does not influence the IL-4 secretion by naïve T cells suggesting that MLT acts only on antigen-activated cells [111]. Melatonin inhibits the inflammatory response in Th1-mediated contact hypersensitivity in mice by inhibiting the production of IFN-γ and IL-12 [113]. Additionally, MLT increases the IL-10 production in an experimental model of septic shock [114] and in acute pancreatitis in rodents [53].

In recent years, certain publications have shown that melatonin affects also T regulatory lymphocytes, which are identified as $CD4^+CD25^+$ cells. Treg lymphocytes are responsible for controlling and inhibiting excessive inflammatory response, hypersensitivity, and hyperactivity reactions, especially in relation to self-antigens [115]. In preliminary clinical trials, MLT decreases the Treg population in cancer patients, whereas no effect in *in vitro* experiments was observed [116]. The authors speculate that MLT acts indirectly by counteracting the macrophage-induced stimulation of Treg lymphocyte production. These assumptions are based on the observation that the generation of T cells *in vivo* is stimulated by several mechanisms, namely by the monocyte-macrophage system [116]. According to Liu et al., MLT downregulates $CD4^+CD25^+$ Treg numbers and Foxp3 expression in lymphocytes infiltrating the tumor tissue, which might be responsible for the antitumor properties of MLT against gastric cancer in *in vivo* animal studies [117]. In turn, MLT increases the number of Treg cells in patients with systemic lupus erythematosus (SLE), where a decrease in the cell number was found in some studies [118].

Although melatonin receptors are expressed by B lymphocytes [18, 119], the role of MLT in modifying the humoral immune response is still undefined [120].

Melatonin inhibits lymphocyte apoptosis [74, 121-123]. One of the described mechanisms of this action is the reduction of Bax activation [74, 123]. Some authors claim that the protective effect of MLT on lymphocyte apoptosis is MT1/MT2 receptor-independent and that the antiapoptotic action is rather from its radical scavenger ability [69, 70]. However Paternoster et al. demonstrated that melatonin reduces stress-induced lymphocyte apoptosis as well as increases the intracellular production of ROS [122], which is in contrast to the above-mentioned data.

POTENTIAL CLINICAL APPLICATIONS OF MELATONIN

Melatonin is a hormone with a very broad spectrum of action and nowadays is available as an over-the-counter (OTC) drug. Treatment of sleep disorders in the elderly, in the blind persons, shift workers and after a swift change of time zones (jet-lag) is the only well documented clinical indication for the use of melatonin in therapy in humans [124]. At present, a strong trend is observed towards other therapeutic possibilities of this hormone application in many diseases; hypertension, gastrointestinal disorders (gastric ulcers, gastroesophageal reflux disease or irritable bowel syndrome), psychiatric, neurological and neurodegenerative disorders (epilepsy, migraine, Alzheimer's disease) with no well proven efficacy [125-130].

Many publications have shown that melatonin modulates the clinical course of acute and chronic inflammation in humans [131] which seems to be a direct result of its ability to inhibit the synthesis of pro-inflammatory cytokines observed in animal models of septic shock [114]. After all, the beneficial effect of MLT is not limited to the suppression of cytokine synthesis, such as IL-12, TNF-α and IFN-γ but the hormone also acts as an immunomodulator, antioxidant and anti-apoptotic agent [114]. MLT revealed anti-oxidative and anti-inflammatory effects in the experimental daytime model of sepsis in humans [132]. When lipopolisacharide (LPS) endotoxin is applied at midnight (nighttime endotoxemia model), MLT does not affect the markers of inflammation and oxidative damage, which is in contrast to daytime endotoxemia [133]. In clinical studies, MLT improves the clinical course of septic newborns and significantly reduces levels of lipid peroxidation products and CRP concentration, which are the markers of oxidative stress and inflammation, respectively [134].

The immunomodulatory effects of MLT were also observed in patients with bronchial asthma in whom the hormone stimulated IL-1, -6 and TNF-α synthesis in zymosan-induced peripheral blood mononuclear cells [135]. Sutherland et al. observed that patients with nocturnal asthma had significantly elevated MLT levels that inversely correlated with the overnight change in forced expiratory volume (FEV), which might explain the worsening of the nocturnal symptoms of asthma [136]. These observations suggest that MLT used in asthma might be unfavorable by aggravating inflammation.

Numerous experimental studies have documented the oncostatic properties of melatonin. The mechanisms through which MLT can exert these effects include anti-proliferation, anti-angiogenic, anti-oxidant effects, the above discussed immunomodulatory properties and its ability to induce apoptosis in cancer cells [137]. In clinical studies, MLT enhances therapeutic efficacy and reduces the toxicity of standard chemotherapy in metastatic non-small cell lung cancer, and colorectal or gastric cancer [106, 138, 139]. These observations suggest that melatonin may be used successfully in clinical oncology as a supportive therapy. Melatonin also presents an oncostatic and anti-proliferative effect in breast cancer [140]. MLT decreases the incidence and size of mammary adenocarcinomas and the incidence of lung metastases in a mouse model of mammary carcinogenesis [141].

The exact effect and mechanism of MLT actions in these disorders are still far from being understood. The majority of studies are often carried out on small groups, and therefore, have limited significance, although the outcomes are promising. As can be seen, investigations related to melatonin are moving in many directions, arousing great hopes and expectations.

CONCLUSION

The knowledge of melatonin is very extensive and the interest of scientists in the immunomodulatory properties and potential clinical applications of this extraordinary hormone is still remarkable. Despite the large number of studies, its impact on immune response is still a matter of controversy. The immunomodulatory effects of melatonin on the chosen cells involved in innate and adaptive responses are summarized in Figure 2.

In conclusion, the influence of MLT on immune functions has not been fully understood and still remains an intriguing research target.

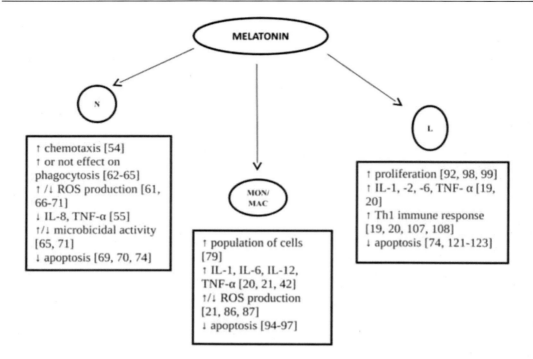

Figure 2. A scheme of melatonin effects on chosen immune cells in basal conditions. (N) neutrophils, (Mon/Mac) monocytes/macrophages, (L) lymphocytes.

REFERENCES

[1] McCord, CP; Allen, FP. Evidence associating pineal gland function with alterations in pigmentation. *J Exp Zool*, 1917, 23, 207–224.

[2] Lerner, AB; Case, JD; Takahashi, Y. Isolation of melatonin and 5-methoxyindole-3-acetic acid from bovine pineal glands. *J Biol Chem*, 1960, 235, 1992–1997.

[3] Pandi-Perumal, SR; Srinivasan, V; Maestroni, GJ; Cardinali, DP; Poeggeler, B; Hardeland, R. Melatonin: Nature's most versatile biological signal? *FEBS J*, 2006, 273, 2813-2838.

[4] Zawilska, JB; Skene, DJ; Arendt, J. Physiology and pharmacology of melatonin in relation to biological rhythms. *Pharmacol Rep*, 2009, 61, 383-410.

[5] Karasek, M; Winczyk, K. Melatonin in humans. *J Physiol Pharmacol*, 2006, 57, 19-39.

[6] Kennaway, DJ; Van Dorp, CF. Free-running rhythms of melatonin, cortisol, electrolytes, and sleep in humans in Antarctica. *Am J Physiol*, 1991, 260, R1137-1144.

[7] Wehr, TA; Moul, DE; Barbato, G; Giesen, HA; Seidel, JA; Barker, C; Bender, C. Conservation of photoperiod-responsive mechanisms in humans. *Am J Physiol*, 1993, 265, R846-857.

[8] Yoneyama, S; Hashimoto, S; Honma, K. Seasonal changes of human circadian rhythms in Antarctica. *Am J Physiol*, 1999, 277, R1091–R1097.

[9] Honma, K; Honma, S; Kohsaka, M; Fukuda, N. Seasonal variation in the human circadian rhythm: dissociation between sleep and temperature rhythm. *Am J Physiol*, 1991, 262, R885-891.

[10] Maldonado, MD; Mora-Santos, M; Naji, L; Carrascosa-Salmoral, MP; Naranjo, MC; Calvo, JR. Evidence of melatonin synthesis and release by mast cells. Possible modulatory role on inflammation. *Pharmacol Res*, 2010, 62, 282-287.

[11] Carrillo-Vico, A; Calvo, JR; Abreu, P; Lardone, PJ; García-Mauriño, S; Reiter, RJ; Guerrero, JM. Evidence of melatonin synthesis by human lymphocytes and its physiological significance: possible role as intracrine, autocrine, and/or paracrine substance. *FASEB J*, 2004, 18, 537-539.

[12] Zawilska, J. Melatonin as a chemical indicator of environmental light-dark cycle. *Acta Neurobiol Exp*, 1996, 56, 757-767.

[13] Reiter, RJ. The melatonin rhythm: both a clock and a calendar. *Experientia*, 1993, 49, 654-664.

[14] Carrillo-Vico, A; Guerrero, JM; Lardone, PJ; Reiter, RJ. A review of multiple actions of melatonin on the immune system. *Endocrine*, 2005, 27, 189-200.

[15] Maestroni, GJ; Conti, A; Pierpaoli, W. Role of the pineal gland in immunity. Circadian synthesis and release of melatonin modulates the antibody response and antagonizes the immunosuppressive effect of corticosterone. *J Neuroimmunol*, 1986, 13, 19-30.

[16] Guerrero, JM; Reiter, RJ. Melatonin-immune system relationships. *Curr Top Med Chem*, 2002, 2, 167-179.

[17] Lopez-Gonzalez, MA; Calvo, JR; Segura, JJ; Guerrero, JM. Characterization of melatonin binding sites in human peripheral blood neutrophils. *Biotechnol Ther*, 1993, 4, 253-262.

[18] Pozo, D; García-Mauriño, S; Guerrero, JM; Calvo, JR. mRNA expression of nuclear receptor RZR/RORα, melatonin membrane receptor MT1, and hydroxindole-O-methyltransferase in different populations of human immune cells. *J Pineal Res*, 2004, 37, 48-54.

[19] García-Mauriño, S; Gonzalez-Haba, MG; Calvo, JR; Rafii-El-Idrissi, M; Sanchez-Margalet, V; Goberna, R; Guerrero, JM. Melatonin enhances IL-2, IL-6, and IFN-gamma production by human circulating CD4+ cells: a possible nuclear receptor-mediated mechanism involving T helper type 1 lymphocytes and monocytes. *J Immunol*, 1997, 159, 574-581.

[20] García-Mauriño, S; Pozo, D; Carrillo-Vico, A; Calvo, JR; Guerrero, JM. Melatonin activates Th1 lymphocytes by increasing IL-12 production. *Life Sci*, 1999, 65, 2143-2150.

[21] Morrey, KM; McLachlan, JA; Serkin, CD; Bakouche, O. Activation of human monocytes by the pineal hormone melatonin. *J Immunol*, 1994, 153, 2671-2680.

[22] Cuzzocrea, S; Zingarelli, B; Costantino, G; Caputi, AP. Protective effect of melatonin in a non-septic shock model induced by zymosan in the rat. *J Pineal Res*, 1998, 25, 24-33.

[23] Mayo, JC; Sainz, RM; Tan, DX; Hardeland, R; Leon, J; Rodriguez, C; Reiter, RJ. Anti-inflammatory actions of melatonin and its metabolites, N1-acetyl-N2-formyl-5-methoxykynuramine (AFMK) and N1-acetyl-5-methoxykynuramine (AMK), in macrophages. *J Neuroimmunol*, 2005, 165, 139-149.

[24] Carrillo-Vico, A; Lardone, PJ; Àlvarez-Sánchez, N; Rodríguez- Rodríguez, A; Guerrero, JM. Melatonin: buffering the immune system. *Int J Mol Sci*, 2013, 14, 8638-8683.

[25] Radogna, F; Diederich, M; Ghibelli, L. Melatonin: A pleiotropic molecule regulating inflammation. *Biochem Pharmacol*, 2010, 80, 1844-1852.

[26] Radogna, F; Sestili, P; Martinelli, C; Paolillo, M; Paternoster, L; Albertini, MC; Accorsi, A; Gualandi, G; Ghibelli, L. Lipoxygenase-mediated pro-radical effect of melatonin via stimulation of arachidonic acid metabolism. *Toxicol Appl Pharmacol*, 2009, 238, 170-177.

[27] Lissoni, P. Is there a role for melatonin in supportive care? *Support Care Cancer*, 2002, 10, 110-116.

[28] Sulli, A; Maestroni, GJ; Villaggio, B; Hertens, E; Craviotto, C; Pizzorni, C; Briata, M; Seriolo, B; Cutolo, M. Melatonin serum levels in rheumatoid arthritis. *Ann NY Acad Sci*, 2002, 966, 276-283.

[29] Melamud, L; Golan, D; Luboshitzky, R; Lavi, I; Miller, A. Melatonin dysregulation, sleep disturbances and fatigue in multiple sclerosis. *J Neurol Sci*, 2012, 314, 37-40.

[30] Srinivasan, V; Pandi-Perumal, SR; Cardinali, DP; Poeggeler, B; Hardeland, R. Melatonin in Alzheimer's disease and other neurodegenerative disorders. *Behav Brain Funct*, 2006, 2, 15.

[31] Karasek, M. Melatonin, human aging, and age-related diseases. *Exp. Gerontol*, 2004, 39, 1723-1729.

[32] Srinivasan, V; Spence, DW; Trakht, I; Pandi-Perumal, SR; Cardinali, DP; Maestroni, GJ. Immunomodulation by melatonin: its significance for seasonally occurring diseases. *Neuroimmunomodulation*, 2008, 15, 93-101.

[33] Davis, S; Mirick, DK; Stevens, RG. Night-shift work, light at night, and risk of breast cancer. *J Natl Cancer Inst*, 2001, 93, 1557–1562.

[34] Bhatti, P; Mirick, DK; Davis, S. Invited commentary: Shift work and cancer. *Am J Epidemiol*, 2012, 176, 760–763.

[35] Lin, GJ; Huang, SH; Chen, SJ; Wang, CH; Chang, DM; Sytwu, HK. Modulation by melatonin of the pathogenesis of inflammatory autoimmune diseases. *Int J Mol Sci*, 2013, 14, 11742-11766.

[36] Srinivasan, V; Spence, DW; Pandi-Perumal, SR; Trakht, I; Cardinali, DP. Therapeutic actions of melatonin in cancer: possible mechanisms. *Integr Cancer Ther*, 2008, 7, 189-203.

[37] Macchi, MM; Bruce, JN. Human pineal physiology and functional significance of melatonin. *Front Neuroendocrinol*, 2004, 25, 177–195.

[38] Dubocovich, ML; Cardinali, DP; Guardiola-Lemaitre, B; Hagan, RM; Krause, DN; Sudgen, D; Vanhoutte, PM; Yocca, FD. The IUPHAR compendium of receptor characterization and classification. *IUPHAR Media*, 1998, 187–193.

[39] Luchetti, F; Canonico, B; Betti, M; Arcangeletti, M; Pilolli, F; Piroddi, M; Canesi, L; Papa, S; Galli, F. Melatonin signaling and cell protection function. *FASEB J*, 2010, 24, 3603–3624.

[40] Vanecek, J. Cellular mechanisms of melatonin action. *Physiol Rev*, 1998, 78, 687–711.

[41] Pandi-Perumal, SR; Trakht, I; Srinivasan, V; Spence, DW; Maestroni, GJ; Zisapel, N; Cardinali, DP. Physiological effects of melatonin: role of melatonin receptors and signal transduction pathways. *Prog Neurobiol*, 2008, 85, 335-353.

[42] Barjavel, MJ; Mamdouh, Z; Raghbate, N; Bakouche, O. Differential expression of the melatonin receptor in human monocytes. *J Immunol*, 1998, 160, 1191-1197.

[43] Blask, DE; Sauer, LA; Dauchy, RT. Melatonin as a chronobiotic/anticancer agent: cellular, biochemical, and molecular mechanisms of action and their implications for circadian-based cancer therapy. *Curr Top Med Chem*, 2002, 2, 113-132.

[44] Reiter, RJ; Tan, DX; Mayo, JC; Sainz, RM; Leon, J; Czaronocki, Z. Melatonin as an antioxidant: biochemical mechanisms and pathophysiological implications in humans. *Acta Biochim Pol*, 2003, 50, 1129-1146.

[45] Tan, DX; Reiter, RJ; Manchester, LC; Yan, MT; El-Sawi, M; Sainz, RM; Mayo, JC; Kohen, R; Allegra, M; Hardeland, R. Chemical and physical properties and potential mechanisms: melatonin as a broad spectrum antioxidant and free radical scavenger. *Curr Top Med Chem*, 2002, 2, 181-197.

[46] Medzhitov, R. Origin and physiological roles of inflammation. *Nature*, 2008, 428-435.

[47] Ren, DL; Li, YJ; Hu, BB; Wang, H; Hu, B. Melatonin regulates the rhythmic migration of neutrophils in live zebrafish. *J Pineal Res*, 2015, 58, 452–460.

[48] Scheiermann, C; Kunisaki, Y; Lucas, D; Chow, A; Jang, JE; Zhang, D; Hashimoto, D; Merad, M; Frenette, PS. Adrenergic nerves govern circadian leukocyte recruitment to tissues. *Immunity*, 2012, 37, 290-301.

[49] Ren, DL; Sun, AA; Li, YJ; Chen, M; Ge, SC; Hu, B. Exogenous melatonin inhibits neutrophil migration through suppression of ERK activation. *J Endocrinol*, 2015, 227, 49-60.

[50] Lin, XJ; Mei, GP; Liu, J; Li, YL; Zuo, D; Liu, SJ; Zhao, TB; Lin, MT. Therapeutic effects of melatonin on heatstroke-induced multiple organ dysfunction syndrome in rats. *J Pineal Res*, 2011, 50, 436-444.

[51] Calvo, JR; Reiter, RJ; García, JJ; Ortiz, GG; Tan, DX; Karbownik, M. Characterization of the protective effects of melatonin and related indoles against alpha-naphthylisothiocyanate-induced liver injury in rats. *J Cell Biochem*, 2001, 80, 461-470.

[52] Wu, WS; Chou, MT; Chao, CM; Chang, CK; Lin, MT; Chang, CP. Melatonin reduces acute lung inflammation, edema, and hemorrhage in heatstroke rats. *Acta Pharmacol Sin*, 2012, 33, 775-782.

[53] Jaworek, J; Szklarczyk, J; Jaworek, AK; Nawrot-Porąbka, K; Leja-Szpak, A; Bonior, J; Kot, M. Protective effect of melatonin on acute pancreatitis. *Int J Inflam*, 2012, 2012, 173675.

[54] Peña, C; Rincon, J; Pedreanez, A; Viera, N; Mosquera, J. Chemotactic effect of melatonin on leukocytes. *J Pineal Res*, 2007, 43, 263-269.

[55] Silva, SO; Rodrigues, MR; Ximenes, VF; Bueno-da-Silva, AE; Amarante-Mendes, GP; Campa, A. Neutrophils as a specific target for melatonin and kynuramines: effects on cytokine release. *J Neuroimmunol*, 2004, 156, 146-152.

[56] Lotufo, CM; Lopes, C; Dubocovich, ML; Farsky, SH; Markus, RP. Melatonin and N-acetylserotonin inhibit leukocyte rolling and adhesion to rat microcirculation. *Eur J Pharmacol*, 2001, 430, 351-357.

[57] Garrelly, L; Bureau, JP; Labreque, G. Temporal study of carrageenan-induced PMN migration in mice. *Agents Actions*, 1991, 33, 225–228.

[58] Tamura, EK; Fernandes, PA; Marçola, M; da Silveira Cruz-Machado, S; Markus, RP. Long-lasting priming of endothelial cells by plasma melatonin levels. *PLoS One*, 2010, 5, e13958. doi: 10.1371/journal. pone.0013958.

[59] Marçola, M; da Silveira Cruz-Machado, S; Fernandes, PA; Monteiro, AW; Markus, RP; Tamura, EK. Endothelial cell adhesiveness is a function of environmental lighting and melatonin level. *J Pineal Res*, 2013, 54, 162–169.

[60] Lotufo, CM; Yamashita, CE; Farsky, SH; Markus, RP. Melatonin effect on endothelial cells reduces vascular permeability increase induced by leukotriene B4. *Eur J Pharmacol*, 2006, 534, 258-263.

[61] Recchioni, R; Marcheselli, F; Moroni, F; Gáspár, R; Damjanovich, S; Pieri, C. Melatonin increases the intensity of respiratory burst and prevents L-selectin shedding in human neutrophils *in vitro*. *Biochem Biophys Res Commun*, 1998, 252, 20-24.

[62] Hriscu, ML. Modulatory factors of circadian phagocytic activity. *Ann NY Acad Sci* 2005, 1057, 403-430.

[63] Rodríguez, AB; Marchena, JM; Nogales, G; Durán, J; Barriga, C. Correlation between the circadian rhythm of melatonin, phagocytosis, and superoxide anion levels in ring dove heterophils. *J Pineal Res*, 1999, 26, 35-42.

[64] Rodríguez, AB; Terrón, MP; Durán, J; Ortega, E; Barriga, C. Physiological concentrations of melatonin and corticosterone affect phagocytosis and oxidative metabolism of ring dove heterophils. *J Pineal Res*, 2001, 31, 31-38.

[65] Silva, SO; Carvalho, SR; Ximenes, VF; Okada, SS; Campa, A. Melatonin and its kynurenin-like oxidation products affect the microbicidal activity of neutrophils. *Microbes Infect*, 2006, 8, 420-425.

[66] Pieri, C; Recchioni, R; Moroni, F; Marcheselli, F; Marra, M; Marinoni, S; Di Primio, R. Melatonin regulates the respiratory burst of human neutrophils and their depolarization. *J Pineal Res*, 1998, 24, 43-49.

[67] Bouhafs, RK; Jarstrand, C. Effects of antioxidants on surfactant peroxidation by stimulated human polymorphonuclear leukocytes. *Free Radic Res*, 2002, 36, 727-734.

[68] Fischer, TW; Scholz, G; Knöll, B; Hipler, UC; Elsner, P. Melatonin reduces UV-induced reactive oxygen species in a dose-dependent manner in IL-3-stimulated leukocytes. *J Pineal Res*, 2001, 31, 39-45.

[69] Espino, J; Bejarano, I; Paredes, SD; Barriga, C; Rodríguez, AB; Pariente, JA. Protective effect of melatonin against human leukocyte apoptosis induced by intracellular calcium overload: relation with its antioxidant actions. *J Pineal Res*, 2011, 51, 195-206.

[70] Espino, J; Rodríguez, AB; Pariente, JA. The inhibition of TNF-α-induced leucocyte apoptosis by melatonin involves membrane receptor MT1/MT2 interaction. *J Pineal Res*, 2013, 54, 442-452.

[71] França, E; Junior, A; de Oliveira, S; Honório-França, A. Chronoimmunomodulation of melatonin on bactericidal activity of human blood phagocytes. *Internet J Microbiol*, 2008, 6, 1-7.

[72] Watson, RW; Rotstein, OD; Nathens, AB; Parodo, J; Marshall, JC. Neutrophil apoptosis is modulated by endothelial transmigration and adhesion molecule engagement. *J Immunol*, 1997, 158, 945–953.

[73] Kerr, JF; Wyllie, AH; Currie, AR. Apoptosis: a basic biological phenomenon with wide-ranging implications in tissue kinetics. *Br J Cancer*, 1972, 26, 239-257.

[74] Espino, J; Bejarano, I; Redondo, PC; Rosado, JA; Barriga, C; Reiter, RJ; Pariente, JA; Rodríguez, AB. Melatonin reduces apoptosis induced by calcium signaling in human leukocytes: evidence for the involvement of mitochondria and bax activation. *J Membrane Biol*, 2010, 233, 105-118.

[75] Chen, HM; Hsu, JT; Chen, JC; Ng, CJ; Chiu, DF; Chen, MF. Delayed neutrophil apoptosis attenuated by melatonin in human acute pancreatitis. *Pancreas*, 2005, 31, 360-364.

[76] Maestroni, GJ; Conti, A. Melatonin and the immune-hematopoietic system therapeutic and adverse pharmacological correlates. *Neuroimmunomodulation*, 1996, 3, 325-332.

[77] Kuci, S; Becker, J; Veit, G; Handgretinger, R; Attanasio, A; Bruchelt, G; Treuner, J; Niethammer, D; Gupta, D. Circadian variations in the immunomodulatory role of the pineal gland. *Neuroendocrinol Lett*, 1988, 10, 65-79.

[78] Haldar, C; Häussler, D; Gupta, D. Effect of the pineal gland on circadian rhythmicity of colony forming units for granulocytes and macrophages (CFU-GM) from rat bone marrow cell cultures. *J Pineal Res*, 1992, 12, 79-83.

[79] Currier, NL; Sun, LZ; Miller, SC. Exogenous melatonin: quantitative enhancement *in vivo* of cells mediating non-specific immunity. *J Neuroimmunol*, 2000, 104, 101-108.

[80] García-Mauriño, S; Pozo, D; Calvo, JR; Guerrero, JM. Correlation between nuclear melatonin receptor expression and enhanced cytokine production in human lymphocytic and monocytic cell lines. *J Pineal Res*, 2000, 29, 129-137.

[81] Xia, MZ; Liang, YL; Wang, H; Chen, X; Huang, YY; Zhang, ZH; Chen, YH; Zhang, C; Zhao M; Xu, DX; Song, LH. Melatonin modulates TLR4-mediated inflammatory genes through MyD88- and TRIF-dependent signaling pathways in lipopolysaccharide-stimulated RAW264.7 cells. *J Pineal Res*, 2012, 53, 325-334.

[82] Huang, SH; Cao, XJ; Wei, W. Melatonin decreases TLR3-mediated inflammatory factor expression via inhibition of NF-kappa B activation in respiratory syncytial virus-infected RAW264.7 macrophages. *J Pineal Res*, 2008, 45, 93–100.

[83] Choi, EY; Jin, JY; Lee, JY; Choi, JI; Choi, IS; Kim, SJ. Melatonin inhibits Prevotella intermedia lipopolysaccharide-induced production of nitric oxide and interleukin-6 in murine macrophages by suppressing NF-κB and STAT1 activity. *J Pineal Res*, 2011, 50, 197–206.

[84] Williams, JG; Bernstein, S; Prager, M. Effect of melatonin on activated macrophage TNF, IL-6, and reactive oxygen intermediates. *Shock*, 1998, 9, 406-411.

[85] Shafer, LL; McNulty, JA; Young, MR. Assessment of melatonin's ability to regulate cytokine production by macrophage and microglia cell types. *J Neuroimmunol*, 2001, 120, 84-93.

[86] Radogna, F; Paternoster, L; De Nicola, M; Cerella, C; Ammendola, S; Bedini, A; Tarzia, G; Aquilano, K; Ciriolo, M; Ghibelli, L. Rapid and transient stimulation of intracellular reactive oxygen species by melatonin in normal and tumor leukocytes. *Toxicol Appl Pharmacol*, 2009, 239, 37-45.

[87] Albertini, MC; Radogna, F; Accorsi, A; Uguccioni, F; Paternoster, L; Cerella, C; De Nicola, M; D'Alessio, M; Bergamaschi, A; Magrini, A; Ghibelli, L. Intracellular pro-oxidant activity of melatonin deprives U937 cells of reduced glutathione without affecting glutathione peroxidase activity. *Ann NY Acad Sci*, 2006, 1091, 10-16.

[88] Islam, SA; Thomas, SY; Hess, C; Medoff, BD; Means, TK; Brander, C; Lilly, CM; Tager, AM; Luster, AD. The leukotriene B$_4$ lipid chemoattractant receptor BLT1 defines antigen-primed T cells in humans. *Blood*, 2006, 107, 444-453.

[89] Gilad, E; Wong, HR; Zingarelli, B; Virag, L; O'Connor, M; Salzman, AL; Szabo, C. Melatonin inhibits expression of the inducible isoform of nitric oxide synthase in murine macrophages: Role of inhibition of NF-κB activation. *FASEB J*, 1998, 12, 685–693.

[90] Bagchi, M; Balmoori, J; Ye, X; Bagchi, D; Ray, SD; Stohs, SJ. Protective effect of melatonin on naphthalene-induced oxidative stress and DNA damage in cultured macrophage J774A.1 cells. *Mol Cell Biochem*, 2001, 221, 49-55.

[91] Seljelid, R; Eskeland, T. The biology of macrophages: I. General principles and properties. *Eur J Haematol*, 1993, 51, 267-275.

[92] Pioli, C; Caroleo, MC; Nistico, G; Doria, G. Melatonin increases antigen presentation and amplifies specific and non specific signals for T-cell proliferation. *Int J Immunopharmacol*, 1993, 15, 463–468.

[93] Parihar, A; Eubank, TD; Doseff, AI. Monocytes and macrophages regulate immunity through dynamic networks of survival and cell death. *J Innate Immun*, 2010, 2, 204-215.

[94] Luchetti, F; Canonico, B; Curci, R; Battistelli, M; Mannello, F; Papa, S; Tarzia, G; Falcieri, E. Melatonin prevents apoptosis induced by UV-B treatment in U937 cell line. *J Pineal Res*, 2006, 40, 158–167.

[95] Radogna, F; Cristofanon, S; Paternoster, L; D'Alessio, M; De Nicola, M; Cerella, C; Dicato, M; Diederich, M; Ghibelli, L. Melatonin antagonizes the intrinsic pathway of apoptosis via mitochondrial targeting bcl-2. *J Pineal Res*, 2008, 44, 316-325.

[96] Radogna, F; Paternoster, L; Albertini, MC; Cerella, C; Accorsi, A; Bucchini, A; Spadoni, G; Diamantini, G; Tarzia, G; Nicola, MD; D'Alessio, M; Ghibelli, L. Melatonin antagonizes apoptosis via receptor interaction in U937 monocytic cells. *J Pineal Res*, 2007, 43, 154–162.

[97] Radogna, F; Albertini, MC; De Nicola, M; Diederich, M; Bejarano, I; Ghibelli, L. Melatonin promotes Bax sequestration to mitochondria reducing cell susceptibility to apoptosis via the lipoxygenase metabolite 5-hydroxyeicosatetraenoic acid. *Mitochondrion*, 2015, 21, 113-121.

[98] Markowska, M; Waloch, M; Skwarło-Sońta, K. Melatonin inhibits PHA-stimulated chicken lymphocyte proliferation *in vitro*. *J Pineal Res*, 2001, 30, 220-226.

[99] Kühlwein, E; Irwin, M. Melatonin modulation of lymphocyte proliferation and Th1/Th2 cytokine expression. *J Neuroimmunol*, 2001, 117, 51–57.

[100] Persengiev, SP; Kyurkchiev, S. Selective effect of melatonin on the proliferation of lymphoid cells. *Int J Biochem*, 1993, 25, 441–443.

[101] Konakchieva, R; Kyurkchiev, S; Kehayov, I; Taushanova, P; Kanchev, L. Selective effect of methoxyindoles on the lymphocyte proliferation and melatonin binding to activated human lymphoid cells. *J Neuroimmunol*, 1995, 63, 125–132.

[102] Raghavendra, V; Singh, V; Shaji, AV; Vohra, H; Kulkarni, SK; Agrewala, JN. Melatonin provides signal 3 to unprimed $CD4^+$ T cells but failed to stimulate LPS primed B cells. *Clin Exp Immunol*, 2001, 124, 414–422.

[103] Carrillo-Vico, A; Lardone, PJ; Fernández-Santos, JM; Martin-Lacave, I; Calvo, JR; Karasek, M; Guerrero, JM. Human lymphocyte-synthesized melatonin is involved in the regulation of the IL-2/IL-2 receptor system. *J Clin Endocrinol Metab*, 2005, 90, 992–1000.

[104] Lissoni, P; Barni, S; Cazzaniga, M; Ardizzoia, A; Rovelli, F; Brivio, F; Tancini, G. Efficacy of the concomitant administration of the pineal hormone melatonin in cancer immunotherapy with low-dose IL-2 in patients with advanced solid tumors who had progressed on IL-2 alone. *Oncology*, 1994, 51, 344-347.

[105] Lissoni, P; Barni, S; Rovelli, F; Brivio, F; Ardizzoia, A; Tancini, G; Conti, A; Maestroni, GJ. Neuroimmunotherapy of advanced solid neoplasms with single evening subcutaneous injection of low-dose interleukin-2 and melatonin: preliminary results. *Eur J Cancer*, 1993, 29, 185-189.

[106] Lissoni, P; Brivio, F; Ardizzoia, A; Tancini, G; Barni, S. Subcutaneous therapy with low-dose interleukin-2 plus the neurohormone melatonin in metastatic gastric cancer patients with low performance status. *Tumori*, 1993, 79, 401-404.

[107] Petrovsky, N; Harrison, L. The chronobiology of human cytokine production. *Int Rev Immunol*, 1998, 16, 635–649.

[108] Nunnari, G; Nigro, L; Palermo, F; Leto, D; Pomerantz, RJ; Cacopardo, B. Reduction of serum melatonin levels in HIV-1-infected individuals' parallel disease progression: correlation with serum interleukin-12 levels. *Infection*, 2003, 31, 379–382.

[109] Maestroni, GJ. T-helper-2 lymphocytes as a peripheral target of melatonin. *J Pineal Res*, 1995, 18, 84–89.

[110] Kelestimur, H; Sahin, Z; Sandal, S; Bulmus, O; Ozdemir, G; Yilmaz, B. Melatonin-related alterations in Th1/Th2 polarisation in primary thymocyte cultures of pinealectomized rats. *Front Neuroendocrinol*, 2006, 27, 103–110.

[111] Shaji, AV; Kulkarni, SK; Agrewala, JN. Regulation of secretion of IL-4 and IgG1 isotype by melatonin-stimulated ovalbumin-specific T cells. *Clin Exp Immunol*, 1998, 111, 181-185.

[112] Raghavendra, V; Singh, V; Kulkarni, SK; Agrewala, JN. Melatonin enhances Th2 cell mediated immune responses: lack of sensitivity to reversal by naltrexone or benzodiazepine receptor antagonists. *Mol Cell Biochem*, 2001, 221, 57–62.

[113] Majewska, M; Zając, K; Zemelka, M; Szczepanik, M. Influence of melatonin and its precursor L-tryptophan on Th1 dependent contact hypersensitivity. *J Physiol Pharmacol*, 2007, 58, 125–132.

[114] Carrillo-Vico, A; Lardone, PJ; Naji, L; Fernández-Santos, JM; Martín-Lacave, I; Guerrero, JM; Calvo, JR. Beneficial pleiotropic actions of melatonin in an experimental model of septic shock in mice: regulation of pro-/anti-inflammatory cytokine network, protection against oxidative damage and anti-apoptotic effects. *J Pineal Res*, 2005, 39, 400–408.

[115] Sakaguchi, S; Yamaguchi, T; Nomura, T; Ono, M. Regulatory T cells and immune tolerance. *Cell*, 2008, 133, 775-787.

[116] Vigoré, L; Messina, G; Brivio, F; Fumagalli, L; Rovelli, F; di Fede, G; Lissoni, P. Psychoneuroendocrine modulation of regulatory T lymphocyte system: *in vivo* and *in vitro* effects of the pineal immunomodulating hormone melatonin. *In Vivo*, 2010, 24, 787–789.

[117] Liu, H; Xu, L; Wei, JE; Xie, MR; Wang, SE; Zhou, RX. Role of CD4+CD25+ regulatory T cells in melatonin-mediated inhibition of murine gastric cancer cell growth *in vivo* and *in vitro*. *Anat Rec (Hoboken)*, 2011, 294, 781–788.

[118] Medrano-Campillo, P; Sarmiento-Soto, H; Álvarez-Sánchez, N; Álvarez-Ríos, AI; Guerrero, JM; Rodríquez-Prieto, I; Castillo-Palma, MJ; Lardone, PJ; Carrillo-Vico, A. Evaluation of the immunomodulatory effect of melatonin on the T-cell response in peripheral blood from systemic lupus erythematosus patients. *J Pineal Res*, 2015, 58, 219-226.

[119] Calvo, JR; Rafii-el-Idrissi, M; Pozo, D; Guerrero, JM. Immunomodulatory role of melatonin: specific binding sites in human and rodent lymphoid cells. *J Pineal Res*, 1995, 18, 119–126.

[120] Demas, GE; Nelson, RJ. Exogenous melatonin enhances cell-mediated, but not humoral, immune function in adult male deer mice (Peromyscus maniculatus) *J Biol Rhythms*, 1998, 13, 245–252.

[121] Yu, Q; Miller, SC; Osmond, DG. Melatonin inhibits apoptosis during early B-cell development in mouse bone marrow. *J Pineal Res*, 2000, 29, 86-93.

[122] Paternoster, L; Radogna, F; Accorsi, A; Cristina Albertini, M; Gualandi, G; Ghibelli, L. Melatonin as a modulator of apoptosis in B-lymphoma cells. *Ann NY Acad Sci*, 2009, 1171, 345-349.

[123] Mohseni, M; Mihandoost, E; Shirazi, A; Sepehrizadeh, Z; Bazzaz, JT; Ghazi-khansari, M. Melatonin may play a role in modulation of bax and bcl-2 expression levels to protect rat peripheral blood lymphocytes from gamma irradiation-induced apoptosis. *Mutat Res*, 2012, 738-739, 19-27.

[124] Dodson, ER; Zee, PC. Therapeutics for Circadian Rhythm Sleep Disorders. *Sleep Med Clin*, 2010, 5, 701-715.

[125] Simko, F; Paulis, L. Melatonin as a potential antihypertensive treatment. *J Pineal Res*, 2007, 42, 319-322.

[126] Bubenik, GA. Gastrointestinal melatonin: localization, function, and clinical relevance. *Dig Dis Sci*, 2002, 47, 2336-2348.

[127] Srinivasan, V; Pandi-Perumal, SR; Maestroni, GJ; Esquifino, AI; Hardeland, R; Cardinali, DP. Role of melatonin in neurodegenerative diseases. *Neurotox Res*, 2005, 7, 293-318.

[128] Maldonado, MD; Pérez-San-Gregorio, MA; Reiter, RJ. The role of melatonin in the immuno-neuro-psychology of mental disorders. *Recent Pat CNS Drug Discov*, 2009, 4, 61–69.

[129] Peres, MF; Masruha, MR; Zukerman, E; Moreira-Filho, CA; Cavalheiro, EA. Potential therapeutic use of melatonin in migraine and other headache disorders. *Expert Opin Investig Drugs*, 2006, 15, 367–375.

[130] Cardinali, DP; Brusco, LI; Liberczuk, C; Furio, AM. The use of melatonin in Alzheimer's disease. *Neuro Endocrinol Lett*, 2002, 23, 20-23.

[131] Cuzzocrea, S; Reiter, RJ. Pharmacological actions of melatonin in acute and chronic inflammation. *Curr Top Med Chem*, 2002, 2, 153-165.

[132] Alamili, M; Bendtzen, K; Lykkesfeldt, J; Rosenberg, J; Gögenur, I. Melatonin supresses markers of inflammation and oxidative damage in a human daytime endotoxemia model. *J Crit Care*, 2014, 29, 184.e9-184.e13.

[133] Alamili, M; Bendtzen, K; Lykkesfeldt, J; Rosenberg, J; Gögenur, I. Effect of melatonin on human nighttime endotoxaemia: randomized, double-blinded, crossover study. *In Vivo*, 2014, 28, 1057-1063.

[134] Gitto, E; Karbownik, M; Reiter, RJ; Tan, DX; Cuzzocrea, S; Chiurazzi, P; Cordaro, S; Corona, G; Trimarchi, G; Barberi, I. Effects of melatonin treatment in septic newborns. *Pediatr Res*, 2001, 50, 756-760.

[135] Sutherland, ER; Martin, RJ; Ellison, MC; Kraft, M. Immunomodulatory effects of melatonin in asthma. *Am J Respir Crit Care Med*, 2002, 166, 1055-1061.

[136] Sutherland, ER; Ellison, MC; Kraft, M; Martin, RJ. Elevated serum melatonin is associated with the nocturnal worsening of asthma. *J Allergy Clin Immunol*, 2003, 112, 513-517.

[137] Cardinali, D; Escames, G; Acuña-Castroviejo, D; Ortiz, F; Fernández-Gil, B; Guerra Librero, A; García-Lòpez, S; Shen, Y; Florido, J. Melatonin-Induced Oncostasis, Mechanisms and Clinical Relevance. *J Integr Oncol*, 2016, S1:006. doi:10.4172/2329-6771.S1-006.

[138] Lissoni, P; Chilelli, M; Villa, S; Cerizza, L; Tancini, G. Five years survival in metastatic non-small cell lung cancer patients treated with chemotherapy alone or chemotherapy and melatonin: a randomized trial. *J Pineal Res*, 2003, 35, 12-15.

[139] Lissoni, P; Barni, S; Tancini, G; Ardizzoia, A; Rovelli, F; Cazzaniga, M; Brivio, F; Piperno, A; Aldeghi, R; Fossati, D. Immunotherapy with subcutaneous low-dose interleukin-2 and the pineal indole melatonin as a new effective therapy in advanced cancers of the digestive tract. *Br J Cancer*, 1993, 67, 1404-1407.

[140] Blask, DE; Hill, SM. Effects of melatonin on cancer: studies on MCF-7 human breast cancer cells in culture. *J Neural Transm Suppl*, 1986, 21, 433-449.

[141] Anisimov, VN; Alimova, IN; Baturin, DA; Popovich, IG; Zabezhinski, MA; Manton, KG; Semenchenko, AV; Yashin, AI. The effect of melatonin treatment regimen on mammary adenocarcinoma development in HER-2/neu transgenic mice. *Int J Cancer*, 2003, 103, 300-305.

In: Melatonin: Medical Uses and Role in Health and Disease ISBN: 978-1-53612-987-8
Editors: Lore Correia and Germaine Mayers © 2018 Nova Science Publishers, Inc.

Chapter 6

HOW MELATONIN COMBATS ISCHEMIC BRAIN INJURY

Eva Ramos[1], PhD, Paloma Patiño[2], BSN,
Javier Egea[3,4], PhD and Alejandro Romero[1,], PhD*

[1]Department of Toxicology and Pharmacology, Faculty of Veterinary Medicine,
Complutense University of Madrid, Madrid, Spain
[2]Paediatric Unit, La Paz University Hospital, Madrid, Spain
[3]Instituto de Investigación Sanitaria, Servicio de Farmacología Clínica, Hospital Universitario de la Princesa, Madrid, Spain
[4]Instituto de I+D del Medicamento Teófilo Hernando (ITH), Facultad de Medicina, Universidad Autónoma de Madrid, Spain

ABSTRACT

Stroke is the second leading cause of death and the main cause of disability worldwide. To date, there is no effective treatment to prevent the brain damage in ischemic stroke. Consequently, there is an obvious need to develop neuroprotective treatments for this pathology. The physiopathological events of the ischemic cascade; oxidative stress, Ca2+ dyshomeostasis, mitochondrial dysfunction, proinflammatory mediators, excitotoxicity and/or programmed neuronal cell death are regulated through multiple signaling pathways. In this chapter, we highlighted the pleiotropic effects of melatonin ameliorating the molecular, tissue and organ damage associated to brain ischemia. Its protective effects encompass from preserving the functional integrity of the blood-brain barrier, inducing neurogenesis and cell proliferation to improving synaptic transmission. This has been lately reinforced by the low cost of melatonin and its reduced toxicity. In fact, melatonin can be highly useful when combined with other therapies.

---

* Corresponding Author: Dr. Alejandro Romero; Department of Toxicology and Pharmacology, Faculty of Veterinary Medicine, Complutense University of Madrid, Avda. Puerta de Hierro s/n 28040 Madrid, Spain; Tel: +34 913943970; Email: manarome@ucm.es.

However, additional investigations are necessary to determine clinical effectiveness, doses and the optimal timing of administration.

Keywords: melatonin, brain ischemia, neuroprotection, free radicals

INTRODUCTION

Taking into account that stroke is the main cause of disability and the second leading cause of death worldwide [1, 2], and that there is no effective treatment to prevent the brain damage in ischemic stroke to date, there is an urgent need to develop neuroprotective treatments for ischemic stroke [3].

The adult human brain constitutes only 2% of the total body weight, while it receives 15% of the total cardiac output and consumes 20% of the total oxygen intake. Neurons need a high metabolic rate to preserve their normal activities: synaptic transmission, macromolecular synthesis and transmembrane ionic gradients [4]. After blockage of a blood vessel, there is a critical reduction in the cerebral blood flow (less than 25%) to the brain, since neurons require a continuous supply of oxygen and glucose. Under deprivation of one and/or another of these substrates, cell death occurs in 2 main phases: first, there is a cell death as a result of anoxia/hypoxia and energy depletion, followed by a phase of reperfusion, where it occurs an increase free radical formation, oxidative stress, excitotoxicity, and nitric oxide production with a final energy failure and delayed cell death [4-6].

The only treatment currently used in therapeutics is the recombinant tissue plasminogen activator (tPA) is used to open obstructed blood vessels, but it has a therapeutic window of 3.5 h [7, 8]. The main objective in this field is to rescue the ischemic penumbra, which is the hypoperfused, non-functional, but yet viable tissue adjacent to the infarcted core. In this sense, several key players in ischemic cell death within the penumbra have been identified, including excitotoxicity and disturbed calcium ion homeostasis, mitochondrial failure, oxidative and nitrosative stress, inflammation and apoptosis [9-13]. However, numerous clinical trials testing drugs that modulate one or more of these mechanisms, have unfortunately failed to show efficacy in patients with acute ischemic stroke, despite promising preclinical data.

Melatonin, is mainly synthesized in the pineal gland and then released into blood and cerebrospinal fluid (CSF) to exert its regulatory actions, controlling circadian rhythms [14, 15]. Additionally, it is also produced in many other extra-pineal tissues such as the immune system cells, brain, airway epithelium, bone marrow, gut, ovary, testes, skin …[16-18]. As a consequence of age, there is a reduction of melatonin levels in serum and CSF [19, 20], this fact may contribute to numerous dysfunctions and pathophysiological changes, such as stroke [21-23]. Furthermore, the pineal gland accumulates calcium deposits with age, leading to a decrease of 6-sulfatoxymelatonin levels in urine, which demonstrates that in calcified glands there is a reduction in the synthesis of this neurohormone [24]. Interestingly, some observations determined that there is a relationship between, a higher pineal gland calcification and intracerebral hemorrhage in stroke patients than in healthy subjects [25, 26]. Thereby, pineal calcification may represent a potential new risk factor for cerebral infarction, however, in another observation study, prevalence and severity of pineal calcification could not been related to stroke incidence [27]. Therefore, further studies are needed to reconcile these discrepancies.

Additionally, melatonin has emerged in the last decade as a potential therapeutic agent to ameliorate the molecular and organ/tissue damage associated to brain ischemia. Herein, we summarize the neuroprotective effect of melatonin counteracting several steps of the ischemic cascade, serving as a free radical (FR) scavenger and reducing excitotoxicity.

Transplantation of stem cells near the ischemic area of stroke patients was first performed in 2000 [28]. To evaluate whether the release of melatonin after intracerebral transplantation of the pineal gland was effective in stroke brains, a middle cerebral artery occlusion (MCAO) rat model was used, the results displayed a significant less motor asymmetry and reduced cerebral infarction compared with control counterparts not transplanted. Furthermore, the higher levels of melatonin in CSF observed were associated to neuroprotection [29].

Moreover, due to a reduction in the levels of circulating melatonin, a relation between pinealectomized animals and higher neurological damage has been detected in stroke models [30, 31]. Whilst, melatonin administration was able to reverse at least partially these effects. [30, 32, 33]. Even more interesting is the fact that reduced endogenous melatonin production has been related to a higher risk of cerebrovascular accidents [25, 26].

Taking all of this into account, we must assume that the experimental design is a critical point to obtain solid translatability from preclinical studies to humans. The selection of the dose, administration route, timing and furthermore the animal model has a direct impact on it.

MELATONIN AND OXIDATIVE STRESS DURING BRAIN ISCHEMIA

Undoubtedly, central nervous system (CNS) is especially sensitive to oxidative stress (OS), given that it consumes a considerable amount of oxygen and it generates more reactive oxygen species (ROS) than any other tissue. In addition, the CNS contains a large amount of polyunsaturated fatty acids and transition metals that participate in the

Haber Weiss/Fenton reactions, while possessing relatively low levels of endogenous antioxidants.

Mitochondria is the first ROS generator and conversely the main target of ROS and reactive nitrogen species (RNS) [34]. The mitochondria synthesizes ATP by coupling the discharge of proton potential generated by the electron transport chain (ETC). The most abundant ROS produced by the ETC is the $O_2^{\cdot-}$, arising from electron leakage into molecular oxygen, leading to the ulterior formation of $HO^{\cdot}$, H_2O_2, $NO^{\bullet}$, and $ONOO^-$, which are even more reactive than $O_2^{\cdot-}$. During ischemia/reperfusion (I/R) conditions, ROS generation is one of the main hallmarks in the pathogenesis of tissue destruction [35], and a high increase of FR associated to OS has been widely documented early during brain ischemia injury [36-38].

In the acute phase of stroke, the protection of neuronal cells from FR-induced injury has been considered as an important therapeutic strategy. Melatonin is a powerful scavenger against ROS/RNS either through non-receptor mediated mechanisms [39-41] or by increasing the activity and expression of the antioxidant enzymes [41-43]. This property, gives melatonin an interesting profile against cerebral I/R damage, considering that the $HO^{\cdot}$ generated from H_2O_2 via the Fenton reaction induces an extensive DNA damage, ATP depletion and aggravates neurotoxicity in brain ischemia [44]. Indeed, preclinical studies show that in stroke animal models pre-administration of melatonin is able to reduce the molecular $HO^{\cdot}$ dependent damage [45]. $O_2^{\cdot-}$ plays an important role in oxidative chain reactions, it is known to initiate the signaling cell death pathways and the damage of nucleic acids, proteins and lipids. Melatonin, exerts its antioxidant properties again at this point when it is administrated in animal models of cerebral ischemia decreasing intensity of this oxidative species [46]. Nox family is a major source of ROS production in the reperfusion-associated brain damage [47], therefore, reducing its expression may be a useful therapeutic strategy. Melatonin pretreatment, in animal models, was able to reduce the increased expression both of Nox2 and Nox4 and also decreased ROS levels, and inhibited cell apoptosis in a cerebral I/R [48].

Even though $NO^{\bullet}$ has a dual action as a physiological messenger and neuromodulator [49], it is also a neurotoxic mediator in the pathophysiology of cerebral ischemia [47, 50]. Exogenous melatonin administration has demonstrated to reduce NO levels after transient ischemia *in vivo*, which shows a new therapeutic characteristic of this molecule [51-54].

However, these effects may not be enough, to explain the high protective potency displayed by melatonin in preclinical studies. Certainly, melatonin has an important indirect function based on upregulation of antioxidant and downregulation of prooxidant enzymes and besides, it contributes to the maintenance of physiological glutathione (GSH) levels, via γ-glutamyl-cystein synthetase, a rate-limiting enzyme in GSH synthesis [42]. Under brain ischemic conditions, reduced GSH levels are observed, which are significantly restored after melatonin administration both in acute and chronic models of cerebral I/R injury [55, 56]. Furthermore, melatonin mitigates the increased brain stress markers such as 8-hydroxy-2-deoxyguanosine [57] and malondialdehyde (MDA), [58] as well as other thiobarbituric acid-reactive substances (TBARS) [59] that lead to lipid peroxidation and DNA damage. In view of these observations, there is, obviously, a need to design more ambitious preclinical and clinical studies to determine whether the efficiency of melatonin therapy is effective for the stroke treatment.

ROLE OF MELATONIN IN EXCITOTOXIC BRAIN INJURY

Glutamate is the most important excitotoxic-mediating neurotransmitter [60], melatonin is able to reduce neuronal damage due to glutamate overexposure, counteracting the activity of PP2A [61], a phosphatase that reduces the Ca^{2+} influx through the NMDA-sensitive glutamate receptors.

The antioxidant actions of melatonin are believed to be responsible of some of its protective features [62-65]. E.g., the presence of kainate, in cell cultures, increase the levels of oxidized GSH making cells more vulnerable to ROS overload, this can be counteracted by melatonin administration [66, 67]. Moreover, tissue viability of rat hippocampal slices exposed to glutamate is preserved by melatonin in a concentration dependent manner [68], similarly extended damage of chick embryo cerebellar cortex is dramatically reduced after melatonin exposure [69].

To clarify the mechanism of action melatonin versus excitotoxicity a series of experiments have been carried out. The excitotoxicity elicited by glutamate is mainly mediated by NMDA receptors given that, NMDA receptor antagonist MK-801 prevents the glutamate harmful effect in rat cortical neurons, while incubation with CNQX, AMPA/kainate receptor antagonist was infective. Conversely, melatonin excitotoxicity protection was not a receptor mediated effect, since, this effect was not dismissed after luzindole administration, its main antagonist. It can be concluded that this effect is independent from either melatonin or NMDA receptors [70]. NO is a mediator of glutamate excitotoxicity, by suppressing nitric oxide synthase (NOS), melatonin inhibits NO synthesis [51], reduces microglial activation [71] and generates a sustained activation of Akt survival pathway [72]. Therefore, melatonin is able to protect against excitotoxicity, which is a characteristic to consider in relation to the excitotoxic events that occur in brain injury.

MELATONIN ANTI-INFLAMMATORY ACTIVITY IN BRAIN ISCHEMIA

It is well known that inflammation plays a crucial role in the development of cerebral ischemia, not only predisposing to brain ischemia, but it can also drive directly to many of the pathological processes in stroke [73]. In fact, patients that suffer concomitant inflammation have the worse clinical outcomes [74, 75]. Immediately after a cerebral infarct, there is a migration of different cell types (lymphocytes, neutrophils, T cells, monocytes and macrophages) to the damage region and especially to the penumbra area. Cytokines, prostaglandins, leukotrienes, eicosanoids, as well as adhesion molecules are dramatically increased in cerebral stroke episodes [76]. Resident glial cells (astrocytes and microglia) are also pivotal in acute brain injuries, microglia cells have a beneficial role phagocytizing neuron debris, releasing trophic factors and promoting tissue repair, participating therefore, in the resolution of inflammation. [77-79]. Conversely, it has been reported that activated microglia may aggravate the lesion and contribute to the tissue damage in stroke by increasing the production of nitrites, myeloperoxidase, prostaglandins, and cytokines such as TNF-α and IL-1β [80]. Within cerebral ischemia this dual role of microglia has been described, microglia exhibit a pro-inflammatory phenotype (a few days after stroke) and eventually acquire an anti-inflammatory phenotype (several days after stroke) [81].

Hence, the inflammatory process is a complex scenario, where melatonin, however, exerts several interesting actions that may counterbalance the inflammation process in stroke. It has been determined that melatonin reduced a 51% cortical macrophage/activated microglial brain infiltration, and furthermore, CD86, a classical marker of microglial activation, was also under-expressed [82, 83]. In several animal models, melatonin strongly reduced white matter inflammation, 8-isoprostanes levels, activated microglia cells, TUNEL-positive cells, IL-1β, TNF-α and GFAP overexpression [84-88]. The findings published to date are consistent with the anti-inflammatory profile of melatonin in stroke models, it has been achieved at a variety of doses ranging from 5 to 20 mg/kg and at different administration routes.

Based on these data, melatonin can be considered an effective inhibitor of the cellular inflammatory response after cerebral ischemia, particularly blocking microglia activation and macrophage infiltration. Despite the promising preclinical evidence, a large number of clinical trials have failed to show the efficacy of drugs that modulate one or more anti-inflammatory mechanisms in animals. To date, there are no human studies related to the use of melatonin in stroke patients. Judging its multiple actions, however, melatonin has proven to be a suitable candidate for the treatment of brain ischemia. Clearly, studies involving these lines should be performed in an attempt to overcome the harmful effects of the ischemic cascade.

MELATONIN AND CEREBRAL EDEMA

In the ischemic process, cerebral edema, the pathological accumulation of fluid in brain, is one of the early symptoms developed. After the ischemic injury, cell swelling by the intracellular accumulation of solutes such as sodium, chloride, etc., together with the extracellular elevation of water content leads to an increase of the intracranial pressure. This reduces cerebral blood flow and adds further ischemia, herniation of intracranial contents, damage of brain structures, neurological injury or even death [89].

To avoid the subsequent ischemic damage, an early reduction of cerebral edema turns up as a potent tool and herein melatonin has demonstrated to be effective. Mainly by reducing approximately the half of the total volume of cerebral edema compared to controls, additionally it seems to be most effective in the cerebral cortex [90, 91].

The preclinical data and the fact that melatonin crosses the BBB, along with the efficacy of this indoleamine as a free radical scavenger and its ability to activate antioxidant enzymes, makes this molecule strategic for a more complete evaluation in clinical trials. Accumulated evidence suggests that it may be as well effective reducing cerebral edema in animals and in humans.

REGULATION OF BLOOD BRAIN BARRIER INTEGRITY BY MELATONIN

During cerebrovascular insults, secretion of proinflammatory mediators such as COX-2, TNF-α, IL-1β, IL-6, an increased protein extravasation, production of reactive oxygen metabolites and interstitial edema are the causes of blood brain barrier (BBB) dysfunction. The BBB acts as a biological and selective interface between blood and brain, it restricts the entry of toxic compounds into the CNS and maintains the optimal brain microenvironment. BBB constituents include neurons, glia, pericytes, endothelial cells and extracellular matrix components, that collectively, form a functional neurovascular unit.

Under brain ischemia conditions, there is a NO overproduction, which plays an important role in the vasodilatation process and the physiological modulation of blood pressure. Additionally, there is an augmented iNOS activity, inducing cellular damage. Taking all together, there is BBB breakdown, which leads to an increase of its permeability [92].

In animal stroke models, melatonin has been able to reduce brain NO levels in the ischemic brain tissue, decrease superoxide production, reduce nitrosative damage, reducing in this way BBB injury and maintaining its integrity [52, 93, 94].

A co-administration of melatonin with tPA in a mice thrombolysis model was able to exert a promising profile preventing thrombolysis-induced hemorrhagic damage, reducing the altered BBB permeability. These findings point melatonin to be considered as an add-on therapy against hemorrhagic transformation [46].

In vitro, melatonin has also demonstrated to preserve functional integrity of the neurovascular unit. After a culture of murine brain endothelial cells were subjected to oxygen and glucose deprivation followed by reoxygenation (OGD/R), melatonin was able to augment cell survival, reduce ROS production and prevent claudin-5 degradation. Claudin-5 is a tight junction protein involved in regulating barrier integrity against OGD/R injury [95]. In a perinatal brain damage model, melatonin as well, showed neuroprotective skills reducing BBB damage and preserving its integrity through tight junctions, moreover, the inflammatory effects were also dismissed by melatonin [96]. These collective findings confirm the neuroprotective signaling pathways of melatonin regulating the functional integrity after OGD/R-exposed brain endothelial cells.

Matrix metalloproteinases (MMPs), expressed in astrocytes and neurons, may increase BBB permeability via tight junctions and extracellular matrix degradation. Specifically, MMP-9 has been proposed as a modulator of BBB hyperpermeability and high MMP-9 levels have been found in cerebrospinal fluid (CSF) samples of traumatic brain injury (TBI) patients [97]. Recently, the role of melatonin as MMP-9 inhibitor has been documented both *in vitro* and *in vivo* studies [98], counteracting BBB hyperpermeability after TBI.

Taking these findings together, melatonin represents a promising profile maintaining BBB integrity after brain ischemia process; therefore, it needs to be more deeply investigated in clinical trials.

MELATONIN AND NEUROGENESIS IN BRAIN ISCHEMIA

Brain ischemia affects a large number of cell populations including neurons, microglia, oligodendrocytes, among them, the neural stem cells (NSCs), have the potential to generate new neuronal populations through proliferation, migration and differentiation. The physiological actions of melatonin in brain are mediated, in part, by stimulation of its two receptors, MT1 and MT2. In humans, MT1 mRNA and protein have been detected in several areas including the hypothalamus, cerebellum, frontal cortex, nucleus accumbens, amygdala and hippocampus [13, 34, 41, 44], and MT2 receptor has been described in the hippocampus [45]. In the adult brain, NSCs, are found in hippocampus: more specifically, at the subgranular zone; in the hippocampal dentate gyrus and the subventricular zone [99]. It has been reported that melatonin induces cell proliferative activities in the dentate gyrus [100, 101]. Although its molecular mechanisms remain unclear, it is known that they involve cellular regeneration, which clearly benefits the brain, after the ischemic injury. In this respect, what it has been described is that melatonin affords neurogenesis and cell proliferation through a MT2 receptor-dependent mechanism [102].

Melatonin increased endogenous neurogenesis and neuronal survival *in vivo*, administered in long treatments, beginning 24 h after stroke onset and continued for 29 days. At the same time, melatonin significantly increased the number of striatum neurons, stimulated cell proliferation and also a motor improvement was described [65, 103].

Mesenchymal stem cells (MSCs) are capable of differentiating into various types of somatic cells. When MSCs are injected into the ischemic brain, after implantation, more than 80% of the grafted cells did not survive [104]. Perhaps, due to the ischemic events and cell death signaling pathways. Recently, melatonin's potential to improve MSCs survival and its mechanism has also been investigated *in vitro* [105], melatonin was able to improve the survival after OGD and H_2O_2 exposure. The melatonin antagonist, luzindole, reversed this effect, which indicates that there is a receptor-mediated mechanism through ERK1/2 signaling pathway, VEGF upregulation, inhibition of apoptosis and promotion of focal angiogenesis and neurogenesis. Briefly, transplantation of MSCs pre-treated with melatonin into the ischemic area significantly increased their survival.

In an attempt to relate the effects of exercise training to neuroprotection and neurogenesis following cerebral ischemia, a combination therapy based on melatonin (10 mg/kg) with and without treadmill exercise (20 m/min, 6 days/week) in MCAO rats enhanced neurogenesis and neuronal cell reconstruction in brain damaged lesions [106]. The benefits of combined exercise plus melatonin in terms of neuro-rehabilitation were also seen in rats after spinal cord injury [107]. Among the multiple actions of melatonin, its role on transplantation of exogenous cells or promoting proliferation and differentiation of endogenous cells represents an important research line for brain health improvement.

THERAPEUTIC POTENTIAL OF MELATONIN IN TRAUMATIC BRAIN INJURY (TBI)

TBI is a leading cause of death and disability worldwide and particularly most common in people aged ≤ 45 [108]. The damage is heterogeneous with complex clinical outcomes and with a short therapeutic window; thus, the treatment of this kind of acute brain injury is difficult [109]. The multifactorial pathogenesis of TBI involves different etiologies that lead to cell death, including excessive ROS, RNS, lipid peroxidation, microcirculation and microcirculation disruption, glutamate excitotoxicity, exacerbated inflammation, sleep-wake cycle impairments and diffuse axonal injury [110-113]. These factors contribute to tissue damage and difficult regeneration, wound healing and tissue repair [114]. In this regard, melatonin has been postulated as a neuroprotective agent against TBI in various *in vitro* and *in vivo* studies [115, 116]. It is well known that melatonin scavenges FR and over expressed antioxidant enzymes. When melatonin is administered in TBI models, Nrf2 protein is translocate from the cytoplasm to the nucleus. Additionally, melatonin reduces intracellular ROS, caspase-3 activation, Ca^{2+} overload, mitochondrial depolarization and brain lipid peroxidation thereby affording the neuroprotection [117-119]. Moreover, a treatment with melatonin reduced brain edema,

BBB permeability and intracranial pressure in other various *in vivo* models of TBI [98, 120-122]. Pinealectomized rats were more vulnerable to brain trauma, as well, but the increased damage could be reduced with melatonin immediately after TBI [123].

Inflammation is crucial in acute brain injuries, pro-inflammatory cytokines and adhesion molecules are dramatically increased after TBI. In this context, melatonin administration prevented the increased of IL-1β, TNF-α, IL-6 and MPP-9 following *in vivo* acute brain injuries [98, 124, 125]. The mRNA expression of NF-κB was markedly upregulated after TBI [126], while the administration of melatonin was able to block NF-κB translocation to the nucleus [122, 125], thereby, reducing the characteristic proinflammatory cascade displayed in acute brain injuries. Several doses of melatonin demonstrated to significantly reduce astrogliosis and neuronal death following TBI [127]. Consequently, melatonin is clearly protective against TBI and this is, at least in part, by blocking inflammatory responses.

Besides, sleep-wake disturbances are highly prevalent [128-130] and melatonin levels in peripheral blood are also reduced in TBI patients [131]. However, little is known about the mechanism involved in this sleep impairment. Interestingly, lower levels of melatonin production (by 42%) as well as delayed melatonin secretion (1.5 h) was recently reported in salivary samples of TBI patients in comparison with healthy controls [128]. Thus, restoring the melatonin concentration imbalance may be a good target to treat these sleep disturbances. Melatonin has some ideal properties for this purpose; it has demonstrated to be safe and well-tolerated in many studies treating sleep disorders after TBI [132].

MELATONIN AS CO-TREATMENT AGAINST BRAIN ISCHEMIA

The complexity of brain ischemia opens a wide and interesting field in the search for new effective treatments in the multiple processes implicated in the ischemic cascade. To date, only tPA, a US Food and Drug Administration (FDA)-approved drug, is effective in the treatment of acute ischemic stroke. Perhaps, the approach *"two better than one"* will lead to a more effective or synergic combination of drugs, with complementary pharmacological/biochemical profiles to treat stroke. In this scenario, literature includes several studies using drugs alone or in combination, showing a better therapeutic performance against cerebral ischemia when combined [133-135].

Regardless of the well-known therapeutic value of melatonin in a large number of diseases, in the last decade, there have been undertaken several studies assaying the coadministration of melatonin with other drug, probing the enhanced melatonin's protective effect as well as the own effect of the supplementary drug. Even if this accompanying drug induces biological damage by itself, melatonin is capable of minimizing the side effects. Herein, we summarize several studies that show the diverse effects of this multifunctional indole, in combination with other pharmaceutical agents used in brain ischemia treatment. As is the case of melatonin administration in combination with tPA, it ameliorates brain damage caused by the thrombolytic agent in animal stroke models [136]. Thus, melatonin combined with thrombolytic therapy may be applied in clinical settings to significantly benefit patients with stroke.

In an acute stage of cerebral H/I melatonin has revealed the capacity to reduce brain damage by reducing OS, promoting mitochondrial function, decreasing excitotoxicity, suppressing inflammation and diminishing apoptosis and cell death.

Disruption of neuronal Ca^{2+} homeostasis represents an important event in the series of biochemical reactions that initiates the ischemic cascade. Cellular Ca^{2+} overload plays a key role in neuronal excitotoxicity and promotes oxidative and nitrosative stress in mitochondria. Animal experiments conducted with the Ca^{2+}-antagonist nimodipine demonstrated to be effective against cerebral ischemia leading to reduced infarct volume [137]. Likewise, the use of nifedipine, a dihydropyridine (DHP) Ca^{2+}- antagonist, in combination with melatonin, may be the "*two better than one*" against brain ischemia. In 2002 [138], melatonin was tested alone and combined with nifedipine. During ischemic conditions, β-hydroxybutyrate (β-HB) content increase and a correlation between reduced brain glucose concentration and augmented lactate dehydrogenase levels is observed. In ischemic rats, melatonin reduced all energy fuels analyzed in the plasma, while nifedipine restored only β-HB and lactate. Of the two antioxidant enzymes evaluated, superoxide dismutase (SOD) and glutathione reductase (GR), the ischemic insult reduced the activity of both. These changes were prevented by the combination of melatonin and nifedipine. So, both compounds had neuroprotective actions and helped to maintain the antioxidant defense system balanced.

Amlodipine, another Ca^{2+} channel blocker, was tested with melatonin in an experimental model of I/R-induced brain damage [139]. Administration of melatonin prior to reperfusion restored all parameters evaluated; it improved clinical signs, reduced lipid peroxidation and myeloperoxidase activity, preserved the Na^+/K^+-ATPase pump, decreased brain edema and maintained BBB integrity. Amlodipine alone, restored all these parameters except for protecting the BBB integrity. This study indicates that both melatonin and amlodipine counteract the behavioral and neurochemical pathological changes in global ischemia.

Inflammatory responses deeply influence the progression of the ischemic cascade. Thus, the administration of a COX-2 inhibitor, meloxicam, in combination with melatonin in a MCAO model of acute ischemic stroke in rat, provided better protection than each drug alone, a wide clinical relevance finding [140]. Combination therapy in a TBI model, based on melatonin and dexamethasone, afforded greater beneficial effects than a single drug to reduce brain edema, to avoid the infiltration of astrocytes to inhibit apoptosis and metalloproteinase and to reduce iNOS expression [141].

As reviewed above, glutamate excitotoxicity represents an important molecular mechanism of ischemic brain damage. Therefore, melatonin, in combination with drugs designed to block AMPA/KA receptors may help to mitigate the neurotoxic processes involved in brain injury. The antiepileptic drug topiramate, alone or in combination with melatonin, was evaluated in a neonatal hypoxic ischemic rat model. When topiramate was given alone or in combination with melatonin, it significantly reduced the percentage of infarct volume and number of TUNEL positive cells. In this case, combined therapy was not more effective than single drug treatment, perhaps due to different routes of administration, experimental design or dosage [142]. The non-competitive NMDA receptor blocker memantine, a known neuroprotective agent, was also evaluated alone or in combination with melatonin in a model of focal cerebral ischemia [143]. After 90 min of ischemia and 24 h of reperfusion, both melatonin and memantine alone or in combination reduced ischemic damage in a significant proportion; they exhibited synergistic actions in the inactivation of stress kinases p21, p38/MAPK and SAPK/JNK1/2 pathways. Moreover, melatonin/memantine combination reduced IgG extravasation relative to either drug alone. Recently, in a TBI model, melatonin and/or memantine attenuated brain injury and DNA fragmentation in mice. These effects were exacerbated in animals treated with the two drugs combination. The melatonin/memantine combination also reduced SAPK/JNK1/2, p38 and ERK-1/2 phosphorylation and iNOS activity [144]. Consequently, all these results make necessary further clinical trials to evaluate the possible synergistic effects of the combined therapy with these two safe drugs, memantine and melatonin, in the post-traumatic application.

Poly(ADP-ribose) polymerases (PARPs) are a family of nuclear enzymes whose overactivation is involved in the pathogenesis of CNS disorders, such as brain ischemia. Testing potential of combination therapy, Gupta et al. [145] found a significantly decrease of edema volume and improved neurological signs when nicotinamide or 3-aminobenzamide, both PARP inhibitors, were combined with melatonin in MCAO focal ischemia model. Again, the combination of drugs provided greater neural protection.

Methazolamide, a cytochrome c inhibitor, exerts its neuroprotective effects by reducing apoptosis [146]. Wang et al. [147] demonstrated that melatonin plus Methazolamide, were neuroprotective in a *in vivo* model of cerebral ischemia, inhibiting cytochrome c release, caspase-7 and 1 activation and IL-1β. This promising combination of drugs requires further evaluation in humans.

Selenium has been recognized as regulator of brain function via selenium-dependent enzymes involved in antioxidant defense and redox balance and its ability to scavenge FR. Knowing the importance of imbalance between oxidants and antioxidants in the ischemic condition, the strategy of synergistic effect based on two antioxidants, selenium and melatonin, could be attractive to counteract free-radical damage. Therefore, Ahmad et al. [148] administered simultaneously sodium selenite and melatonin in a transient focal cerebral ischemia model. This treatment ameliorated OS parameters, inhibited iNOS expression and the inflammatory response, and prevented apoptosis.

Acupuncture has been proposed to be effective inhibiting the inflammatory reaction, reducing OS injury and edema, and promoting neural regeneration. In this way, the effects of melatonin pre-treatment (single dose) in combination with electroacupuncture in a rat MCAO model [149] significantly improved the neurological function including a decrease in the infarct volume, brain inflammation and TUNEL-positive cells.

CONCLUSION

There is an annual increase of published melatonin studies that examine its protective effects counteracting the toxic processes that generate free radicals and associated reactants. There are so many that it becomes almost impossible to summarize here all available information documenting the relevant and beneficial aspects of melatonin counteracting numerous desadaptative phenotypes.

Here in, we summarized numerous preclinical studies in different animal models, doses, administration routes and both acute or long-term studies. To achieve rigorous translational data for clinical studies is necessary to take into account STAIR or RIGOR guidelines for preclinical stroke research [150, 151]. Melatonin, in fact, has been found to be beneficial to treat many animal diseases models, but correlative clinical studies are lacking, possibly due to the fact that it is not a patentable molecule.

Melatonin is an exceptionally pleiotropy neurohormone with beneficial effects, such as a direct free radical scavenger activity, indirect antioxidant properties, anti-inflammatory, neuromodulatory and antiexcitotoxic effects, which have attracted many scientists in the last years, dedicating their efforts to investigate the molecular mechanisms stopping the beginning and progression of the ischemic cascade. Melatonin, in this sense, has been described as a versatile modulator of multiple signaling pathways involved in neuronal tissue damage, promoting cell proliferation and neurogenesis, maintaining the integrity and permeability of BBB, as well as improving the electrophysiological recovery either alone or in combination therapy with other neuroprotective drugs.

Considering melatonin's beneficial properties, low toxicity and high efficacy, it makes this molecule a suitable treatment against brain ischemia injury and its multiple pathophysiological side-effects, however, a more deeply clinical profile should be investigated.

CONFLICT OF INTEREST

The authors declare that there is no conflict of interests regarding the publication of this chapter.

REFERENCES

[1] Flynn RW, MacWalter RS, Doney AS. The cost of cerebral ischaemia. *Neuropharmacology*. 2008;55(3):250-256.

[2] Mathers CD, Boerma T, Ma Fat D. Global and regional causes of death. *Br Med Bul*l. 2009;92:7-32.

[3] O'Collins VE, Macleod MR, Donnan GA, Horky LL, van der Worp BH, Howells DW. 1,026 experimental treatments in acute stroke. *Ann Neurol.* 2006;59(3):467-477.

[4] Hossmann KA. Viability thresholds and the penumbra of focal ischemia. *Ann Neurol.* 1994;36(4):557-565.

[5] Choi DW. Ischemia-induced neuronal apoptosis. *Curr Opin Neurobiol.* 1996;6(5):667-672.

[6] Lee JM, Zipfel GJ, Choi DW. The changing landscape of ischaemic brain injury mechanisms. *Nature.* 1999;399(6738 Suppl):A7-14.

[7] Tissue plasminogen activator for acute ischemic stroke. The National Institute of Neurological Disorders and Stroke rt-PA Stroke Study Group. *N Engl J Med.* 1995;333(24):1581-1587.

[8] Zivin JA, Fisher M, DeGirolami U, Hemenway CC, Stashak JA. Tissue plasminogen activator reduces neurological damage after cerebral embolism. *Science.* 1985;230(4731):1289-1292.

[9] Lo EH, Moskowitz MA, Jacobs TP. Exciting, radical, suicidal: how brain cells die after stroke. *Stroke.* 2005;36(2):189-192.

[10] Paschen W. Role of calcium in neuronal cell injury: which subcellular compartment is involved? *Brain Res Bull.* 2000;53(4):409-413.

[11] Chan PH. Reactive oxygen radicals in signaling and damage in the ischemic brain. *J Cereb Blood Flow Metab.* 2001;21(1):2-14.

[12] Niizuma K, Yoshioka H, Chen H, Kim GS, Jung JE, Katsu M, et al. Mitochondrial and apoptotic neuronal death signaling pathways in cerebral ischemia. *Biochim Biophys Acta.* 2010;1802(1):92-99.

[13] Iadecola C, Alexander M. Cerebral ischemia and inflammation. *Curr Opin Neurol.* 2001;14(1):89-94.

[14] Hardeland R. Melatonin: signaling mechanisms of a pleiotropic agent. *Biofactors.* 2009;35(2):183-192.

[15] Vriend J, Reiter RJ. Melatonin feedback on clock genes: a theory involving the proteasome. *J Pineal Res.* 2015;58(1):1-11.

[16] Menendez-Pelaez A, Reiter RJ. Distribution of melatonin in mammalian tissues: the relative importance of nuclear versus cytosolic localization. *J Pineal Res.* 1993; 15(2):59-69.

[17] Venegas C, Garcia JA, Escames G, Ortiz F, Lopez A, Doerrier C, et al. Extrapineal melatonin: analysis of its subcellular distribution and daily fluctuations. *J Pineal Res.* 2012;52(2):217-27.

[18] Acuna-Castroviejo D, Escames G, Venegas C, Diaz-Casado ME, Lima-Cabello E, Lopez LC, et al. Extrapineal melatonin: sources, regulation, and potential functions. *Cell Mol Life Sci.* 2014;71(16):2997-3025.

[19] Reiter RJ, Richardson BA, Johnson LY, Ferguson BN, Dinh DT. Pineal melatonin rhythm: reduction in aging Syrian hamsters. *Science.* 1980;210(4476):1372-1373.

[20] Reiter RJ, Craft CM, Johnson JE, Jr., King TS, Richardson BA, Vaughan GM, et al. Age-associated reduction in nocturnal pineal melatonin levels in female rats. *Endocrinology.* 1981;109(4):1295-1297.

[21] Fiorina P, Lattuada G, Ponari O, Silvestrini C, DallAglio P. Impaired nocturnal melatonin excretion and changes of immunological status in ischaemic stroke patients. *Lancet.* 1996;347(9002):692-693.

[22] Ritzenthaler T, Nighoghossian N, Berthiller J, Schott AM, Cho TH, Derex L, et al. Nocturnal urine melatonin and 6-sulphatoxymelatonin excretion at the acute stage of ischaemic stroke. *J Pineal Res.* 2009;46(3):349-352.

[23] Reiter RJ, Tan DX, Kim SJ, Cruz MH. Delivery of pineal melatonin to the brain and SCN: role of canaliculi, cerebrospinal fluid, tanycytes and Virchow-Robin perivascular spaces. *Brain Struct Funct.* 2014;219(6):1873-1887.

[24] Uduma FU, Fokam P, Okere PCN, Motah M. Incidence of physiological pineal gland and choroid plexus calcifications in cranio-cerebral computed tomograms in Douala, Cameroon. *Global Journal of Medical Research.* 2011;11:5-11.

[25] Kitkhuandee A, Sawanyawisuth K, Johns J, Kanpittaya J, Tuntapakul S, Johns NP. Pineal calcification is a novel risk factor for symptomatic intracerebral hemorrhage. *Clin Neurol Neurosurg.* 2014;121:51-54.

[26] Kitkhuandee A, Sawanyawisuth K, Johns NP, Kanpittaya J, Johns J. Pineal calcification is associated with symptomatic cerebral infarction. *J Stroke Cerebrovasc Dis.* 2014;23(2):249-253.

[27] Del Brutto OH, Mera RM, Lama J, Zambrano M. Stroke and pineal gland calcification: lack of association. Results from a population-based study (The Atahualpa Project). *Clin Neurol Neurosurg.* 2015;130:91-94.

[28] Kondziolka D, Wechsler L, Goldstein S, Meltzer C, Thulborn KR, Gebel J, et al. Transplantation of cultured human neuronal cells for patients with stroke. *Neurology.* 2000;55(4):565-569.

[29] Borlongan CV, Sumaya I, Moss D, Kumazaki M, Sakurai T, Hida H, et al. Melatonin-secreting pineal gland: a novel tissue source for neural transplantation therapy in stroke. *Cell Transplant.* 2003;12(3):225-234.

[30] Kilic E, Ozdemir YG, Bolay H, Kelestimur H, Dalkara T. Pinealectomy aggravates and melatonin administration attenuates brain damage in focal ischemia. *J Cereb Blood Flow Metab.* 1999;19(5):511-516.

[31] De Butte M, Fortin T, Pappas BA. Pinealectomy: behavioral and neuropathological consequences in a chronic cerebral hypoperfusion model. *Neurobiol Aging.* 2002;23(2):309-317.

[32] Manev H, Uz T, Kharlamov A, Joo JY. Increased brain damage after stroke or excitotoxic seizures in melatonin-deficient rats. *FASEB J.* 1996;10(13):1546-1551.

[33] Joo JY, Uz T, Manev H. Opposite effects of pinealectomy and melatonin administration on brain damage following cerebral focal ischemia in rat. *Restor Neurol Neurosci.* 1998;13(3-4):185-191.

[34] Acuna Castroviejo D, Lopez LC, Escames G, Lopez A, Garcia JA, Reiter RJ. Melatonin-mitochondria interplay in health and disease. *Curr Top Med Chem.* 2011;11(2):221-40.

[35] Sanderson TH, Reynolds CA, Kumar R, Przyklenk K, Huttemann M. Molecular mechanisms of ischemia-reperfusion injury in brain: pivotal role of the mitochondrial membrane potential in reactive oxygen species generation. *Mol Neurobiol.* 2013;47(1):9-23.

[36] Lipton P. Ischemic cell death in brain neurons. *Physiol Rev.* 1999;79(4):1431-568.

[37] Agardh CD, Zhang H, Smith ML, Siesjo BK. Free radical production and ischemic brain damage: influence of postischemic oxygen tension. *Int J Dev Neurosci.* 1991;9(2):127-138.

[38] Flamm ES, Demopoulos HB, Seligman ML, Poser RG, Ransohoff J. Free radicals in cerebral ischemia. *Stroke.* 1978;9(5):445-447.

[39] Reiter RJ, Paredes SD, Manchester LC, Tan DX. Reducing oxidative/nitrosative stress: a newly-discovered genre for melatonin. *Crit Rev Biochem Mol Biol.* 2009;44(4):175-200.

[40] Manchester LC, Coto-Montes A, Boga JA, Andersen LP, Zhou Z, Galano A, et al. Melatonin: an ancient molecule that makes oxygen metabolically tolerable. *J Pineal Res.* 2015;59(4):403-419.

[41] Barlow-Walden LR, Reiter RJ, Abe M, Pablos M, Menendez-Pelaez A, Chen LD, et al. Melatonin stimulates brain glutathione peroxidase activity. *Neurochem Int.* 1995;26(5):497-502.

[42] Rodriguez C, Mayo JC, Sainz RM, Antolin I, Herrera F, Martin V, et al. Regulation of antioxidant enzymes: a significant role for melatonin. *J Pineal Res.* 2004; 36(1):1-9.

[43] Pablos MI, Reiter RJ, Ortiz GG, Guerrero JM, Agapito MT, Chuang JI, et al. Rhythms of glutathione peroxidase and glutathione reductase in brain of chick and their inhibition by light. *Neurochem Int.* 1998;32(1):69-75.

[44] Saito A, Maier CM, Narasimhan P, Nishi T, Song YS, Yu F, et al. Oxidative stress and neuronal death/survival signaling in cerebral ischemia. *Mol Neurobiol.* 2005;31(1-3):105-116.

[45] Li XJ, Zhang LM, Gu J, Zhang AZ, Sun FY. Melatonin decreases production of hydroxyl radical during cerebral ischemia-reperfusion. *Zhongguo Yao Li Xue Bao.* 1997;18(5):394-396.

[46] Chen TY, Lee MY, Chen HY, Kuo YL, Lin SC, Wu TS, et al. Melatonin attenuates the postischemic increase in blood-brain barrier permeability and decreases hemorrhagic transformation of tissue-plasminogen activator therapy following ischemic stroke in mice. *J Pineal Res.* 2006;40(3):242-250.

[47] Bolanos JP, Almeida A. Roles of nitric oxide in brain hypoxia-ischemia. *Biochim Biophys Acta.* 1999;1411(2-3):415-436.

[48] Li H, Wang Y, Feng D, Liu Y, Xu M, Gao A, et al. Alterations in the time course of expression of the Nox family in the brain in a rat experimental cerebral ischemia and reperfusion model: effects of melatonin. *J Pineal Res.* 2014;57(1):110-119.

[49] Luque Contreras D, Vargas Robles H, Romo E, Rios A, Escalante B. The role of nitric oxide in the post-ischemic revascularization process. *Pharmacol Ther.* 2006;112(2):553-563.

[50] Liu H, Li J, Zhao F, Wang H, Qu Y, Mu D. Nitric oxide synthase in hypoxic or ischemic brain injury. *Rev Neurosci.* 2015;26(1):105-117.

[51] Guerrero JM, Reiter RJ, Ortiz GG, Pablos MI, Sewerynek E, Chuang JI. Melatonin prevents increases in neural nitric oxide and cyclic GMP production after transient brain ischemia and reperfusion in the Mongolian gerbil (Meriones unguiculatus). *J Pineal Res.* 1997;23(1):24-31.

[52] Pei Z, Fung PC, Cheung RT. Melatonin reduces nitric oxide level during ischemia but not blood-brain barrier breakdown during reperfusion in a rat middle cerebral artery occlusion stroke model. *J Pineal Res.* 2003;34(2):110-118.

[53] Pei Z, Pang SF, Cheung RT. Administration of melatonin after onset of ischemia reduces the volume of cerebral infarction in a rat middle cerebral artery occlusion stroke model. *Stroke.* 2003;34(3):770-775.

[54] Koh PO. Melatonin regulates nitric oxide synthase expression in ischemic brain injury. *J Vet Med Sci.* 2008;70(7):747-750.

[55] Sinha K, Degaonkar MN, Jagannathan NR, Gupta YK. Effect of melatonin on ischemia reperfusion injury induced by middle cerebral artery occlusion in rats. *Eur J Pharmacol.* 2001;428(2):185-192.

[56] Ozacmak VH, Barut F, Ozacmak HS. Melatonin provides neuroprotection by reducing oxidative stress and HSP70 expression during chronic cerebral hypoperfusion in ovariectomized rats. *J Pineal Res.* 2009;47(2):156-163.

[57] Wakatsuki A, Okatani Y, Izumiya C, Ikenoue N. Melatonin protects against ischemia and reperfusion-induced oxidative lipid and DNA damage in fetal rat brain. *J Pineal Re*s. 1999;26(3):147-152.

[58] Cuzzocrea S, Costantino G, Gitto E, Mazzon E, Fulia F, Serraino I, et al. Protective effects of melatonin in ischemic brain injury. *J Pineal Res.* 2000;29(4):217-227.

[59] Wakatsuki A, Okatani Y, Shinohara K, Ikenoue N, Fukaya T. Melatonin protects against ischemia/reperfusion-induced oxidative damage to mitochondria in fetal rat brain. *J Pineal Res.* 2001;31(2):167-172.

[60] Ma OK, Sucher NJ. Molecular interaction of NMDA receptor subunit NR3A with protein phosphatase 2A. *Neuroreport.* 2004;15(9):1447-1450.

[61] Koh PO. Melatonin attenuates decrease of protein phosphatase 2A subunit B in ischemic brain injury. *J Pineal Res.* 2012;52(1):57-61.

[62] Uz T, Giusti P, Franceschini D, Kharlamov A, Manev H. Protective effect of melatonin against hippocampal DNA damage induced by intraperitoneal administration of kainate to rats. *Neuroscience.* 1996;73(3):631-636.

[63] Giusti P, Lipartiti M, Franceschini D, Schiavo N, Floreani M, Manev H. Neuroprotection by melatonin from kainate-induced excitotoxicity in rats. *FASEB J.* 1996;10(8):891-6.

[64] Giusti P, Franceschini D, Petrone M, Manev H, Floreani M. In vitro and in vivo protection against kainate-induced excitotoxicity by melatonin. *J Pineal Res.* 1996;20(4):226-231.

[65] Juan WS, Huang SY, Chang CC, Hung YC, Lin YW, Chen TY, et al. Melatonin improves neuroplasticity by upregulating the growth-associated protein-43 (GAP-43) and NMDAR postsynaptic density-95 (PSD-95) proteins in cultured neurons exposed to glutamate excitotoxicity and in rats subjected to transient focal cerebral ischemia even during a long-term recovery period. *J Pineal Res.* 2014;56(2):213-223.

[66] Floreani M, Skaper SD, Facci L, Lipartiti M, Giusti P. Melatonin maintains glutathione homeostasis in kainic acid-exposed rat brain tissues. *FASEB J.* 1997;11(14):1309-1315.

[67] Lee YK, Lee SR, Kim CY. Melatonin attenuates the changes in polyamine levels induced by systemic kainate administration in rat brains. *J Neurol Sc*i. 2000;178(2):124-131.

[68] Lorrio S, Romero A, Gonzalez-Lafuente L, Lajarin-Cuesta R, Martinez-Sanz FJ, Estrada M, et al. PP2A ligand ITH12246 protects against memory impairment and focal cerebral ischemia in mice. *ACS Chem Neurosci.* 2013;4(9):1267-1277.

[69] Espinar A, Garcia-Oliva A, Isorna EM, Quesada A, Prada FA, Guerrero JM. Neuroprotection by melatonin from glutamate-induced excitotoxicity during development of the cerebellum in the chick embryo. *J Pineal Res.* 2000;28(2):81-88.

[70] Cazevieille C, Safa R, Osborne NN. Melatonin protects primary cultures of rat cortical neurones from NMDA excitotoxicity and hypoxia/reoxygenation. *Brain Res.* 1997;768(1-2):120-124.

[71] Chung SY, Han SH. Melatonin attenuates kainic acid-induced hippocampal neurodegeneration and oxidative stress through microglial inhibition. *J Pineal Res.* 2003;34(2):95-102.

[72] Lee SH, Chun W, Kong PJ, Han JA, Cho BP, Kwon OY, et al. Sustained activation of Akt by melatonin contributes to the protection against kainic acid-induced neuronal death in hippocampus. *J Pineal Res.* 2006;40(1):79-85.

[73] Jin R, Yang G, Li G. Inflammatory mechanisms in ischemic stroke: role of inflammatory cells. *J Leukoc Biol.* 2010;87(5):779-789.

[74] Emsley HC, Hopkins SJ. Acute ischaemic stroke and infection: recent and emerging concepts. *Lancet Neurol.* 2008;7(4):341-353.

[75] McColl BW, Allan SM, Rothwell NJ. Systemic infection, inflammation and acute ischemic stroke. *Neuroscience.* 2009;158(3):1049-1061.

[76] Barone FC, Feuerstein GZ. Inflammatory mediators and stroke: new opportunities for novel therapeutics. *J Cereb Blood Flow Metab.* 1999;19(8):819-834.

[77] Hanisch UK, Kettenmann H. Microglia: active sensor and versatile effector cells in the normal and pathologic brain. *Nat Neurosci.* 2007;10(11):1387-1394.

[78] Morgan SC, Taylor DL, Pocock JM. Microglia release activators of neuronal proliferation mediated by activation of mitogen-activated protein kinase, phosphatidylinositol-3-kinase/Akt and delta-Notch signalling cascades. *J Neurochem.* 2004;90(1):89-101.

[79] Muller FJ, Snyder EY, Loring JF. Gene therapy: can neural stem cells deliver? *Nat Rev Neurosci.* 2006;7(1):75-84.

[80] Taylor RA, Sansing LH. Microglial responses after ischemic stroke and intracerebral hemorrhage. *Clin Dev Immunol.* 2013;2013:746068.

[81] Hu X, Li P, Guo Y, Wang H, Leak RK, Chen S, et al. Microglia/macrophage polarization dynamics reveal novel mechanism of injury expansion after focal cerebral ischemia. *Stroke.* 2012;43(11):3063-3070.

[82] Robertson NJ, Faulkner S, Fleiss B, Bainbridge A, Andorka C, Price D, et al. Melatonin augments hypothermic neuroprotection in a perinatal asphyxia model. *Brain.* 2013;136(Pt 1):90-105.

[83] Lee MY, Kuan YH, Chen HY, Chen TY, Chen ST, Huang CC, et al. Intravenous administration of melatonin reduces the intracerebral cellular inflammatory

response following transient focal cerebral ischemia in rats. *J Pineal Res.* 2007;42(3):297-309.

[84] Villapol S, Fau S, Renolleau S, Biran V, Charriaut-Marlangue C, Baud O. Melatonin promotes myelination by decreasing white matter inflammation after neonatal stroke. *Pediatr Res.* 2011;69(1):51-55.

[85] Welin AK, Svedin P, Lapatto R, Sultan B, Hagberg H, Gressens P, et al. Melatonin reduces inflammation and cell death in white matter in the mid-gestation fetal sheep following umbilical cord occlusion. *Pediatr Res.* 2007;61(2):153-158.

[86] Paredes SD, Rancan L, Kireev R, Gonzalez A, Louzao P, Gonzalez P, et al. Melatonin Counteracts at a Transcriptional Level the Inflammatory and Apoptotic Response Secondary to Ischemic Brain Injury Induced by Middle Cerebral Artery Blockade in Aging Rats. *Biores Open Access.* 2015;4(1):407-416.

[87] Hutton LC, Abbass M, Dickinson H, Ireland Z, Walker DW. Neuroprotective properties of melatonin in a model of birth asphyxia in the spiny mouse (Acomys cahirinus). *Dev Neurosci.* 2009;31(5):437-451.

[88] Pei Z, Cheung RT. Pretreatment with melatonin exerts anti-inflammatory effects against ischemia/reperfusion injury in a rat middle cerebral artery occlusion stroke model. *J Pineal Res.* 2004;37(2):85-91.

[89] Kahle KT, Simard JM, Staley KJ, Nahed BV, Jones PS, Sun D. Molecular mechanisms of ischemic cerebral edema: role of electroneutral ion transport. *Physiology (Bethesda).* 2009;24:257-265.

[90] Kondoh T, Uneyama H, Nishino H, Torii K. Melatonin reduces cerebral edema formation caused by transient forebrain ischemia in rats. *Life Sci.* 2002;72(4-5):583-590.

[91] Torii K, Uneyama H, Nishino H, Kondoh T. Melatonin suppresses cerebral edema caused by middle cerebral artery occlusion/reperfusion in rats assessed by magnetic resonance imaging. *J Pineal Res.* 2004;36(1):18-24.

[92] Chi OZ, Wei HM, Sinha AK, Weiss HR. Effects of inhibition of nitric oxide synthase on blood-brain barrier transport in focal cerebral ischemia. *Pharmacology.* 1994;48(6):367-373.

[93] Chen HY, Chen TY, Lee MY, Chen ST, Hsu YS, Kuo YL, et al. Melatonin decreases neurovascular oxidative/nitrosative damage and protects against early increases in the blood-brain barrier permeability after transient focal cerebral ischemia in mice. *J Pineal Res.* 2006;41(2):175-182.

[94] Han F, Chen YX, Lu YM, Huang JY, Zhang GS, Tao RR, et al. Regulation of the ischemia-induced autophagy-lysosome processes by nitrosative stress in endothelial cells. *J Pineal Res.* 2011;51(1):124-135.

[95] Song J, Kang SM, Lee WT, Park KA, Lee KM, Lee JE. The beneficial effect of melatonin in brain endothelial cells against oxygen-glucose deprivation followed by reperfusion-induced injury. *Oxid Med Cell Longev.* 2014;2014:639531.

[96] Moretti R, Zanin A, Pansiot J, Spiri D, Manganozzi L, Kratzer I, et al. Melatonin reduces excitotoxic blood-brain barrier breakdown in neonatal rats. *Neuroscience.* 2015;311:382-397.

[97] Grossetete M, Phelps J, Arko L, Yonas H, Rosenberg GA. Elevation of matrix metalloproteinases 3 and 9 in cerebrospinal fluid and blood in patients with severe traumatic brain injury. *Neurosurgery.* 2009;65(4):702-708.

[98] Alluri H, Wilson RL, Anasooya Shaji C, Wiggins-Dohlvik K, Patel S, Liu Y, et al. Melatonin Preserves Blood-Brain Barrier Integrity and Permeability via Matrix Metalloproteinase-9 Inhibition. *PLoS One.* 2016;11(5):e0154427.

[99] Zhao C, Deng W, Gage FH. Mechanisms and functional implications of adult neurogenesis. *Cell.* 2008;132(4):645-660.

[100] Kim MJ, Kim HK, Kim BS, Yim SV. Melatonin increases cell proliferation in the dentate gyrus of maternally separated rats. *J Pineal Res.* 2004;37(3):193-197.

[101] Ayao MS, Olaleyer O, Ihunwo AO. Melatonin Potentiates Cells Proliferation in the Dentate Gyrus Following Ischemic Brain Injury in Adult Rats. *Journal of Animal and Veterinary Advances.* 2010;9(11):1633-1638.

[102] Chern CM, Liao JF, Wang YH, Shen YC. Melatonin ameliorates neural function by promoting endogenous neurogenesis through the MT2 melatonin receptor in ischemic-stroke mice. *Free Radic Biol Med.* 2012;52(9):1634-1647.

[103] Kilic E, Kilic U, Bacigaluppi M, Guo Z, Abdallah NB, Wolfer DP, et al. Delayed melatonin administration promotes neuronal survival, neurogenesis and motor recovery, and attenuates hyperactivity and anxiety after mild focal cerebral ischemia in mice. *J Pineal Res.* 2008;45(2):142-148.

[104] Roh JK, Jung KH, Chu K. Adult stem cell transplantation in stroke: its limitations and prospects. *Curr Stem Cell Res Ther.* 2008;3(3):185-196.

[105] Tang Y, Cai B, Yuan F, He X, Lin X, Wang J, et al. Melatonin pretreatment improves the survival and function of transplanted mesenchymal stem cells after focal cerebral ischemia. *Cell Transplant.* 2014;23(10):1279-1291.

[106] Lee M, Lee S, Hong Y. Melatonin plus treadmill exercise synergistically promotes neurogenesis and reduce apoptosis in focal cerebral ischemic rats (877.17). *The FASEB Journal.* 2014;28(1 Supplement).

[107] Hong Y, Palaksha KJ, Park K, Park S, Kim HD, Reiter RJ, et al. Melatonin plus exercise-based neurorehabilitative therapy for spinal cord injury. *J Pineal Res.* 2010;49(3):201-9.

[108] Roozenbeek B, Maas AI, Menon DK. Changing patterns in the epidemiology of traumatic brain injury. *Nat Rev Neurol.* 2013;9(4):231-236.

[109] Werner C, Engelhard K. Pathophysiology of traumatic brain injury. *Br J Anaesth*. 2007;99(1):4-9.

[110] Lucke-Wold BP, Smith KE, Nguyen L, Turner RC, Logsdon AF, Jackson GJ, et al. Sleep disruption and the sequelae associated with traumatic brain injury. *Neurosci Biobehav Rev*. 2015;55:68-77.

[111] Greve MW, Zink BJ. Pathophysiology of traumatic brain injury. *Mt Sinai J Med*. 2009;76(2):97-104.

[112] Cederberg D, Siesjo P. What has inflammation to do with traumatic brain injury? *Childs Nerv Syst*. 2010;26(2):221-226.

[113] Browne KD, Chen XH, Meaney DF, Smith DH. Mild traumatic brain injury and diffuse axonal injury in swine. *J Neurotrauma*. 2011;28(9):1747-1755.

[114] Aertker BM, Bedi S, Cox CS, Jr. Strategies for CNS repair following TBI. *Exp Neurol*. 2016;275 Pt 3:411-426.

[115] Tsai MC, Chen WJ, Tsai MS, Ching CH, Chuang JI. Melatonin attenuates brain contusion-induced oxidative insult, inactivation of signal transducers and activators of transcription 1, and upregulation of suppressor of cytokine signaling-3 in rats. *J Pineal Res*. 2011;51(2):233-245.

[116] Naseem M, Parvez S. Role of melatonin in traumatic brain injury and spinal cord injury. *Scientific World Journal*. 2014;2014:586270.

[117] Ding K, Wang H, Xu J, Li T, Zhang L, Ding Y, et al. Melatonin stimulates antioxidant enzymes and reduces oxidative stress in experimental traumatic brain injury: the Nrf2-ARE signaling pathway as a potential mechanism. *Free Radic Biol Med*. 2014;73:1-11.

[118] Yuruker V, Naziroglu M, Senol N. Reduction in traumatic brain injury-induced oxidative stress, apoptosis, and calcium entry in rat hippocampus by melatonin: Possible involvement of TRPM2 channels. *Metab Brain Dis*. 2015;30(1):223-231.

[119] Ozdemir D, Uysal N, Gonenc S, Acikgoz O, Sonmez A, Topcu A, et al. Effect of melatonin on brain oxidative damage induced by traumatic brain injury in immature rats. *Physiol Res*. 2005;54(6):631-637.

[120] Dehghan F, Khaksari Hadad M, Asadikram G, Najafipour H, Shahrokhi N. Effect of melatonin on intracranial pressure and brain edema following traumatic brain injury: role of oxidative stresses. *Arch Med Res*. 2013;44(4):251-258.

[121] Kabadi SV, Maher TJ. Posttreatment with uridine and melatonin following traumatic brain injury reduces edema in various brain regions in rats. *Ann N Y Acad Sci*. 2010;1199:105-113.

[122] Beni SM, Kohen R, Reiter RJ, Tan DX, Shohami E. Melatonin-induced neuroprotection after closed head injury is associated with increased brain antioxidants and attenuated late-phase activation of NF-kappaB and AP-1. *FASEB J*. 2004;18(1):149-151.

[123] Ates O, Cayli S, Gurses I, Yucel N, Iraz M, Altinoz E, et al. Effect of pinealectomy and melatonin replacement on morphological and biochemical recovery after traumatic brain injury. *Int J Dev Neurosci.* 2006;24(6):357-363.

[124] Ding K, Wang H, Xu J, Lu X, Zhang L, Zhu L. Melatonin reduced microglial activation and alleviated neuroinflammation induced neuron degeneration in experimental traumatic brain injury: Possible involvement of mTOR pathway. *Neurochem Int.* 2014;76:23-31.

[125] Wang Z, Wu L, You W, Ji C, Chen G. Melatonin alleviates secondary brain damage and neurobehavioral dysfunction after experimental subarachnoid hemorrhage: possible involvement of TLR4-mediated inflammatory pathway. *J Pineal Res.* 2013;55(4):399-408.

[126] Chen G, Zhang S, Shi J, Ai J, Qi M, Hang C. Simvastatin reduces secondary brain injury caused by cortical contusion in rats: possible involvement of TLR4/NF-kappaB pathway. *Exp Neurol.* 2009;216(2):398-406.

[127] Babaee A, Eftekhar-Vaghefi SH, Asadi-Shekaari M, Shahrokhi N, Soltani SD, Malekpour-Afshar R, et al. Melatonin treatment reduces astrogliosis and apoptosis in rats with traumatic brain injury. *Iran J Basic Med Sci.* 2015;18(9):867-872.

[128] Grima NA, Ponsford JL, St Hilaire MA, Mansfield D, Rajaratnam SM. Circadian Melatonin Rhythm Following Traumatic Brain Injury. *Neurorehabil Neural Repair.* 2016.

[129] Andrabi SS, Parvez S, Tabassum H. Melatonin and Ischemic Stroke: Mechanistic Roles and Action. *Adv Pharmacol Sci.* 2015;2015:384750.

[130] Vermaelen J, Greiffenstein P, deBoisblanc BP. Sleep in traumatic brain injury. *Crit Care Clin.* 2015;31(3):551-561.

[131] Paparrigopoulos T, Melissaki A, Tsekou H, Efthymiou A, Kribeni G, Baziotis N, et al. Melatonin secretion after head injury: a pilot study. *Brain Inj.* 2006;20(8):873-878.

[132] Barlow KM, Brooks BL, MacMaster FP, Kirton A, Seeger T, Esser M, et al. A double-blind, placebo-controlled intervention trial of 3 and 10 mg sublingual melatonin for post-concussion syndrome in youths (PLAYGAME): study protocol for a randomized controlled trial. *Trials.* 2014;15:271.

[133] Yang Y, Li Q, Shuaib A. Enhanced neuroprotection and reduced hemorrhagic incidence in focal cerebral ischemia of rat by low dose combination therapy of urokinase and topiramate. *Neuropharmacology.* 2000;39(5):881-888.

[134] Schmid-Elsaesser R, Hungerhuber E, Zausinger S, Baethmann A, Reulen HJ. Neuroprotective efficacy of combination therapy with two different antioxidants in rats subjected to transient focal ischemia. *Brain Res.* 1999;816(2):471-479.

[135] Steiner T, Hacke W. Combination therapy with neuroprotectants and thrombolytics in acute ischaemic stroke. *Eur Neurol.* 1998;40(1):1-8.

[136] Chen TY, Lee MY, Chen HY, Kuo YL, Lin SC, Wu TS, et al. Melatonin attenuates the postischemic increase in blood-brain barrier permeability and decreases hemorrhagic transformation of tissue-plasminogen activator therapy following ischemic stroke in mice. *J Pineal Res.* 2006;40(3):242-250.

[137] Gelmers HJ, Gorter K, de Weerdt CJ, Wiezer HJ. A controlled trial of nimodipine in acute ischemic stroke. *N Engl J Med.* 1988;318(4):203-207.

[138] El-Abhar HS, Shaalan M, Barakat M, El-Denshary ES. Effect of melatonin and nifedipine on some antioxidant enzymes and different energy fuels in the blood and brain of global ischemic rats. *J Pineal Res.* 2002;33(2):87-94.

[139] Toklu H, Deniz M, Keyer-Uysal M, Sener G. The protective effect of melatonin and amlodipine against cerebral ischemia/reperfusion-induced oxidative brain injury in rats. *Marmara Medical Journal.* 2009;22:34-44.

[140] Gupta YK, Chaudhary G, Sinha K. Enhanced protection by melatonin and meloxicam combination in a middle cerebral artery occlusion model of acute ischemic stroke in rat. *Can J Physiol Pharmacol.* 2002;80(3):210-217.

[141] Campolo M, Ahmad A, Crupi R, Impellizzeri D, Morabito R, Esposito E, et al. Combination therapy with melatonin and dexamethasone in a mouse model of traumatic brain injury. *J Endocrinol.* 2013;217(3):291-301.

[142] Ozyener F, Cetinkaya M, Alkan T, Goren B, Kafa IM, Kurt MA, et al. Neuroprotective effects of melatonin administered alone or in combination with topiramate in neonatal hypoxic-ischemic rat model. *Restor Neurol Neurosci.* 2012;30(5):435-444.

[143] Kilic U, Yilmaz B, Reiter RJ, Yuksel A, Kilic E. Effects of memantine and melatonin on signal transduction pathways vascular leakage and brain injury after focal cerebral ischemia in mice. *Neuroscience.* 2013;237:268-276.

[144] Kelestemur T, Yulug B, Caglayan AB, Beker MC, Kilic U, Caglayan B, et al. Targeting different pathophysiological events after traumatic brain injury in mice: Role of melatonin and memantine. *Neurosci Lett.* 2016;612:92-97.

[145] Gupta S, Kaul CL, Sharma SS. Neuroprotective effect of combination of poly (ADP-ribose) polymerase inhibitor and antioxidant in middle cerebral artery occlusion induced focal ischemia in rats. *Neurol Res.* 2004;26(1):103-107.

[146] Wang X, Zhu S, Pei Z, Drozda M, Stavrovskaya IG, Del Signore SJ, et al. Inhibitors of cytochrome c release with therapeutic potential for Huntington's disease. *J Neurosci.* 2008;28(38):9473-9485.

[147] Wang X, Figueroa BE, Stavrovskaya IG, Zhang Y, Sirianni AC, Zhu S, et al. Methazolamide and melatonin inhibit mitochondrial cytochrome C release and are neuroprotective in experimental models of ischemic injury. *Stroke.* 2009; 40(5):1877-1885.

[148] Ahmad A, Khan MM, Ishrat T, Khan MB, Khuwaja G, Raza SS, et al. Synergistic effect of selenium and melatonin on neuroprotection in cerebral ischemia in rats. *Biol Trace Elem Res*. 2011;139(1):81-96.

[149] Liu L, Cheung RT. Effects of pretreatment with a combination of melatonin and electroacupuncture in a rat model of transient focal cerebral ischemia. *Evid Based Complement Alternat Med.* 2013;2013:953162.

[150] Fisher M, Feuerstein G, Howells DW, Hurn PD, Kent TA, Savitz SI, et al. Update of the stroke therapy academic industry roundtable preclinical recommendations. *Stroke.* 2009;40(6):2244-2250.

[151] Lapchak PA, Zhang JH, Noble-Haeusslein LJ. RIGOR guidelines: escalating STAIR and STEPS for effective translational research. *Transl Stroke Res.* 2013;4(3):279-285.

In: Melatonin: Medical Uses and Role in Health and Disease ISBN: 978-1-53612-987-8
Editors: Lore Correia and Germaine Mayers © 2018 Nova Science Publishers, Inc.

Chapter 7

PARTICIPATION OF MELATONIN RECEPTORS IN NEUROLOGICAL DISORDERS

Anna Karynna A. A. Rocha[1,*],
Edilson D. da Silva Júnior[2], *Fernanda Amaral*[3]
and Débora Amado[3]

[1]Brain Institute, Department of Neuroscience,
[2]Department of Biophysics and Pharmacology,
Federal University of Rio Grande do Norte, Natal, Brazil
[3]Department of Neurology and Neurosurgery,
Federal University of São Paulo, São Paulo, Brazil

ABSTRACT

Melatonin regulates a variety of neurophysiological and neuroendocrine functions through the activation of two G protein-coupled melatonin receptors that have been identified in mammals as MT_1 and MT_2. Melatonin receptors are located in areas of the central nervous system, including the cerebral cortex, suprachiasmatic nucleus, cerebellum, thalamus, and hippocampus. On one hand, evidence has shown that MT_1 receptors can inhibit calcium influx, inducing inhibitory effects on neuronal electrical activity. On the other hand, MT_2 receptors are able to (a) decrease cAMP and cGMP accumulation, (b) mediate circadian rhythm, and (c) inhibit long-term potentiation and increase the activity of PKC. Some studies have demonstrated changes in MT_1 and MT_2 receptor expression in several neurological disorders such as epilepsy, Alzheimer's and Parkinson's diseases, and traumatic brain injury. The role of melatonin or melatonin receptor agonists as potent neuroprotective agents in these diseases has been considered both in humans and rodents. In addition, progress in understanding the role of melatonin receptors in the modulation of neurological disorders has led to the discovery of novel classes of drugs for treating several brain-related disorders including insomnia and mood

---

[*] Corresponding Author Email: rocha.anna2@gmail.com.

disorders. In this review we will show the participation of MT_1 and MT_2 receptors in neurological disorders and the future therapeutic prospects of melatonin receptors agonists/antagonists on various central nervous system diseases.

Keywords: melatonin receptors, melatonin, neurological disorders

INTRODUCTION

Melatonin (N-acetyl-5-methoxytrypamine) was isolated and characterized from bovine pineal gland more than 50 years ago by Lerner et al. [1] who found that this indolamine caused bleaching of frogs' skin.

Most of the body's melatonin is synthesized in the pineal gland [2]. From fundamental experiences made during the first half of the last century, it is now known that the pineal gland is more than a phylogenetic rudiment of the so-called "third eye", or the parietal eye of lower vertebrates but plays a role in all vertebrates as a neuroendocrine transducer. Nowadays, it is known that the pineal gland can directly or indirectly convert ambient lighting conditions into a neurohormonal message: the synthesis of nocturnal melatonin [3]. This process is species-dependent.

Melatonin is produced at night in all of the species that have been investigated. Therefore, during the day, melatonin synthesis and sympathetic activity in the pineal gland are reduced [4]. After entering the circulation, melatonin acts as an endocrine factor and chemical messenger for the light/dark cycle (a circadian and circannual pacemaker) [5].

There is evidence that melatonin is produced in various extra-pineal organs, including the brain, retina, pigmented epithelium of the retina, gastrointestinal tract, bone marrow, lymphocytes, and skin [6, 7]. In these organs, melatonin may act as an autocrine or paracrine molecule, which includes the possibility that it acts through intracellular receptors expressed by the same cells that produce it. In addition, melatonin may protect the cell from free radical damage when it is produced locally [8]. In adult subjects, circulating levels of melatonin are extremely low (<20 pmol/L). These levels increase approximately at 9:00 p.m., (peak at 03:00 a.m.), and return to lower levels at approximately 9:00 a.m. [9]. Once synthesized, melatonin is released quickly and passively into the bloodstream [2]. It enters the blood circulatory system and travels to different regions of the body to obtain desirable physiological responses.

NEUROPROTECTIVE EFFECTS OF MELATONIN

Although melatonin is produced in several different tissues and organs, its main action is on the central nervous system (CNS) [10]. It is known that melatonin's actions are predominantly inhibitory [11, 12]. In addition to its well-known properties of sleep promotion humans, melatonin has several other sedative, anti-excitatory, and anticonvulsant effects, which are involved in different actions such as facilitating gamma-aminobutyric acid (GABA) transmission, modulating glutamate receptors, decreasing transport of cytosolic Ca^{2+} through GABAc and/or mGlu3 metabotropic receptors, interference with neuronal nitric oxide synthase (nNOS), and changes in K^+ currents [13].

Melatonin inhibits the activity of constitutive NOS (NOSc) through complex formation with calmodulin, thereby reducing nitric oxide (NO) production in various regions of the brain [14, 15]. In addition, melatonin controls mitochondrial NOS isoform, regulates mitochondrial and bioenergetic functions, and protects mitochondria from excess NO [16].

Melatonin also has well-described antioxidant properties. It has the capacity to be a powerful "cleaner" of both reactive oxygen and nitrogen species (ROS and RNS, respectively). It can stimulate the activity of antioxidant enzymes such as glutathione peroxidase (GPx), superoxide dismutase (SOD), and glutathione reductase (GRd) [17, 18]. It can reduce electron leakage from the mitochondria, thus reducing the generation of free radicals [19]. In addition, melatonin may act synergistically with other antioxidants such as vitamin E and ascorbic acid [20].

As an antioxidant, melatonin is not only effective in protecting against oxidative damage in nuclear DNA, membrane lipids and cytosolic proteins but also acts by altering the activity of enzymes that enhance the body's total antioxidant defense capacity [21]. In addition, melatonin decreases electron leakage from the mitochondria, thereby decreasing the generation of free radicals [19]. All melatonin antioxidant functions are facilitated by its ability to cross morphophysiological barriers such as the blood-brain barrier and intracellular and subcellular barriers [22].

In order to present a free radical scavenger activity, melatonin must be at sufficiently high concentrations [23–25]. However, in physiological concentrations, melatonin acts as an excitatory process suppressor [26]. Melatonin has been associated with protection of various cells, tissues, and organs by repairing the damage caused by a variety of free radical-generating processes such as cyanide poisoning, glutathione depletion, ischemia-reperfusion, and cainic acid excitotoxicity [27, 28].

Some authors have shown that melatonin exerts anti-inflammatory effects via inhibition of neutrophil adhesion and rolling in endothelial cells [29], reducing the synthesis of the inducible nitric oxide synthase (iNOS) and inducible cyclooxygenase (COX2) [20]. Melatonin's anti-inflammatory effects appear to be directly related to its ability to block the activation of the nuclear kappa factor (NFK), which is the common pathway in inducing iNOS and COX2 in inflammatory processes [30].

In rat striatal neurons, melatonin was able to inhibit calcium influx into neurons and incorporation into the calcium-calmodulin complex, thereby inhibiting nNOS activity and nitric oxide (NO) production, which leads to reduction in the excitatory effect of the

N-methyl_D-aspartate (NMDA) receptor [31]. According to Acuña-Castroviejo et al. [32], melatonin specifically inhibits the NMDA subtype of the excitatory glutamatergic receptors. In addition, melatonin increases the concentration of GABA in the brain and its affinity for its receptor [33] in addition to potentiating inhibitory transmission through gabaergic synapses [34].

Moreover, melatonin promotes neurogenesis in the dentate gyrus of pinealectomized rats [35] and when acting *in vitro*, increases the number of new neurons derived from adult hippocampal precursor neuronal cells by promoting cell survival. This effect was partially dependent on activation of melatonin receptors since it could be blocked by the application of the melatonin receptor antagonist, luzindole [36].

MELATONIN RECEPTORS

Many of the pineal melatonin effects are mediated via MT_1 and MT_2 melatonin receptors, which are found in several CNS structures. Exogenous melatonin also modulates many processes and responses in the CNS via activation of the MT_1 and/or MT_2 melatonin receptors [37]. Besides its intrinsic free-radical scavenging characteristics, melatonin's antioxidant effects may also occur by triggering the MT_1/MT_2-dependent induction of enzymes involved in the removal of ROS and RNS, a pathway(s) that requires much lower ligand concentrations [38].

Melatonin receptors are members of the seven transmembrane G protein-coupled receptors (GPCRs) superfamily [39]. Activation of MT_1 and MT_2 melatonin receptors leads to dissociation of heterotrimeric G proteins, which involves Gα subunit separation from the Gβγ complex and subsequent interaction with several effectors (Figure 1). An example is the regulation of the rhythmic expression of the biological clock gene via the adenylyl cyclase/cAMP/CREB pathway [40]. Phosphorylated cyclic adenosine monophosphate response element binding protein (CREB) can bind to the response element (CRE) site on the circadian rhythm-associated mPER1 promoter and increase the expression of the clock gene [41], and the melatonin-induced signaling cascade may then be able to modulate the circadian rhythm in the suprachiasmatic nucleus (SCN) [42].

In addition to cAMP-dependent signaling, Gi-coupled melatonin receptors may recruit Gs, Gz, and G16 proteins for signal propagation [43]. MT_1 and MT_2 receptor-activation of the phospholipase C pathway, which leads to an increase in inositol triphosphate (IP3) and 1, 2-diacylglycerol (DAG) levels, has been also demonstrated [44]. This activation is found to be linked to the pertussis toxin-insensitive Gq/11 protein. Thus, the MT_1/MT_2 heterodimer also shows augmented activation for these pathways, indicating that there are positive allosteric interactions between these two receptors [45].

Furthermore, melatonin acts as a ligand of the retinoic acid receptor-related orphan receptors (ROR receptors) [46]. The subfamily RZR/ROR consists of three subtypes (α, β, and γ) and four splicing variants of subtype α with RORα1 and α2 receptors involved in immune modulation. It has been suggested that they regulate immune system cell-associated cytokine production after melatonin binds in addition to participating in antioxidant enzyme gene expression [47, 48]. RORα receptors are found in several brain regions such as the hypothalamus, pituitary, hippocampus, thalamus, and neocortex [49].

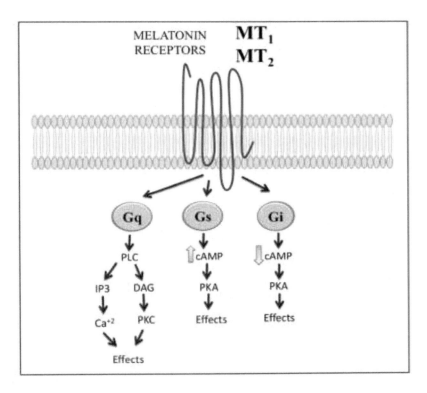

Figure 1. Putative signaling pathways activated by MT₁ and MT₂ melatonin receptors. Multiple signaling pathways for MT1 and MT2 melatonin receptors coupled to Gαi, Gαq/11 and Gs.

Physiological concentrations of nocturnal melatonin are approximately 0.1–0.4 nM. In the picomolar range, melatonin binds with high affinity to the membrane receptors, MT₁ and MT₂, and in a low nanomolar range it binds to the RORα nuclear receptor in addition to calmodulin. Melatonin's effects are dependent on the dose and the type of receptor expression in the cell. The relative distribution of such receptors in different areas of the brain and in different tissues in conjunction with melatonin levels may directly affect their signaling pathways [50].

Recently, the immunoreactive approach complemented with polymerase chain reaction analysis and *in situ* hybridization has become the most popular method for investigating melatonin receptor expression patterns in different areas of the CNS [51–55]. All current studies show melatonin receptors are widely distributed in various regions of the encephalon in several species. Melatonin receptors are found in the cerebral cortex, hippocampus, reticular thalamic nucleus, habenula, hypothalamus, pituitary gland, periaqueductal gray, dorsal raphe nucleus, midbrain, and cerebellum [56]. MT_1 and MT_2 receptor functions have been established using mice with genetic deletions of either the MT_1 or MT_2 melatonin receptors [57, 43] or by using pharmacological approaches through the use of prototype competitive melatonin receptor antagonists such as luzindole, a nonselective ligand with 15- to 25-fold higher affinity for the MT_2 melatonin receptors and 4-phenyl-2-propionamidotetralin (4P-PDOT), a selective MT_2 ligand with 300- to 1,500-fold higher affinity for MT_2 receptors [58, 59]. Luzindole and 4P-PDOT competitively block MT_1 melatonin receptors at concentrations of ≥ 300 nM, and both act as inverse agonists in systems endowed with constitutively active MT_1 receptors [60]. They are considered the gold standard for pharmacological characterization of melatonin receptors.

MT_1

MT_1 receptor is a protein with 350 amino acids and a molecular weight of 39,374 Da [61]. It is characterized as a receptor linked to a pertussis toxin-sensitive guanine nucleotide binding protein (G protein) that mediates the inhibition of adenylyl cyclase in native tissues [62]. Through the use of (125 I) 2-iodomelatonin, the MT_1 receptor has been found to be expressed in various regions of the brain such as the pars tuberal, hypothalamus, and peripheral tissues, including the cardiovascular system, gastrointestinal tract, and immune system [63].

Functionally, melatonin MT_1 receptors are related to reproduction, metabolism, and vasoconstriction, whereas MT_2 receptors are involved in the control of circadian rhythms and the release of retinal dopamine in addition to vasodilation and other types of actions [64]. Activation of MT_1 receptors by melatonin leads to different responses, most of them depending on the decrease in cAMP levels through the inhibitory actions of Gi proteins, and intracellular calcium increases through actions of Gq/11. Melatonin binding to MT_2 receptors lead to a decrease in cAMP and cGMP levels. MT_2 receptors can also use phospholipase C pathways [65].

It is well established that MT_1 melatonin receptor signaling can couple to both Gi and Gq/11 [66]. In addition, these receptors can also regulate ionic fluxes through specific ion channels. Activation of endogenous MT_1 receptors in ovine pars tuberalis cells was shown to increase intracellular calcium via the Gq protein. Furthermore, melatonin acting through MT_1-associated Gi proteins can inhibit calcium influx into neonatal rat pituitary cells and into a pituitary adenoma cell line [65].

In mice with a systematic elimination of MT_1 receptor signaling, the acute inhibitory effects of melatonin on suprachiasmatic nucleus (SCN) activity was shown to be completely abolished while the phase shift responses to melatonin (at physiological concentrations) were normal. In *pars tuberalis* (PT), the MT_1 receptor appears to be essential for rhythmic synthesis of the biological clock gene product, mPER1 [42].

In addition, a recent study has shown that in HEK293 cells transfected with the MT_1 melatonin receptor, recruitment of Gq proteins occurred after exposure to melatonin, and the subsequent dissociation of βγ subunits activated the PI3K/PDK1/PKC pathway, leading to ERK1/2 phosphorylation at later times (≥5min). Melatonin MT_1 receptors can also activate ERK1/2 via the Gs/cAMP/PKA pathway (≤2Min) [65]. Thus, available data indicate that the activation of MT_1 receptors culminate in a variety of tissue-dependent response signaling.

MT_2

The MT_2 receptor was cloned only one year after the MT_1 receptor [67]. The MT_2 receptor consists of 362 amino acids with a molecular weight of 40,188 Da, and it shares 60% homology with the MT_1 receptor [68]. The MT_2 receptor is a Gi-protein coupled receptor that is able to inhibit cAMP and GMPc formation [67]. In the SCN, MT_2 is involved in the induction of circadian rhythm phase changes by producing an increase in PKC activity. In rat hippocampal slices, melatonin inhibits long-term potentiation (LTP) in a concentration-dependent manner (100 nM IC_{50}), and excitability of Schaffer collateral pathways in the CA1 dendritic layer has been shown to occur via MT_2 receptors [69].

Evaluation of the pharmacology and function of melatonin receptors is challenging since their native binding site densities in animal tissues are low or undetectable, particularly for the MT_2 receptors [57]. Melatonin responses, mediated by the activation of the MT_1 and MT_2 melatonin receptors, are dependent on circadian time, duration, and mode of exposure to endogenous or exogenous melatonin in addition to functional receptor sensitivity. Thus, it is important to carefully link each type of melatonin receptor to specific functional responses in target tissues to facilitate the design and development of novel therapeutic agents [64].

Many of the melatonin receptor ligands have been used as therapeutic agents in various types of neurological diseases. The nonselective melatonin agonist, agomelatine (S 20098), has been studied extensively using *in vitro* and *in vivo* models [70]. Agomelatine is a high-affinity MT_1/MT_2 melatonin receptor agonist is also a 5-HT_{2C} serotonin receptor antagonist. It is a novel and efficacious antidepressant, shows a unique pharmacological profile with high efficacy on sleep, circadian rhythm dysfunction, and depression [69, 71].

Another widely studied agent is Ramelteon. It is a novel, high-affinity, nonselective MT_1/MT_2 receptor agonist with little affinity for other GPCRs. It was developed for the treatment of insomnia and circadian sleep disorders [72].

In addition to agonists, many studies use specific antagonists to prove the presence and function of melatonin receptors. 4P-PDOT was the first available selective MT_2 melatonin receptor antagonist, and it has been used in many studies as a pharmacological tool to demonstrate the involvement of the MT_2 receptor type in physiological functions.

The role of melatonin receptors agonists as potent neuroprotective agents in many diseases has been elucidated both in humans and rodents. In fact, progress in understanding the role of melatonin receptors in neurological disorder modulation has led to the discovery of novel classes of melatonin agonists for treating insomnia, mood disorders, epilepsy, Parkinson's disease (PD), and others.

MELATONIN RECEPTORS AND NEUROLOGICAL DISORDERS

Melatonin's neuroprotective effects on neurodegeneration, apoptosis, and ischemia/reperfusion injury are often attributed to its free radical scavenging properties [73]. However, recent evidence suggests that activation of MT_1 and/or MT_2 melatonin receptors are involved in the induction of antioxidant genes such as superoxide dismutase and catalase, acting via receptor-mediated transcriptional signaling [74]. Thus, melatonin receptors may be viable targets for novel agents capable of counteracting oxidative stress components that commonly occur in ischemia.

The relationship of ischemic events to melatonin receptors has been demonstrated by studies showing changes in the expression of these receptors after hypoxic-ischemic injury. Some authors have shown a significant down-regulation of MT_1 receptors following hypoxic-ischemic brain injury in a model of newborn hypoxia-ischemia. They demonstrated that there was a graduated reduction of melatonin receptors in brain of newborn mice at 12, 24, and 48 h following injury. However, after melatonin treatment, the expression of MT_1 receptors increased. They also showed high mortality in MT_1 knockout mice in this model of hypoxic-ischemic brain injury [75].

Participation of the melatonin MT_2 receptor has also been demonstrated in ischemic events. In hippocampal slice cultures deprived of oxygen and glucose, melatonin reduces ROS to nearby basal levels, which is an effect blocked by luzindole, a melatonin receptor antagonist with higher affinity for MT_2 melatonin receptors [74]. Another study showed that melatonin treatment significantly improved survival rates and neural functioning with modestly prolonged life span in stroke mice subject to transient middle cerebral ischemic/reperfusional injury. The protective effects of melatonin were reversed by pretreatment with the MT_2 melatonin receptor antagonists, 4P-PDOT and luzindole, showing that the MT_2 receptor also has an important role in neuroprotection after ischemic events. In addition, melatonin promotes neurogenesis and cell proliferation through an MT_2 receptor-dependent mechanism [76].

MT₂ immunoreactivity and protein levels were increased in the cortical area (CA)1 after ischemic damage [77]. Activation of the MT₂ melatonin receptor in the CA1 after melatonin treatment may be involved in the melatonin-associated neuroprotective effects after ischemic injury. This relationship between ischemic events and melatonin receptors demonstrates that treatment with specific agonists may be a way of reducing the consequences of ischemia.

In this context, melatonin receptor agonists can be used for ischemia/stroke treatment. Agomelatine is an agonist of the melatonergic MT₁/MT₂ receptors and an antagonist of the serotonergic 5-HT receptors. Its actions mimic antioxidative and anti-inflammatory melatonin effects. Agomelatine treatment significantly decreased apoptosis with subsequent decreases in Bax and cleaved caspase-3 and an increase in Bcl-X$_L$ along with a decrease in apoptotic neuronal cells in cerebral ischemic/reperfusion injury rats. Moreover, agomelatine was also found to significantly increase the expression of HO-1, antioxidative enzymes, and the activity of Nrf2-mediated superoxide dismutase (SOD) [78].

Ramelteon, a melatonin receptor agonist, is also able to treat delirium in elderly stroke patients with insomnia. Ohta et al. [79] used Ramelton and other drugs to treat elderly patients with delirium and insomnia after acute stroke. They showed that all patients treated with Ramelteon had a significant improvement in their conditions within a week. No patient experienced oversedation, neurologic deterioration, or any other adverse Ramelton-associated effects. This study indicates that melatonin receptor agonists may be effective for delirium treatment in elderly patients with acute stroke.

In conjunction with impaired production of melatonin, some neurodegenerative diseases also present altered expression of melatonin receptors. Several studies have shown that Parkinson's disease (PD) patients exhibit changes in melatonin production and in melatoninergic receptors MT₁ and MT₂ expressions in the amygdala and SCN [51, 80–82]. This reduction in endogenous melatonin production in PD patients along with the discovery of melatonin's antioxidant activity has led to increasing interest in using meltonin for neuroprotection in PD.

Melatonin and its receptors (MT₁ and MT₂), are impaired in Alzheimer's disease (AD) with severe effects on neuropathological and clinical symptoms. A study performed in the human hippocampus of AD patients showed that MT₂ melatonin receptor is present in both control and AD patients. In control patients, MT₂ was localized in pyramidal neurons of the hippocampal subfields CA1–4 and in some granular neurons of the *stratum granulosum*. The overall intensity of the MT₂ staining had distinctly decreased in AD cases showing that MT₂ expression in the human hippocampus is heavily impaired in AD patients [83]. In addition, MT2 melatonin receptors are present in human retinas but are decreased in retinas of AD patients [84]. Other studies showed a decrease in MT₁ and MT₂ expressions in the cortex and pineal gland of AD patients, which paralleled degenerative tissue changes [54]. AD patients' brains tissues also showed a decrease in

MT_1 expression in the SCN [85]. Moreover, β-amyloid directly binds to MT_1 and MT_2 receptors, which results in a decrease in MT_1 binding sites and activation of downstream pathways triggered by both melatonin receptors subtypes [86].

Age-related dementia such as AD causes both diminished memory and cognition. In this context, understanding the mechanisms of melatonin receptors in the memory seems to be a way to search for therapeutic agents capable of reducing the consequences of this disease. It is known that there is a reduced baseline level of long-term potentiation (LTP) in hippocampi from MT_2 knockout C3H/HeN mice compared to wild-type mice [87]. Furthermore, melatonin inhibits LTP in the CA1 dendritic layer of the Schaffer collaterals of mouse hippocampal brain slices from MT_2 knockout but not MT_1 knockout mice revealing the requirement of the MT_2 receptor for this response [88]. Indeed, the melatonin receptor antagonists, luzindole and K-185, reversed long-term memory formation suppression in zebrafish [89]. These data show that melatonin and its receptors have a role in memory formation, indicating that melatonin receptor agonists may also be important agents in the treatment of this disease.

Melatonin (via melatonin receptors) provides neuroprotection in genetic models of Huntington's disease (HD). In a study using the–R6/2 mouse model, melatonin slows disease progression and blocks the mitochondrial death pathway. Lower levels of MT_1 (but not the melatonin MT_2 receptor) are seen in the brains of R6/2 mice; this is a deficit that can be partially reversed by treatment with melatonin. Moreover, it has been demonstrated that melatonin inhibits mutant huntington (htt)-mediated toxicity caspase activation and preserves the MT_1 receptor. It has also been demonstrated that MT_1 receptor knockdown by informational ribonucleic acid makes cells more vulnerable to cell death, enhances neuronal vulnerability, and potentially accelerates the neurodegenerative process. Conversely, MT_1 receptor overexpression increases resistance to cell death. In addition, melatonin inhibits cell death in primary cerebrocortical and primary striatal neuronal cultures. This inhibition is reversed by luzindole showing that the MT_2 receptor participates in melatonin-mediated neuroprotection. These findings imply that the development of MT_1/MT_2 selective receptor agonists may lead to neuroprotective therapy capable of treating patients with Huntington's disease (HD) [90].

Via a number of mechanisms, traumatic brain injury (TBI) is a devastating and costly acquired condition that affects individuals of all ages, races, and locations. The effects of TBI on melatonin receptors (MT_1 and MT_2) were studied in an experimental model of TBI. This study showed that melatonin receptor levels were reduced in a brain region- and time point- dependent manner. Both MT_1 and MT_2 were reduced in the frontal cortex at 24h and in the hippocampus at both 6 and 24h after injury [91].

Clinical and experimental findings show that melatonin may be used as an adjuvant to the treatment for epilepsy-related complications by alleviating epilepsy-associated sleep disturbances, circadian alterations, and attenuates seizures alone or in combination with antiepileptic drugs. In addition, it has been observed that there is a circadian component to seizures, which cause changes in circadian system and in melatonin production. Nevertheless, the dynamic changes of the melatonin membrane receptors (MT$_1$ and MT$_2$) in the natural course of temporal lobe epilepsy was studied. MT$_1$ and MT$_2$ melatonin receptor mRNA expression levels were increased in rat hippocampi a few hours after epilepsy induction. They found a reduction in both melatonin receptors in the chronic phase of model epilepsy showing that melatonin receptors MT$_1$ and MT$_2$ are changed by epilepsy [92]. Further evidence shows that valproate, an antiepileptic drug used in epilepsy, was able to induce upregulation of both MT$_1$ and MT$_2$ receptors in rat brain [33], suggesting that the melatonergic system is an epigenetic target for this neuropsychotropic agent and other epi-drugs.

In fact, recent studies about the novel antidepressant agomelatine, which is a mixed MT$_1$ and MT$_2$ melatonin receptor agonist and 5HT2C serotonin receptor antagonist, has anticonvulsant and neuroprotective actions in an acute penetylenetetrazol- and pilocarpine-induced seizure model in rats, suggest that it may have the potential to function against epileptogenetic- and epilepsy-induced memory impairment [93]. In order to evaluate the effects on epileptogenesis after *status epilepticus* (SE) induced by kainate acid (KA), chronic agomelatine treatment was used by Tchekalarova et al. [94]. Agomelatine treatment started one hour after SE and continued up to 10 weeks in Wistar rats significantly decreased the latency for spontaneous motor seizures onset and increased seizure frequency during the 2nd and the 3rd weeks of treatment. Agomelatine exacerbated KA-induced hyperlocomotion and impulsive behavior, and it was unable to prevent spatial memory impairment in epileptic rats. In this study, agomelatine also induced neuroprotection in the dorsal hippocampus, specifically in the CA1, septal CA2, and partially in the CA3 region, the hilus of the dentate gyrus, piriform cortex, and septo-temporal and temporal basolateral amygdala. These findings suggest that agomelatine protects against neuronal damage without preventing epileptogenesis in the KA model of temporal lobe epilepsy [94].

In contrast, Ramelteon, a synthetic and selective melatonin receptor agonist, might possess anticonvulsant and/or antiepileptogenic properties. Ramelton's effects were evaluated in two epileptic animal models. In a rat rapid kindling model, Ramelteon reversed kindling-induced hippocampal excitability although it did not modify the baseline after discharge properties and progression and establishment of the kindled state in the rapid kindling model. However, in a spontaneously epileptic Kcna1-null mouse model, Ramelteon significantly attenuated seizure periodicity and frequency and improved circadian rest-activity rhythms compared with control animals [95]. These findings show that Ramelteon possesses anticonvulsant properties in a chronic epilepsy model and provide further support for melatonin receptors as potential novel targets for anticonvulsant drug development.

Current treatments to major depressive disorder primarily include tricyclic antidepressants (TCAs) and selective serotonin reuptake inhibitors (SSRIs). These antidepressants, however, lead to unwanted side effects such as sexual dysfunction, weight gain, and a cluster of cognitive, autonomic, and motor signs. They also do not treat sleep disturbances or circadian and/or seasonal rhythm dysfunctions associated with depressive disorders (sleep patterns and circadian and/or seasonal rhythm entrainment [96]). Thus, novel medications and co-adjuvant therapies that reduce the side effects of conventional antidepressants is highly needed.

Melatonin receptors (MT_1 and MT_2) are important targets for the development of novel antidepressants. Anti-depressant effects of melatonin and its agonists on learned helplessness rodent models indicate that melatonin receptors are involved in depression. Melatonin-mediated antidepressant-like effects in the forced swim test were blocked by luzindole, suggesting the potential involvement of MT_1 and MT_2 receptors [97]. Other studies showed that mice with MT_1 receptor deletion have an increased depressive-like behaviors in the forced swim test [98, 99]. The role of the MT_1 melatonin receptor in depression has also been shown in patients with upregulation in immunoreactivity of these receptors, suggesting this receptor type is a potential target for treating some depression symptoms [100]. Thus, apparently MT_1 melatonin receptors are the agonist targets for the treatment of clinical manifestations of depressive disorders.

Agomelatine is a novel antidepressant, which acts through synergistic actions at the melatonin receptors and 5-HT2C receptor. Clinical studies have demonstrated that this agonist is effective and also better tolerated than SSRIs and serotonin/norepinephrine reuptake inhibitors [101]. In clinical studies, agomelatine is as effective and better tolerated than selective serotonin reuptake inhibitors (SSRIs); this group includes paroxetine, fluoxetine, and sertraline and serotonin/norepinephrine reuptake inhibitors such as Venlafaxine [102] for classic treatment of the depression. Agomelatine has shown great promise for the treatment of depression because it not only relieves depressed moods, anxiety, neurochemical imbalance, and neuronal atrophy (Soumier et al., 2009) but also improves disturbed sleep patterns and circadian/seasonal rhythm entrainment [103].

The most well-known and popular function of melatonin and its agonists is related to sleep disturbances and in jet-leg dysfunction. Insomnia or sleep-wake disorder is defined as difficulty in initiating and/or maintaining sleep or as non-passive sleep and is usually associated with daytime disability or suffering. Insomnia is related to other comorbid disorders, particularly mood and circadian rhythm disorders. Current treatments for insomnia include benzodiazepine- and non-benzodiazepine-related medications, but they have considerable side effects resulting in impaired cognitive and psychomotor skills [49]. Melatonin and synthetic melatonin agonists generally do not have the common side effects often seen with sleep medications (such as impairment of learning, memory, or motor function). For this reason, molecules with improved safety profiles and melatonin receptor agonists (such as Ramelteon, tasimelteon, and agomelatine) have been widely used in the treatment of sleep disturbances [101, 104, 105].

Activation of melatonin receptors in the SCN plays a key role in coordinating the phase and amplitude of circadian rhythms throughout the body [65]. The phase shift of circadian rhythms caused by melatonin is mediated by MT_1 melatonin receptor activation as demonstrated by mice with a genetic deletion of the MT_1 receptor in a circadian reentrant model [64]. Understanding of this physiology is important for creating strategies for the use of melatonin receptor agonists and promoting efficient treatment for sleep induction.

Ramelteon, the high affinity MT_1 and MT_2 melatonin receptor agonist widely discussed in this chapter, promotes sleep in several mammalian species, including humans, without learning, memory, or impairment of motor functions or induction of reward [106]. In fact, in individuals with primary chronic insomnia, studies show that Ramelteon shows a significant reduction in latency for persistent sleep and an increase in total sleep time with no apparent residual effects on the following day [107], which demonstrates that this MT_1 and MT_2 agonist has excellent prospects for being a coadjuvant in sleep disturbance treatment.

CONCLUSION

Besides the vital role of melatonin in the GPCR-associated signaling pathways, its small molecular size, high lipophilicity, excellent blood–brain barrier permeability, and minimal side effects in humans makes it an attractive option for neuroprotection. Discovery of MT_1 or MT_2 melatonin receptor–selective drugs may improve efficacy compared to nonselective ligands and lead to the discovery of new therapeutic targets. Additionally, progress in understanding the role of melatonin receptors in the modulation of neurological disorders has led to the discovery of novel classes of melatonin agonists for treating hypoxic-ischemic events, stroke, PD, AD, HD, epilepsy, major depressive disorder, insomnia, and others. The main advantage of treatment with these agents is the incredible efficiency, zero toxicity, and reduction of the side effects seen with conventional treatments. In this review we showed the participation of MT_1 and MT_2 receptors in neurological disorders and the future therapeutic prospects of melatonin receptors agonists/antagonists on various central nervous system diseases.

REFERENCES

[1] Lerner, A. B., Case, J. D., Mori, W., & Wright, M. R. (1959). Melatonin in peripheral nerve. *Nature*, *183*(4678), 1821-1821.

[2] Reiter, R. J. (1991). Pineal melatonin: cell biology of its synthesis and of its physiological interactions. *Endocrine reviews*, *12*(2), 151-180.

[3] Maronde, E., & Stehle, J. H. (2007). The mammalian pineal gland: known facts, unknown facets. *Trends in Endocrinology & Metabolism*, *18*(4), 142-149.

[4] Arendt, J., Deacon, S., English, J., Hampton, S., & Morgan, L. (1995). Melatonin and adjustment to phase shift. *Journal of sleep research*, *4*(s2), 74-79.

[5] Reiter, R. J. (1993). The melatonin rhythm: both a clock and a calendar. *Experientia*, *49*(8), 654-664.

[6] Pandi-Perumal, S. R., Trakht, I., Srinivasan, V., Spence, D. W., Maestroni, G. J., Zisapel, N., & Cardinali, D. P. (2008). Physiological effects of melatonin: role of melatonin receptors and signal transduction pathways. *Progress in neurobiology*, *85*(3), 335-353.

[7] Slominski, A., Tobin, D. J., Zmijewski, M. A., Wortsman, J., & Paus, R. (2008). Melatonin in the skin: synthesis, metabolism and functions. *Trends in Endocrinology & Metabolism*, *19*(1), 17-24.

[8] Slominski, R. M., Reiter, R. J., Schlabritz-Loutsevitch, N., Ostrom, R. S., & Slominski, A. T. (2012). Melatonin membrane receptors in peripheral tissues: distribution and functions. *Molecular and cellular endocrinology*, *351*(2), 152-166.

[9] Kennaway, D. J., & Rowe, S. A. (2000). Effect of stimulation of endogenous melatonin secretion during constant light exposure on 6-sulphatoxymelatonin rhythmicity in rats. *Journal of pineal research*, *28*(1), 16-25.

[10] Banach, M., Gurdziel, E., Jędrych, M., & Borowicz, K. K. (2011). Melatonin in experimental seizures and epilepsy. *Pharmacological Reports*, *63*(1), 1-11.

[11] Ayar, A., Martin, D. J., Ozcan, M., & Kelestimur, H. (2001). Melatonin inhibits high voltage activated calcium currents in cultured rat dorsal root ganglion neurones. *Neuroscience letters*, *313*(1), 73-77.

[12] Baydas, G., Kutlu, S., Naziroglu, M., Canpolat, S., Sandal, S., Ozcan, M., & Kelestimur, H. (2003). Inhibitory effects of melatonin on neural lipid peroxidation induced by intracerebroventricularly administered homocysteine. *Journal of pineal research*, *34*(1), 36-39.

[13] Hardeland, R., & Poeggeler, B. (2008). Melatonin beyond its classical functions.

[14] Bettahi, I., Pozo, D., Osuna, C., Reiter, R. J., Acuña-Castroviejo, D., & Guerrero, J. M. (1996). Melatonin reduces nitric oxide synthase activity in rat hypothalamus. *Journal of pineal research*, *20*(4), 205-210.

[15] Pozo, D., Reiter, R. J., Calvo, J. R., & Guerrero, J. M. (1997). Inhibition of cerebellar nitric oxide synthase and cyclic GMP production by melatonin via complex formation with calmodulin. *Journal of cellular biochemistry*, *65*(3), 430-442.

[16] Acuña-Castroviejo, D., Escames, G., López, L. C., Hitos, A. B., & León, J. (2005). Melatonin and nitric oxide. *Endocrine*, *27*(2), 159-168.

[17] Reiter, R. J., Tan, D. X., Manchester, L. C., Terron, M. P., Flores, L. J., & Koppisepi, S. (2007). Medical implications of melatonin: receptor-mediated and receptor-independent actions. *Advances in Medical Sciences (De Gruyter Open)*, *52*.

[18] Hardeland, R. (2005). Atioxidative protection by melatonin. *Endocrine*, *27*(2), 119-130.

[19] Kabuto, H., Yokoi, I., & Ogawa, N. (1998). Melatonin Inhibits Iron-Induced Epileptic Discharges in Rats by Suppressing Peroxidation. *Epilepsia*, *39*(3), 237-243.

[20] Cuzzocrea, S., & Reiter, R. J. (2001). Pharmacological action of melatonin in shock, inflammation and ischemia/reperfusion injury. *European journal of pharmacology*, *426*(1), 1-10.

[21] Antolín, I. S. A. A. C., Rodríguez, C. A. R. M. E. N., Saínz, R. M., Mayo, J. C., Uría, H. I. G. I. N. I. O., Kotler, M. L., ... & Menendez-Pelaez, A. (1996). Neurohormone melatonin prevents cell damage: effect on gene expression for antioxidant enzymes. *The FASEB Journal*, *10*(8), 882-890.

[22] Reiter, R. J., Tan, D. X., Cabrera, J., & D'Arpa, D. (1999). Melatonin and tryptophan derivatives as free radical scavengers and antioxidants. In *Tryptophan, Serotonin, and Melatonin* (pp. 379-387). Springer US.

[23] Poeggeler, B., Reiter, R. J., Tan, D. X., Chen, L. D., & Manchester, L. C. (1993). Melatonin: A potent, endogenous hydroxyl radical scavenger. *J. Pineal Res*, *14*, 57-60.

[24] Reiter, R. J., Acuña-Castroviejo, D. A. R. I. O., Tan, D. X., & Burkhardt, S. (2001). Free radical-mediated molecular damage. *Annals of the New York Academy of Sciences*, *939*(1), 200-215.

[25] Reiter, R. J., Tan, D. X., Mayo, J. C., Sainz, R. M., Leon, J., & Czarnocki, Z. (2003). Melatonin as an antioxidant: biochemical mechanisms and pathophysiological implications in humans. *Acta Biochimica Polonica-English Edition-*, *50*(4), 1129-1146.

[26] Hardeland, R., Cardinali, D. P., Srinivasan, V., Spence, D. W., Brown, G. M., & Pandi-Perumal, S. R. (2011). Melatonin—a pleiotropic, orchestrating regulator molecule. *Progress in neurobiology*, *93*(3), 350-384.

[27] Giusti, P., Lipartiti, M., Franceschini, D. A. V. I. D. E., Schiavo, N., Floreani, M., & Manev, H. (1996). Neuroprotection by melatonin from kainate-induced excitotoxicity in rats. *The FASEB journal*, *10*(8), 891-896.

[28] Floreani, M., Skaper, S. D., Facci, L. A. U. R. A., Lipartiti, M. A. R. I. A., & Giusti, P. I. E. T. R. O. (1997). Melatonin maintains glutathione homeostasis in kainic acid-exposed rat brain tissues. *The FASEB journal*, *11*(14), 1309-1315.

[29] Lotufo, C. M., Lopes, C., Dubocovich, M. L., Farsky, S. H., & Markus, R. P. (2001). Melatonin and N-acetylserotonin inhibit leukocyte rolling and adhesion to rat microcirculation. *European journal of pharmacology*, *430*(2), 351-357.

[30] Mohan, N., Sadeghi, K., Reiter, R. J., & Meltz, M. L. (1995). The neurohormone melatonin inhibits cytokine, mitogen and ionizing radiation induced NF-kappa B. *Biochemistry and molecular biology international*, *37*(6), 1063-1070.

[31] León, J., Macías, M., Escames, G., Camacho, E., Khaldy, H., Martín, M., ... & Acuña-Castroviejo, D. (2000). Structure-related inhibition of calmodulin-dependent neuronal nitric-oxide synthase activity by melatonin and synthetic kynurenines. *Molecular pharmacology*, *58*(5), 967-975.

[32] Acuña-Castroviejo, D., Escames, G., Macks, M., Hoyos, A. M., Carballo, A. M., Arauzo, M., ... & Vives, F. (1995). Minireview: cell protective role of melatonin in the brain. *Journal of pineal research*, *19*(2), 57-63.

[33] Niles, L. P., Sathiyapalan, A., Bahna, S., Kang, N. H., & Pan, Y. (2012). Valproic acid up-regulates melatonin MT1 and MT2 receptors and neurotrophic factors CDNF and MANF in the rat brain. *International Journal of Neuro psycho pharmacology*, *15*(9), 1343-1350.

[34] Stewart, L. S., & Leung, L. S. (2005). Hippocampal melatonin receptors modulate seizure threshold. *Epilepsia*, *46*(4), 473-480.

[35] Rennie, K., De Butte, M., & Pappas, B. A. (2009). Melatonin promotes neurogenesis in dentate gyrus in the pinealectomized rat. *Journal of pineal research*, *47*(4), 313-317.

[36] Ramírez-Rodríguez, G., Klempin, F., Babu, H., Benítez-King, G., & Kempermann, G. (2009). Melatonin modulates cell survival of new neurons in the hippocampus of adult mice. *Neuropsycho-pharmacology*, *34*(9), 2180-2191.

[37] Liu, J., Clough, S. J., Hutchinson, A. J., Adamah-Biassi, E. B., Popovska-Gorevski, M., & Dubocovich, M. L. (2016). MT1 and MT2 melatonin receptors: a therapeutic perspective. *Annual review of pharmacology and toxicology*, *56*, 361-383.

[38] Carrillo-Vico, A., Lardone, P. J., Álvarez-Sánchez, N., Rodríguez-Rodríguez, A., & Guerrero, J. M. (2013). Melatonin: buffering the immune system. *International Journal of Molecular Sciences*, *14*(4), 8638-8683.

[39] Witt-Enderby, P. A., Bennett, J., Jarzynka, M. J., Firestine, S., & Melan, M. A. (2003). Melatonin receptors and their regulation: biochemical and structural mechanisms. *Life sciences*, *72*(20), 2183-2198.

[40] Masana, M. I., Witt-Enderby, P. A., & Dubocovich, M. L. (2003). Melatonin differentially modulates the expression and function of the hMT 1 and hMT 2 melatonin receptors upon prolonged withdrawal. *Biochemical pharmacology*, *65*(5), 731-739.

[41] Travnickova-Bendova, Z., Cermakian, N., Reppert, S. M., & Sassone-Corsi, P. (2002). Bimodal regulation of mPeriod promoters by CREB-dependent signaling and CLOCK/BMAL1 activity. *Proceedings of the National Academy of Sciences*, *99*(11), 7728-7733.

[42] Jin, X., Von Gall, C., Pieschl, R. L., Gribkoff, V. K., Stehle, J. H., Reppert, S. M., & Weaver, D. R. (2003). Targeted disruption of the mouse Mel1b melatonin receptor. *Molecular and cellular biology*, *23*(3), 1054-1060.

[43] Chan, A. S., Lai, F. P., Lo, R. K., Voyno-Yasenetskaya, T. A., Stanbridge, E. J., & Wong, Y. H. (2002). Melatonin mt1 and MT2 receptors stimulate c-Jun N-terminal kinase via pertussis toxin-sensitive and-insensitive G proteins. *Cellular signalling*, *14*(3), 249-257.

[44] Sharkey, J., & Olcese, J. (2007). Transcriptional inhibition of oxytocin receptor expression in human myometrial cells by melatonin involves protein kinase C signaling. *The Journal of Clinical Endocrinology & Metabolism*, *92*(10), 4015-4019.

[45] Jockers, R., Maurice, P., Boutin, J. A., & Delagrange, P. (2008). Melatonin receptors, heterodimerization, signal transduction and binding sites: what's new?. *British journal of pharmacology*, *154*(6), 1182-1195.

[47] Skwarlo-Sonta, K., Majewski, P., Markowska, M., Oblap, R., & Olszanska, B. (2003). Bidirectional communication between the pineal gland and the immune system. *Canadian journal of physiology and pharmacology*, *81*(4), 342-349.

[48] Smirnov, A. N. (2001). Nuclear melatonin receptors. *Biochemistry (Moscow)*, *66*(1), 19-26.

[49] Laudon, M., & Frydman-Marom, A. (2014). Therapeutic effects of melatonin receptor agonists on sleep and comorbid disorders. *International journal of molecular sciences*, *15*(9), 15924-15950.

[50] Acuna-Castroviejo, D., Escames, G., Rodriguez, M. I., & Lopez, L. C. (2007). Melatonin role in the mitochondrial function. *Front Biosci*, *12*, 947-963.

[51] Adi, N., Mash, D. C., Ali, Y., Singer, C., Shehadeh, L., & Papapetropoulos, S. (2010). Melatonin MT1 and MT2 receptor expression in Parkinson's disease. *Medical science monitor*, *16*(2), BR61-BR67.

[52] Klosen, P., Bienvenu, C., Demarteau, O., Dardente, H., Guerrero, H., Pévet, P., & Masson-Pévet, M. (2002). The mt1 melatonin receptor and RORβ receptor are co-localized in specific TSH-immunoreactive cells in the pars tuberalis of the rat pituitary. *Journal of Histochemistry & Cytochemistry*, *50*(12), 1647-1657.

[53] Musshoff, U., Riewenherm, D., Berger, E., Fauteck, J. D., & Speckmann, E. J. (2002). Melatonin receptors in rat hippocampus: molecular and functional investigations. *Hippocampus*, *12*(2), 165-173.

[54] Brunner, P., Sozer-Topcular, N., Jockers, R., Ravid, R., Angeloni, D., & Fraschini, F. (2006). Pineal and cortical melatonin receptors MT1 and MT2 are decreased in Alzheimer's disease. *European journal of histochemistry: EJH, 50*(4), 311.

[55] Waly, N. E., & Hallworth, R. (2015). Circadian pattern of melatonin MT1 and MT2 receptor localization in the rat suprachiasmatic nucleus. *Journal of circadian rhythms, 13*.

[56] Ng, K. Y., Leong, M. K., Liang, H., & Paxinos, G. (2017). Melatonin receptors: distribution in mammalian brain and their respective putative functions. *Brain Structure and Function*, 1-19.

[57] Liu, C., Weaver, D. R., Jin, X., Shearman, L. P., Pieschl, R. L., Gribkoff, V. K., & Reppert, S. M. (1997). Molecular dissection of two distinct actions of melatonin on the suprachiasmatic circadian clock. *Neuron, 19*(1), 91-102.

[58] Dubocovich, M. L. (1988). Luzindole (N-0774): a novel melatonin receptor antagonist. *Journal of Pharmacology and Experimental Therapeutics, 246*(3), 902-910.

[59] Dubocovich, M. L., Masana, M. I., Iacob, S., & Sauri, D. M. (1997). Melatonin receptor antagonists that differentiate between the human Mel1a and Mel1b recombinant subtypes are used to assess the pharmacological profile of the rabbit retina ML1 presynaptic heteroreceptor. *Naunyn-Schmiedeberg's archives of pharmacology, 355*(3), 365-375.

[60] Browning, C., Beresford, I., Fraser, N., & Giles, H. (2000). Pharmacological characterization of human recombinant melatonin mt1 and MT2 receptors. *British journal of pharmacology, 129*(5), 877-886.

[61] Song, Y., Chan, C. W., Brown, G. M., Pang, S. F., & Silverman, M. (1997). Studies of the renal action of melatonin: evidence that the effects are mediated by 37 kDa receptors of the Mel1a subtype localized primarily to the basolateral membrane of the proximal tubule. *The FASEB journal, 11*(1), 93-100.

[62] Carlson, L. L., Weaver, D. R., & Reppert, S. M. (1989). Melatonin signal transduction in hamster brain: inhibition of adenylyl cyclase by a pertussis toxin-sensitive G protein. *Endocrinology, 125*(5), 2670-2676.

[63] Wu, Y. H., Ursinus, J., Zhou, J. N., Scheer, F. A., Ai-Min, B., Jockers, R., ... & Swaab, D. F. (2013). Alterations of melatonin receptors MT1 and MT2 in the hypothalamic suprachiasmatic nucleus during depression. *Journal of affective disorders, 148*(2), 357-367.

[64] Dubocovich, M. L., & Markowska, M. (2005). Functional MT1 and MT2 melatonin receptors in mammals. *Endocrine, 27*(2), 101-110.

[65] Dubocovich, M. L. (2007). Melatonin receptors: role on sleep and circadian rhythm regulation. *Sleep medicine, 8*, 34-42.

[66] Dubocovich, M. L., Hudson, R. L., Sumaya, I. C., Masana, M. I., & Manna, E. (2005). Effect of MT1 melatonin receptor deletion on melatonin-mediated phase shift of circadian rhythms in the C57BL/6 mouse. *Journal of pineal research*, *39*(2), 113-120.

[67] Reppert, S. M., Godson, C., Mahle, C. D., Weaver, D. R., Slaugenhaupt, S. A., & Gusella, J. F. (1995). Molecular characterization of a second melatonin receptor expressed in human retina and brain: the Mel1b melatonin receptor. *Proceedings of the National Academy of Sciences*, *92*(19), 8734-8738.

[68] Reppert, S. M., Weaver, D. R., Ebisawa, T., Mahle, C. D., & Kolakowski, L. F. (1996). Cloning of a melatonin-related receptor from human pituitary. *FEBS letters*, *386*(2-3), 219-224.

[69] Dubocovich, M. L., Delagrange, P., Krause, D. N., Sugden, D., Cardinali, D. P., & Olcese, J. (2010). International Union of Basic and Clinical Pharmacology. LXXV. Nomenclature, classification, and pharmacology of G protein-coupled melatonin receptors. *Pharmacological reviews*, *62*(3), 343-380.

[70] Racagni, G., Riva, M. A., & Popoli, M. (2007). The interaction between the internal clock and antidepressant efficacy. *International clinical psycho pharmacology*, *22*, S9-S14.

[71] Olié, J. P., & Kasper, S. (2007). Efficacy of agomelatine, a MT 1/MT 2 receptor agonist with 5-HT 2C antagonistic properties, in major depressive disorder. *The The International Journal of Neuropsychopharmacology*, *10*(5), 661-673.

[72] Miyamoto, M. (2009). Pharmacology of ramelteon, a selective MT1/MT2 receptor agonist: a novel therapeutic drug for sleep disorders. *CNS neuroscience & therapeutics*, *15*(1), 32-51.

[73] Tan, D. X., Manchester, L. C., Sainz, R. M., Mayo, J. C., León, J., & Reiter, R. J. (2005). Physiological ischemia/reperfusion phenomena and their relation to endogenous melatonin production. *Endocrine*, *27*(2), 149-157.

[74] Parada, E., Buendia, I., León, R., Negredo, P., Romero, A., Cuadrado, A., ... & Egea, J. (2014). Neuroprotective effect of melatonin against ischemia is partially mediated by alpha-7 nicotinic receptor modulation and HO-1 overexpression. *Journal of pineal research*, *56*(2), 204-212.

[75] Sinha, B., Wu, Q., Li, W., Tu, Y., Sirianni, A. C., Chen, Y., ... & Reiter, R. J. (2017). Protection of melatonin in experimental models of newborn hypoxic-ischemic brain injury through MT1 receptor. *Journal of pineal research*.

[76] Chern, C. M., Liao, J. F., Wang, Y. H., & Shen, Y. C. (2012). Melatonin ameliorates neural function by promoting endogenous neurogenesis through the MT2 melatonin receptor in ischemic-stroke mice. *Free Radical Biology and Medicine*, *52*(9), 1634-1647.

[77] Lee, C. H., Yoo, K. Y., Choi, J. H., Park, O. K., Hwang, I. K., Kwon, Y. G., ... & Won, M. H. (2010). Melatonin's protective action against ischemic neuronal damage is associated with up-regulation of the MT2 melatonin receptor. *Journal of neuroscience research*, *88*(12), 2630-2640.

[78] Chumboatong, W., Thummayot, S., Govitrapong, P., Tocharus, C., Jittiwat, J., & Tocharus, J. (2017). Neuroprotection of agomelatine against cerebral ischemia/reperfusion injury through an antiapoptotic pathway in rat. *Neurochemistry international*, *102*, 114-122.

[79] Ohta, T., Murao, K., Miyake, K., & Takemoto, K. (2013). Melatonin receptor agonists for treating delirium in elderly patients with acute stroke. *Journal of Stroke and Cerebrovascular Diseases*, *22*(7), 1107-1110.

[80] Bordet, R., Devos, D., Brique, S., Touitou, Y., Guieu, J. D., Libersa, C., & Destee, A. (2003). Study of circadian melatonin secretion pattern at different stages of Parkinson's disease. *Clinical neuropharmacology*, *26*(2), 65-72.

[81] Videnovic, A., Noble, C., Reid, K. J., Peng, J., Turek, F. W., Marconi, A., ... & Zee, P. C. (2014). Circadian melatonin rhythm and excessive daytime sleepiness in Parkinson disease. *JAMA neurology*, *71*(4), 463-469.

[82] Mack, J. M., Schamne, M. G., Sampaio, T. B., Pértile, R. A. N., Fernandes, P. A. C. M., Markus, R. P., & Prediger, R. D. (2016). Melatoninergic system in Parkinson's disease: from neuroprotection to the management of motor and nonmotor symptoms. *Oxidative medicine and cellular longevity*, 2016.

[83] Savaskan, E., Ayoub, M. A., Ravid, R., Angeloni, D., Fraschini, F., Meier, F., ... & Jockers, R. (2005). Reduced hippocampal MT2 melatonin receptor expression in Alzheimer's disease. *Journal of pineal research*, *38*(1), 10-16.

[84] Savaskan, E., Olivieri, G., Meier, F., Brydon, L., Jockers, R., Ravid, R., ... & Müller-Spahn, F. (2002). Increased melatonin 1a-receptor immunoreactivity in the hippocampus of Alzheimer's disease patients. *Journal of pineal research*, *32*(1), 59-62.

[85] Wu, Y. H., Zhou, J. N., Van Heerikhuize, J., Jockers, R., & Swaab, D. F. (2007). Decreased MT1 melatonin receptor expression in the suprachiasmatic nucleus in aging and Alzheimer's disease. *Neurobiology of aging*, *28*(8), 1239-1247.

[86] Cecon, E., Chen, M., Marçola, M., Fernandes, P. A., Jockers, R., & Markus, R. P. (2015). Amyloid β peptide directly impairs pineal gland melatonin synthesis and melatonin receptor signaling through the ERK pathway. *The FASEB Journal*, *29*(6), 2566-2582.

[87] Larson, J., Jessen, R. E., Uz, T., Arslan, A. D., Kurtuncu, M., Imbesi, M., & Manev, H. (2006). Impaired hippocampal long-term potentiation in melatonin MT 2 receptor-deficient mice. *Neuroscience letters*, *393*(1), 23-26.

[88] Wang, L. M., Suthana, N. A., Chaudhury, D., Weaver, D. R., & Colwell, C. S. (2005). Melatonin inhibits hippocampal long-term potentiation. *European Journal of Neuroscience*, *22*(9), 2231-2237.

[89] Rawashdeh, O., de Borsetti, N. H., Roman, G., & Cahill, G. M. (2007). Melatonin suppresses nighttime memory formation in zebrafish. *Science*, *318*(5853), 1144-1146.

[90] Wang, X., Sirianni, A., Pei, Z., Cormier, K., Smith, K., Jiang, J., ... & Kim, J. E. (2011). The melatonin MT1 receptor axis modulates mutant Huntingtin-mediated toxicity. *Journal of Neuroscience*, *31*(41), 14496-14507.

[91] Osier, N. D., Pham, L., Pugh, B. J., Puccio, A., Ren, D., Conley, Y. P., ... & Dixon, C. E. (2017). Brain injury results in lower levels of melatonin receptors subtypes MT1 and MT2. *Neuroscience Letters*, *650*, 18-24.

[92] de Alencar Rocha, A. K. A., de Lima, E., Amaral, F., Peres, R., Cipolla-Neto, J., & Amado, D. (2017). Altered MT1 and MT2 melatonin receptors expression in the hippocampus of pilocarpine-induced epileptic rats. *Epilepsy & Behavior*, *71*, 23-34.

[93] Aguiar, C. C. T., Almeida, A. B., Araújo, P. V. P., Vasconcelos, G. S., Chaves, E. M. C., do Vale, O. C., ... & Vasconcelos, S. M. M. (2012). Anticonvulsant effects of agomelatine in mice. *Epilepsy & Behavior*, *24*(3), 324-328.

[94] Tchekalarova, J., Atanasova, D., Nenchovska, Z., Atanasova, M., Kortenska, L., Gesheva, R., & Lazarov, N. (2017). Agomelatine protects against neuronal damage without preventing epileptogenesis in the kainate model of temporal lobe epilepsy. *Neurobiology of Disease*, *104*, 1-14.

[95] Fenoglio-Simeone, K., Mazarati, A., Sefidvash-Hockley, S., Shin, D., Wilke, J., Milligan, H., ... & Maganti, R. (2009). Anticonvulsant effects of the selective melatonin receptor agonist ramelteon. *Epilepsy & Behavior*, *16*(1), 52-57.

[96] Williams, J. W., Mulrow, C. D., Chiquette, E., Noël, P. H., Aguilar, C., & Cornell, J. (2000). A systematic review of newer pharmacotherapies for depression in adults: evidence report summary: clinical guideline, part 2. *Annals of internal medicine*, *132*(9), 743-756.

[97] Micale, V., Arezzi, A., Rampello, L., & Drago, F. (2006). Melatonin affects the immobility time of rats in the forced swim test: the role of serotonin neurotransmission. *European Neuropsychopharmacology*, *16*(7), 538-545.

[98] Weil, Z. M., Hotchkiss, A. K., Gatien, M. L., Pieke-Dahl, S., & Nelson, R. J. (2006). Melatonin receptor (MT1) knockout mice display depression-like behaviors and deficits in sensorimotor gating. *Brain research bulletin*, *68*(6), 425-429.

[99] Adamah-Biassi, E. B., Hudson, R. L., & Dubocovich, M. L. (2014). Genetic deletion of MT 1 melatonin receptors alters spontaneous behavioral rhythms in male and female C57BL/6 mice. *Hormones and behavior*, *66*(4), 619-627.

[100] Wu, Y. H., Ursinus, J., Zhou, J. N., Scheer, F. A., Ai-Min, B., Jockers, R., ... & Swaab, D. F. (2013). Alterations of melatonin receptors MT1 and MT2 in the hypothalamic suprachiasmatic nucleus during depression. *Journal of affective disorders*, *148*(2), 357-367.

[101] De Bodinat, C., Guardiola-Lemaitre, B., Mocaër, E., Renard, P., Muñoz, C., & Millan, M. J. (2010). Agomelatine, the first melatonergic antidepressant: discovery, characterization and development. *Nature reviews. Drug discovery*, *9*(8), 628.

[102] Hickie, I. B., & Rogers, N. L. (2011). Novel melatonin-based therapies: potential advances in the treatment of major depression. *The Lancet*, *378*(9791), 621-631.

[103] Millan, M. J., Gobert, A., Lejeune, F., Dekeyne, A., Newman-Tancredi, A., Pasteau, V., ... & Cussac, D. (2003). The novel melatonin agonist agomelatine (S20098) is an antagonist at 5-hydroxytryptamine2C receptors, blockade of which enhances the activity of frontocortical dopaminergic and adrenergic pathways. *Journal of Pharmacology and Experimental Therapeutics*, *306*(3), 954-964.

[104] Mini, L. J., Wang-Weigand, S., & Zhang, J. (2007). Self-reported efficacy and tolerability of ramelteon 8 mg in older adults experiencing severe sleep-onset difficulty. *The American journal of geriatric pharmacotherapy*, *5*(3), 177-184.

[105] Rajaratnam, S. M., Polymeropoulos, M. H., Fisher, D. M., Roth, T., Scott, C., Birznieks, G., & Klerman, E. B. (2009). Melatonin agonist tasimelteon (VEC-162) for transient insomnia after sleep-time shift: two randomised controlled multicentre trials. *The Lancet*, *373*(9662), 482-491.

[106] Miyamoto, M., Nishikawa, H., Doken, Y., Hirai, K., Uchikawa, O., & Ohkawa, S. (2004). The sleep-promoting action of ramelteon (TAK-375) in freely moving cats. *Sleep*, *27*(7), 1319-1325.

[107] Roth, T., Stubbs, C., & Walsh, J. K. (2005). Ramelteon (TAK-375), a selective MT1/MT2-receptor agonist, reduces latency to persistent sleep in a model of transient insomnia related to a novel sleep environment. *Sleep*, *28*(3), 303-307.

Chapter 8

EPILEPSY: NEUROPROTECTIVE, ANTI-INFLAMMATORY, AND ANTICONVULSANT EFFECTS OF MELATONIN

Anna Karynna A. A. Rocha[1,], José Cipolla-Neto[2] and Débora Amado[3]*

[1]Brain Institute, Department of Neuroscience,
Federal University of Rio Grande do Norte, Natal, Brazil
[2]Department of Physiology and Biophysics,
University of São Paulo, São Paulo, Brazil
[3]Department of Neurology and Neurosurgery,
Federal University of São Paulo, São Paulo, Brazil

ABSTRACT

The International League Against Epilepsy (ILAE) defined a seizure as "a transient occurrence of signs and/or symptoms due to abnormal excessive or synchronous neuronal activity in the brain." The neuropathology of epilepsy involves glutamate excitotoxicity, oxidative stress, and high inflammatory response. Anti-epileptic drugs (AEDs) are the main form of treatment. However, approximately 30% to 40% of individuals are unresponsive to them. Moreover, AEDs can produce undesirable symptoms, such as moderate disturbances of the central nervous system (CNS), liver failure, and bone health impairment. Melatonin has anti-inflammatory, antioxidant, anticonvulsant, and GABAergic activity on the CNS, and low toxicity at pharmacological doses. Melatonin may be a potential adjuvant for epilepsy treatment, as several studies have shown its administration can reduce oxidative stress and high inflammatory responses from occurring. In addition, melatonin has been safe and effective in decreasing seizure frequency, improving sleep quality, regulating the light-dark cycle, improving attention, memory, language, and anxiety in these patients. All of this demonstrates that melatonin

---

[*] Corresponding Author Email: rocha.anna2@gmail.com.

has been effective in the treatment of epilepsy in experimental models as well as in patients. We present recent data from clinical and experimental studies, which show the neuroprotective, anti-inflammatory, and antioxidant effects of melatonin in many types of epilepsy; this should include positive future prospects for treatment.

Keywords: Melatonin, epilepsy, neuroprotection

INTRODUCTION

Epilepsy is a chronic neurological disease that is included among the most common neurological disorders and affects up to 1% to 2% of the population [1]. Epilepsy was defined in 2005 by the International League Against Epilepsy (ILAE) [2] as a brain disorder characterized by a long-standing predisposition to epileptic seizures with neurobiological, cognitive, psychological, and social consequences. This concept of epilepsy requires the occurrence of at least one epileptic seizure, which is characterized as a transient occurrence of signs and/or symptoms due to abnormal excessive or synchronous neuronal activity in the brain. Since 2014, the ILAE has adopted a new definition of epilepsy of which the approach is operational (practical) in contrast with the definition of 2005, which is considered only conceptual. Furthermore, the ILAE and the International Bureau for Epilepsy have recently agreed that epilepsy is best considered a disease [3]. Therefore, by the new definition, a person is considered to have epilepsy if they meet any of the following conditions: 1) at least two unprovoked (or reflex) seizures occurring >24 h apart; 2) one unprovoked (or reflex) seizure and the probability of further seizures similar to the general recurrence risk (at least 60%) after two unprovoked seizures, occurring over the next 10 years; 3) diagnosis of an epilepsy syndrome. Epilepsy is considered to be resolved for individuals who had an age-dependent epilepsy syndrome but are now past the applicable age or those who have remained seizure-free for the last 10 years with no seizure medicines for the last five years [3].

Approximately 50 million people worldwide currently live with epilepsy [4]. The prevalence of epilepsy in South America is 934 per 100,000 followed by Africa at 863 per 100,000 persons. The prevalence of epilepsy in North America is 680 per 100,000 persons, 495 per 100,000 persons in Asia, and 457 per 100,000 persons in Europe [5], showing that in low- and middle-income countries the proportion is much higher. In addition, approximately 2.4 million people worldwide are diagnosed with epilepsy each year. In high-income countries, new cases are between 30 and 50 per 100,000 people in the general population each year. In low- and middle-income countries, this figure can be up to two times higher [4]. In fact, the incidence rate of the epilepsy is highest in Africa at 215.00 per 100,000 person-years followed by South America at 162.45 per 100,000 person-years. The incidence of epilepsy in North America is 23.29 per 100,000 person-years and 42.63 per 100,000 person-years in Europe [5].

In developed countries, the incidence of epilepsy tends to present a U-shaped curve with higher rates in the elderly and children. This pattern is not found in developing countries, in which the incidence of epilepsy demonstrates the incidence high point in early adulthood [6]. A range of etiologic groups with emphasis on those that have implications for treatment has been recognized. The etiology of epilepsy could be structural, genetic, infectious, metabolic, and immune in addition to a group of unknown-etiologies [7].

Temporal lobe epilepsy (TLE) has been studied the most with respect to its histopathology, and the histological lesions are thus best understood in relation to epilepsy due to the high refractoriness and wide range of experimental models. TLE is the major epilepsy disorder that affects >40 million people worldwide [8].

NEUROPATHOLOGY OF EPILEPSY

Currently, there is no universally accepted definition for epileptogenesis. The term "epileptogenesis" is defined as a neurobiological process that leads to the occurrence of the first spontaneous seizure and post-brain injury recurrent epileptiform events, which includes traumatic brain injuries, stroke, status epilepticus (SE), or infectious triggers of brain diseases, or genetic or inflammatory factors [1]. The latency period refers to pre-epileptic seizure-free periods between brain insult and the occurrence of the first spontaneous seizure [9, 10].

The process of epileptogenesis leads to the occurrence of spontaneous convulsions and eventually to an epilepsy diagnosis [11]. Abnormal and excessive electrical discharges are seen in brain during epileptic seizure development. Unbalanced neuronal excitability appears to be the cause of these electrical discharges [12].

Aberrant neuronal excitability and excitotoxic damage are characteristics of epilepsy and are a consequences of elevations in extracellular glutamate levels, the predominant excitatory neurotransmitter in the adult mammalian brain. Epilepsy-generated chronic desynchronized network activity has been shown to induce extraneous neuronal firing and pathological alterations in signaling that may arise from multiple pathways [12].

The hypothesis of neuronal hyperexcitability in epilepsy is due to an imbalance between Glu-mediated excitation/inhibition and gamma-aminobutyric acid (GABA) [13]. The glutamatergic mechanisms that are involved in epilepsy initiation and progression include the upregulation of glutamate receptors [14, 15], elevation of extracellular glutamate concentration [16], and alterations in glutamatergic transporters [17].

On the other hand, GABA is recognized as the main inhibitory neurotransmitter because it generates potential presynaptic inhibitors via neuronal hyperpolarization [18]. The GABAergic system plays an important role in the balance of neuronal excitation and consequent suppression of epileptic discharges [19]. It is assumed that the reduction or loss of GABAergic inhibition may increase the probability of generating excitatory post-synaptic potentials and synchronize discharges, thus inducing epileptogenesis [20]. GABAergic mechanisms that may be involved in epilepsy include impairment of GABA release [21], decrease in GABA synthesis [21], receptor modifications [22], and loss of GABAergic neurons [23].

Seizure activity from an epileptic brain readily induces an inflammatory response, including the activation of microglia and production of pro-inflammatory cytokines [24]. Inflammatory mediators may initiate or trigger early seizures, preceding the onset of diagnosed epilepsy. In fact, IL-6 and IL-1 receptors (IL-1Rs) upregulation occurs in epileptic patients [25]. High levels of cytokines, including IL-1β, have also been identified in neurons and glia of surgically resected epileptic tissue, thus demonstrating the important presence of inflammatory components during the epileptogenic process [26].

Many studies have demonstrated that oxidative stress is regarded as a possible mechanism in epilepsy pathogenesis [27]. Kainic acid (KA) and pilocarpine experimental models of epilepsy have been shown to cause an increase in reactive oxygen species (ROS) production, producing considerable amounts of $O_2^{-\bullet}$ and overloading endogenous protection mechanisms (glutathione peroxidase [GPx], superoxide dismutase [SOD], and catalase [CAT]). This overload results in oxidative damage to proteins, phospholipids, and mitochondrial DNA [28].

Epilepsy Treatment

The current clinical treatment of epilepsy is usually done with antiepileptic drugs (AEDs). The choice of AED is based on the type of epileptic seizure or preferably, on the type of epilepsy or epileptic syndrome that is presented. In addition, patient data such as age, gender, physical condition, and socioeconomic status should be evaluated and taken into consideration [29].

The most well-known traditional AEDs are carbamazepine, phenytoin, phenobarbital, and valproate. Some new drugs such as Oxcarbazepine, Topiramate, Lamotrigine, and Gabapentin have also been used. For newly diagnosed epilepsy, the main medications to be considered are carbamazepine, oxcarbazepine, phenytoin, lamotrigine, and topiramate [30, 31]. With the first appropriate treatment, approximately 40% of the patients were free of seizures [32]. However, in approximately 30% of the patients, epileptic seizures have been shown to be refractory to treatment with monotherapy.

When epilepsy is refractory to monotherapy, polytherapy may be attempted with the knowledge that only a low percentage of these patients (3%–10%) will be seizure-free [32]. Nevertheless, polytherapy often results in a number of unwanted effects, including neurological disturbances (somnolence, ataxia, dizziness), psychiatric and behavioral symptoms, and metabolic alteration (osteoporosis, inducement or inhibition of hepatic enzymes.). If there is no success with polytherapy, an evaluation for surgical treatment can be considered [29]. The need for better tolerated AEDs is even more urgent in this group of people. Thus, there is a great need to explore candidate treatments to improve the conditions of those who live with epilepsy.

NEUROPROTECTIVE, ANTI-INFLAMMATORY AND ANTIOXIDANT PROPERTIES OF MELATONIN

Characteristics of melatonin as an anti-inflammatory, antioxidant, and central nervous system inhibitory molecule, its effects on several pathologies, including epilepsy, have been studied. It has been shown to be safe even at pharmacological doses (above the nanomolar range) [33], in which melatonin functions to eliminate free radicals [34].

Many studies have shown that melatonin plays a neuroprotective role due to its ability to reduce neuronal excitability. Melatonin (at concentrations corresponding to its nocturnal peak) may inhibit neuronal calcium influx neurons and bind to the calcium-calmodulin complex, thus inhibiting neuronal nitric oxide synthase (nNOS) activity, diminishing NO production, and subsequently reduce the excitatory effect of N-methyl-D-aspartate [35].

In hypothalamic slices, melatonin potentiates GABAa currents in neurons of the suprachiasmatic nucleus (SCN) via the MT1 receptor [36]. In addition, melatonin increases the amplitude and frequency of GABAergic mIPSCs, indicating that melatonin enhances gabaergic inhibitory transmission in cultured rat hippocampal neurons. Thus, it is evident that melatonin has a potentiating effect on the GABAergic system, which may indicate a potential pathway for its neuroprotective effects [37].

A neuroprotective role for melatonin has been demonstrated in experimental models of neurodegeneration in which this indole acts as a potent antioxidant and protects neurons and glial cells from free radical-induced damage [38]. In fact, melatonin's antioxidant role has been extensively shown in various pathological central nervous system (CNS) conditions. Melatonin increases brain glutathione levels and significantly reduces neural tissue lipid peroxidation while constant light (which blocks melatonin production) significantly enhanced the breakdown of lipids in the brain, suggesting its potential protection of neurons and glial cells from free radicals [39]. In addition, melatonin promotes neurogenesis in the dentate gyrus in the pinealectomized rat [40] and modulates cell survival of new neurons in adult mice hippocampi [41].

Melatonin also stimulates antioxidant enzymes and reduces oxidative stress in an experimental traumatic brain injury [42] and attenuates manganese-induced neurotoxicity by means of its antioxidant properties [43]. Another study shows that treatment with melatonin effectively prevents glutamate excitotoxicity and oxidative damage in C6 astroglial cells [44].

In a study conducted in the rat brain, it has been shown that dexamethasone was able to increase prooxidant enzyme production and reduce the production of antioxidant enzymes. In addition, dexamethasone produced DNA fragmentation in brain tissue. Treatment with melatonin was able to revert all of these neurotoxic processes. Oxidative stress inhibition via stimulation of the antioxidant enzymes by melatonin pretreatment was shown to play a central protective role in neurotoxicity modulation [45].

In addition to the antioxidant effects and reduction of neuronal excitability, melatonin demonstrated significant anti-inflammatory power. Several groups have shown that melatonin reverses chronic and acute inflammation [46, 47].

Hung et al. [48] showed that there was a decrease in inflammatory mediator expressions and an elevation in antioxidant enzyme expressions in rats with hippocampal injury that were injected with melatonin versus vehicle-treated animals.

In addition, the immunomodulatory properties of melatonin are well known; it acts on the immune system by regulating cytokine production of immunocompetent cells [49, 50]. The hormone acts on immunocompetent cells (monocytes, B-lymphocytes, natural killer lymphocytes, T-helper lymphocytes, cytotoxic T-lymphocytes) and enhances cytokine production/secretion, cell proliferation, and oncostasis [51]. Experimental and clinical data show that melatonin reduces adhesion molecules and pro-inflammatory cytokines including IL-6, IL-8, and tumor necrosis factor (TNF)-α, and modified serum inflammatory parameters [50]. As a consequence, melatonin appears to improve the clinical course of illnesses which have an inflammatory etiology (Reiter et al. 1998).

These data indicate that melatonin attenuates oxidative stress through its antioxidant and anti-inflammatory properties in addition to reducing CNS excitability. All of these properties indicate that the melatonin is a strong candidate for treatment reduction of epilepsy-associated inflammation, excitotoxicity, and oxidative stress.

MELATONIN AND EPILEPSY

Although AED therapy reduces the frequency of recurrent seizures, nearly all available treatments have cognitive side effects, which significantly increase with polypharmacy and higher dosing, especially in children [52]. For this reason, melatonin's effects on various types of epilepsy have been studied for decades both in humans and in experimental models of seizures.

Melatonin and melatonin receptor agonists may be suitable for use as add-on treatments with routine AEDs. Many studies have demonstrated the beneficial effects of melatonin in epilepsy. Studies showing proconvulsant effects of melatonin are scarce. Indeed, melatonin has been shown to alleviate sleep disturbances related to different etiologies in epileptic children and adults [53, 54]. The improvement in sleep parameters is accompanied by seizure activity attenuation in young patients with intractable epilepsy

[55] and in patients on valproate [56]. Furthermore, clinical reports have shown melatonin's beneficial effects on seizure activity during the day [57] and the night [54] in patients with intractable epilepsy, a pathology characterized by an increased post-seizure salivary melatonin levels [58].

Comorbidity of a sleep disorder with epilepsy creates difficulties in choosing an AED [59]. Indeed, while lamotrigine, levetiracetam, valproate, and gabapentin have been reported to lack or even exert few negative effects, other agents such as benzodiazepines, barbiturates, and phenytoin aggravate sleep architecture in epileptic individuals [58]. In addition, seizures produced at night profoundly affect sleep structures, increase sleep disruption, and decrease the amount of REM sleep. Daytime seizures have a similar effect but with less profound results. Patients suffering from intractable epileptic seizures, in addition to being affected by their critical manifestations, experience sleep disturbances and in most cases post ictal lethargy, which can last up to several days after even a very mild episode.

Furthermore, melatonin has been shown to be an excellent adjunct for the treatment of epilepsy-correlated sleep disorders. The relationship between sleep and epilepsy is so strong that changes in sleep pattern affect the electroencephalographic (EEG) readings, triggering paroxysmal abnormalities when they are present and producing morphological changes, both quantitative and qualitative, in the paroxysmal activity of a subject in response to epileptic seizures. Thus, it has long been believed that sleep deprivation has a significant influence on epileptogenic activity [60]. For this reason, it is widely believed that melatonin administration may regulate sleep and consequently, reduce epileptic seizures.

Studies have reported that melatonin's primary beneficial effect on seizure activity is due to its ability to synchronize disturbed circadian rhythms in epileptic patients [54]. Melatonin may alleviate sleep disorders in young patients although without producing any effects on seizures. In addition, melatonin has been shown to be effective in regulating the sleep-wake cycle in pediatric patients [60] and in patients with mental retardation and epileptic seizures [53].

Another study treated children with difficult-to-control epilepsy and high rates of sleep problems before going to bed. Children with difficult-to-control epilepsy had higher scores for each category of sleepwalking, forced grinding teeth, and sleep apnea. At the end of the therapeutic trial, patients with refractory epilepsy showed significant improvement in bedtime resistance, sleep duration, sleep latency, frequent nocturnal awakenings, sleepwalking incidence, excessive daytime sleepiness, nocturnal enuresis, forced grinding teeth, apnea, and Epworth Sleepiness scores. There was also a significant reduction in the seizure severity. Thus, the use of melatonin in patients with refractory epilepsy was associated with the improvement in both the many sleep-related phenomena and seizure severity [55].

Studies have demonstrated that melatonin participates in seizure control, suggesting that there is an increased melatonin concentration during seizures and that patients with seizures of different origins show a change in melatonin rhythm. In fact, Bazil et al. [58] observed that in adult patients with temporal lobe epilepsy, interictal levels of melatonin are significantly decreased when compared to controls. On the other hand, after a seizure, an increase of this hormone is observed for a period of up to 24 h. In patients with complex partial epilepsy a decrease in melatonin levels compared to the control group was also demonstrated. Another study showed that the endogenous melatonin secretion in idiopathic generalized epilepsy tends to have a late circadian phase [61].

An important study performed in 54 children with febrile seizures and epilepsy showed that serum melatonin levels increased during the seizure with normal recovery values of 1 h later. Values were maintained for up to 24 h when the circadian variation of melatonin returns to the normal acrophase [62]. Other researchers have shown that plasma melatonin levels increased during seizures. However, post-seizure plasma melatonin levels were significantly lower in the epilepsy group than in the control group, indicating that although seizures increase serum melatonin levels, patients with epilepsy present lower circulating values than healthy individuals [63]. In this same study, plasma melatonin levels did not differ between patients with afebrile seizures who had or had not used AEDs. Daytime (8 AM to 8 PM) and nighttime (8 PM to 8 AM) post-seizure melatonin levels were not significantly different indicating that only discharges from the seizures are capable of altering melatonin production. These results support the hypothesis that stimulation of melatonin production as a result of seizures may contribute to the body's response to seizures.

In addition to beneficial effects on sleep and circadian alterations, melatonin has been studied due to its anticonvulsant effects. Due to its antioxidant action and the ability to inhibit excitotoxic phenomena, melatonin acts as a cellular protector as it directly degrades important and toxic free radicals [64]. These free radicals are generated during seizures and are capable of inducing neurotoxicity and cell death. Excitotoxicity can be observed with increases in the stimulation of brain glutamate receptors, which in many cases, leads to neuronal death. On the other hand, nitric oxide (NO) is metabolized to peroxynitrite, a toxic oxide formed by the reaction of superoxide and NO under inflammatory and oxidative stress conditions and which is a class of compound capable of inducing lipid peroxidation, with consequent neuronal death. Some studies have shown the ability of melatonin to degrade the peroxynitrite anion, a finding that supports the anti-inflammatory, antioxidant, and consequently the anticonvulsant actions of this hormone [65].

Anton-Tay (1974) described an anticonvulsive action of melatonin after administering it as adjunctive therapy to conventional anticonvulsant treatments for six patients. Another study demonstrated an increase in the activity of antioxidant enzymes (such as GPx and GRd) in children with epilepsy treated with melatonin compared to controls [66]. Other authors suggest that melatonin may suppress neuronal epileptic activity through specific melatonin receptors (MT$_1$ and MT$_2$) [67]. In addition, melatonin increases the release of prolactin, which in turn increases GABAergic neurotransmission [67].

It is known that melatonin potentiates the effect of several AEDs such as carbamazepine and phenobarbital [68, 69]. Clinical evidence has demonstrated that melatonin in addition to antiepileptic therapies play an anticonvulsive role by reducing spicules in the electroencephalographic tract and also the frequency of seizures in epileptic patients [67]. In line with this, Goldberg-Stern et al. [57] demonstrated that melatonin is effective and safe for decreasing the frequency of seizures during the day in patients with difficult-to-control epilepsy.

In order to observe the properties of melatonin as an antioxidant molecule, Gupta et al. [66] administered melatonin to children undergoing valproate monotherapy and observed an improvement in the quality of life of these patients with respect to appetite, improvement in sleep quality, and improvement in attention, memory, language, and anxiety when compared to the placebo group. The authors attributed this improvement to the properties of melatonin as an anticonvulsive and antioxidant molecule and to its favorable effects on sleep.

Scientific evidence from clinical studies indicates that melatonin is a strong candidate for reductions of the AED- and epilepsy-related adverse effects. Animal models of epilepsy, however, play a fundamental role in our understanding of the basic physiological and behavioral changes associated with human epilepsy and its relationship with melatonin. Much of the available literature suggests melatonin's anticonvulsant, antiinflamatory, antioxidant and neuroprotective effects on many types of experimental epilepsies.

The anticonvulsant effects of melatonin have been demonstrated in several different experimental seizure models. In 1996, Uz et al. [70] tested the *in vivo* efficacy of melatonin in preventing KA-induced DNA damage in adult rats' hippocampi. They suggested that melatonin may reduce the extent of cell injury that is mediated through kainate receptors. Other authors have also observed that concurrent intraperitoneal administration of melatonin (20 mg/kg) completely abolishes kainate-induced seizures and mitochondrial DNA damage in the rat cerebral cortex [71]. Melatonin also exerts anticonvulsant effects in a rat model of hyperthermia and achieved its optimum efficacy at a dose of 80 mg/kg when administered 15 min prior to seizure induction [72]. Melatonin's anticonvulsant effects are thought to be mediated by melatonergic MT$_1$ and MT$_2$ receptors in conjunction with GABAergic and serotonergic modulatory mechanisms [73].

In this context, Yildirim and Marangoz [74] observed melatonin's anticonvulsant effects on penicillin-induced epileptiform activity in rats. Melatonin given intracerebroventricularly (ICV) at 40 and 80 ug prolonged the latency of epileptiform activity as analyzed by electrocorticography. In addition, melatonin significantly decreased the frequency of spike-wave activity while the spike amplitude remained unchanged.

A study showed that melatonin is neuroprotective via attenuation of KA-induced status epilepticus (SE)-associated loss of autophagy and mitochondria in the hippocampus of rats [75]. Melatonin also increased the latency in the appearance of spontaneous recurrent seizures (SRSs) and decreased their frequency during the treatment period. The behavioral alterations associated with hyperactivity, depression-like behavior during the light phase, and deficits in hippocampal-dependent working memory were positively affected by melatonin treatment in epileptic rats. Melatonin reduced the neuronal damage in the CA1 area of the hippocampus and piriform cortex and reverse the decrease in hippocampal serotonin (5-HT) level in epileptic rats. Taken together, long-term melatonin treatment after SE was unable to suppress the development of epileptogenesis [76].

A pinealectomy is surgery for the total removal of the pineal gland, the mammalian source of circulating melatonin. Janjoppi et al. [77] showed that pinealectomized animals require a reduced number of stimuli to reach stage 5 in the amygdala model of amygdala. On the other hand, treatment with melatonin in animals subject to amygdala abrasion has shown an anticonvulsant effect by causing a reduction in the susceptibility to seizure induction [78].

Rats that were also pinealectomized and subject experimental pilocarpine-induced epilepsy presented a decrease in the silent period, an expressive increase in the frequency of seizures in the chronic phase, and a greater number of lesions in the hippocampal CA1 and CA3 regions. It was also verified that these animals presented higher numbers of apoptotic cells 48 h after SE [79]. In addition, it was observed that animals pretreated with melatonin presented higher latency for SE, decreased frequency of seizures in the chronic period, and 100% survival in contrast to 40% in rats that did not receive any previous treatment; these findings demonstrated melatonin's neuroprotective effects [80]. These experimental data show that the absence of normal melatonin production by the pineal gland can facilitate epileptogenesis. In contrast, melatonin addition promoted important neuroprotection against epilepsy-associated damage.

Chronic treatment with melatonin after induction of *status epilepticus* by lithium-pilocarpine significantly alleviated seizure severity, reduced neuronal death in the hippocampal CA1 region, improved spatial learning, and reversed long-term potentiation (LTP) impairment. The authors also found that melatonin rescued the decreased GluR2 surface levels in the CA1 region, which may reduce calcium permeability, rescue neurons from death, and rescue epilepsy-induced LTP impairments. These data may indicate the mechanism of melatonin's neuroprotection in alleviating cognitive dysfunction in chronic phase epilepsy [81].

Both in vitro and in vivo studies have suggested an antiepileptic activity of melatonin [82, 67], mediated by an antioxidant effect [82], an increase in gamma-aminobutyric acid (GABA) concentration [83] and GABA receptor affinity, or a reduction of the N-methyl-D-aspartate (NMDA) excitatory effect [84]. Although studies investigating melatonin's long-term effects are still lacking, high melatonin doses have so far proved to be safe [85].

Different melatonin doses (50–100 mg/kg intraperitoneally) were tested for their ability to suppress complex partial seizures in the amygdala kindling rat model. Thirty minutes after melatonin injection, the current threshold necessary to elicit epileptic changes after discharge was significantly increased by about 200%–250%. [78]. Another study demonstrated that melatonin exerts anticonvulsant effects in a rat model of hyperthermic seizure and achieved optimum efficacy at a dose of 80 mg/kg when administered 15 min prior to seizure induction of a seizure [72].

All of these experimental models have demonstrated that melatonin is capable of reversing or reducing pathophysiological lesions in addition to frequency of seizures. However, many melatonin effects are mediated by G-protein-coupled receptors (MT_1 and MT_2) that are found in several structures of the central nervous system (CNS). Evidence has shown that MT_1 and MT_2 receptors have a potential inhibitory effect on the CNS. In this context, MT_1 and MT_2 receptor agonists could be able to decrease seizures in epilepsy. Therefore, in addition to melatonin, several melatonin agonist receptors have also been widely used in several experimental models. In fact, many of them have shown beneficial effects for epilepsy treatment.

Agomelatine is a novel antidepressant drug. Recently, it was hypothesized to be a potential drug for the treatment of epilepsy and its complications primarily due to dual action at $MT_{1/2}$ and 5HT2c receptors coupled with its strong neuroprotective and possible antioxidant properties [86]. Agomelatine was recently reported to have antiepileptic effects in a pilocarpine-induced seizure model and pentylenetetrazole (PTZ)-induced seizures in mice [28, 87].

Agomelatine effectively delayed development of kindling seizure severity, and decreased the incidence of epilepsy. Agomelatine also showed antioxidant effects that could partially contribute to its anticonvulsant actions. In addition, it alleviated PTZ-kindling-associated behavioral despair and favorably modulated liver enzymes. Its effects on improvement of kindling-associated spatial memory could possibly be related to its effects on locomotor activity. Thus, agomelatine, could be explored as an adjunct to AED drugs used for seizure control and alleviating epilepsy-associated depression [88].

In the pilocarpine-induced seizure model, only a high dose of agomelatine demonstrated an increase in latency to convulsion when compared to the control. On the other hand, latency to death was increased in animals treated with two doses of agomelatine when compared to control animals. All animals died and 100% exhibited convulsions with the exception of the group that was pretreated with agomelatine [28].

Recently, both agomelatine and melatonin were demonstrated to reduce spike-and-wave discharges in the model of genetic absence epilepsy in WAG/Rij rats [89].

Another study showed that luzindole, a melatonin receptor antagonist, significantly blocked the protective action of agomelatine on PTZ-induced kindling. This suggests a role for the melatonergic pathway in mediating agomelatine's anticonvulsant effects of on PTZ-induced kindling. Agomelatine pretreatment significantly reduced cortical and hippocampal malondialdehyde (MDA) levels and elevated glutathione (GSH) levels in kindled mice. Thus, this study indicates that melatonin's anticonvulsant effects may be at least partially due to its antioxidant actions and melatonergic pathways [88].

Ramelteon (RozeremTM) has been approved by the United States Food and Drug Administration for long-term use in insomnia treatment. Ramelteon is a sleep agent that selectively binds to MT_1 and MT_2 receptors. It was able to reverse kindling-induced hippocampal excitability. In Kcna1 null mice, Ramelteon significantly attenuated seizure periodicity and frequency and improved circadian rest activity rhythms compared to control animals. In this case, use of the selective melatonin receptor agonist Ramelteon provided further support for melatonin receptors as potential novel targets for anticonvulsant drug development because it shows anticonvulsant properties in a chronic epilepsy model [90].

The integrity of the melatonergic system is important so that the treatment can be efficient. However, almost no experimental data has shown the effects of epilepsy on the pineal gland, melatonin production pathways, or melatonin receptors. A study recently undertaken by our group showed that pilocarpine-induced SE was able to change melatonin receptor MT_1 and MT_2 proteins and mRNA expression levels in rat hippocampi a few hours after SE in addition to in silent and chronic phases [91]. Observation of the epileptic lesions that occur in the melatonergic system may be important for understanding many of the signs and symptoms involving epileptic patients and why treatment often does not have the expected efficiency.

It is known that, the AEDs can produce undesirable symptoms such as moderate disturbances of the CNS, liver failure, and even bone health impairments. As discussed in this review, in the CNS, melatonin has been studied in several neurological pathologies and may be a potential adjuvant for epilepsy treatment since it has anti-inflammatory, antioxidant, anticonvulsant, and GABAergic activities in addition to low toxicity even at pharmacological doses. This has been shown in several studies in humans and in experimental epilepsy models, which show that the administration of melatonin and its agonists is able to reduce oxidative stress and the high inflammatory responses occurring during epilepsy. Melatonin is also safe and effective for decreasing of seizures frequency or even abolishing them, improving sleep quality, regulating the light-dark cycle, improving attention, memory, language and anxiety in patients with epilepsy. As an adjunct to antiepileptic treatment, melatonin can help to improve seizure management, sleep parameters, and lead to an improvement in the quality of life. The data supporting a proconvulsant role of melatonin are limited and controversial. Therefore, it could be useful in the treatment of seizures and to reduce the collateral effects produced by antiepileptic drugs.

CONCLUSION

All of this scientific evidence demonstrates that melatonin is quite effective in the epilepsy treatment in both experimental models and patients. In this review, we presented recent data from clinical and experimental model that show the neuroprotective, anti-inflammatory, and antioxidant effects of melatonin on many types of epilepsy in addition to the future prospects for treatment with this hormone. However, although data on human research and considerable work with animals suggest that melatonin may have antiepileptic properties, one must ask appropriate questions. Perhaps the best question should be: "Under what circumstances can melatonin improve or worsen seizure control?" It would be very useful for clinicians to know which factors may be associated with an improvement or deterioration in melatonin selective control, so that it can be added in the daily routine for treating people with epilepsy and thus improve the quality of life of these people.

REFERENCES

[1] Alyu, F., & Dikmen, M. (2017). Inflammatory aspects of epileptogenesis: Contribution of molecular inflammatory mechanisms. *Acta Neuropsychiatrica*, *29*(1), 1-16.

[2] Fisher, R. S., Boas, W. V. E., Blume, W., Elger, C., Genton, P., Lee, P., & Engel, J. (2005). Epileptic seizures and epilepsy: definitions proposed by the International League against Epilepsy (ILAE) and the International Bureau for Epilepsy (IBE). *Epilepsia*, *46*(4), 470-472.

[3] Fisher, R. S., Acevedo, C., Arzimanoglou, A., Bogacz, A., Cross, J. H., Elger, C. E., & Hesdorffer, D. C. (2014). ILAE official report: a practical clinical definition of epilepsy. *Epilepsia*, *55*(4), 475-482.

[4] WHO 2017. *Epilepsy*. February 2017. Available from: http://www.who.int/mediacentre/factsheets/fs999/en/.

[5] Pringsheim, T., Fiest, K., & Jette, N. (2014). The international incidence and prevalence of neurologic conditions How common are they?. *Neurology, 83*(18), 1661-1664.

[6] Téllez-Zenteno, J. F., & Hernández-Ronquillo, L. (2011). A review of the epidemiology of temporal lobe epilepsy. *Epilepsy research and treatment, 2012*.

[7] Scheffer, I. E., Berkovic, S., Capovilla, G., Connolly, M. B., French, J., Guilhoto, L., & Nordli, D. R. (2017). ILAE classification of the epilepsies: Position paper of the ILAE Commission for Classification and Terminology. *Epilepsia*, *58*(4), 512-521.

[8] De Lanerolle, N. C., Lee, T. S., & Spencer, D. D. (2012). Histopathology of human epilepsy. *Jasper's Basic Mechanisms of the Epilepsies*, *4*.

[9] Pitkänen, A., & Lukasiuk, K. (2011). Mechanisms of epileptogenesis and potential treatment targets. *The Lancet Neurology*, *10*(2), 173-186.

[10] Sloviter, R. S., & Bumanglag, A. V. (2013). Defining "epileptogenesis" and identifying "antiepileptogenic targets" in animal models of acquired temporal lobe epilepsy is not as simple as it might seem. *Neuropharmacology*, *69*, 3-15.

[11] Bovolenta, R., Zucchini, S., Paradiso, B., Rodi, D., Merigo, F., Mora, G. N., & Fabene, P. F. (2010). Hippocampal FGF-2 and BDNF overexpression attenuates epileptogenesis-associated neuro-inflammation and reduces spontaneous recurrent seizures. *Journal of neuroinflammation*, *7*(1), 81.

[12] Barker-Haliski, M., & White, H. S. (2015). Glutamatergic mechanisms associated with seizures and epilepsy. *Cold Spring Harbor perspectives in medicine*, *5*(8), a022863.

[13] Aroniadou-Anderjaska, V., Fritsch, B., Qashu, F., & Braga, M. F. (2008). Pathology and pathophysiology of the amygdala in epileptogenesis and epilepsy. *Epilepsy research*, *78*(2), 102-116.

[14] Aronica, E., Van Vliet, E. A., Mayboroda, O. A., Troost, D., Da Silva, F. H. L., & Gorter, J. A. (2000). Upregulation of metabotropic glutamate receptor subtype mGluR3 and mGluR5 in reactive astrocytes in a rat model of mesial temporal lobe epilepsy. *European Journal of Neuroscience*, *12*(7), 2333-2344.

[15] Debanne, D., Thompson, S. M., & Gähwiler, B. H. (2006). A brief period of epileptiform activity strengthens excitatory synapses in the rat hippocampus in vitro. *Epilepsia*, *47*(2), 247-256.

[16] Yi, J. H., & Hazell, A. S. (2006). Excitotoxic mechanisms and the role of astrocytic glutamate transporters in traumatic brain injury. *Neurochemistry international*, *48*(5), 394-403.

[17] Touret, M., Parrot, S., Denoroy, L., Belin, M. F., & Didier-Bazes, M. (2007). Glutamatergic alterations in the cortex of genetic absence epilepsy rats. *BMC neuroscience*, *8*(1), 69.

[18] Treiman, D. M. (2001). GABAergic mechanisms in epilepsy. *Epilepsia*, *42*(s3), 8-12.

[19] Morimoto, K., Fahnestock, M., & Racine, R. J. (2004). Kindling and status epilepticus models of epilepsy: rewiring the brain. *Progress in neurobiology*, *73*(1), 1-60.

[20] Acharya, J. N. (2002). Recent advances in epileptogenesis. *Current science*, 679-688.
[21] Buzzi, A., Chikhladze, M., Falcicchia, C., Paradiso, B., Lanza, G., Soukupova, M., & Simonato, M. (2012). Loss of cortical GABA terminals in Unverricht–Lundborg disease. *Neurobiology of disease, 47*(2), 216-224.
[22] Dinkel, K., Meinck, H. M., Jury, K. M., Karges, W., & Richter, W. (1998). Inhibition of γ- aminobutyric acid synthesis by glutamic acid decarboxylase autoantibodies in stiff- man syndrome. *Annals of neurology*, *44*(2), 194-201.
[23] Brooks-Kayal, A. R., Shumate, M. D., Jin, H., Rikhter, T. Y., & Coulter, D. A. (1998). Selective changes in single cell GABA A receptor subunit expression and function in temporal lobe epilepsy. *Nature medicine*, *4*(10).
[24] Knopp, A., Frahm, C., Fidzinski, P., Witte, O. W., & Behr, J. (2008). Loss of GABAergic neurons in the subiculum and its functional implications in temporal lobe epilepsy. *Brain, 131*(6), 1516-1527.
[25] Ravizza, T., Balosso, S., & Vezzani, A. (2011). Inflammation and prevention of epileptogenesis. *Neuroscience letters*, *497*(3), 223-230.
[26] Peltola, J., Laaksonen, J., Haapala, A. M., Hurme, M., Rainesalo, S., & Keränen, T. (2002). Indicators of inflammation after recent tonic–clonic epileptic seizures correlate with plasma interleukin-6 levels. *Seizure*, *11*(1), 44-46.
[27] Crespel, A., Coubes, P., Rousset, M. C., Brana, C., Rougier, A., Rondouin, G., & Lerner-Natoli, M. (2002). Inflammatory reactions in human medial temporal lobe epilepsy with hippocampal sclerosis. *Brain research*, *952*(2), 159-169.
[28] Chang, S. J., & Yu, B. C. (2010). Mitochondrial matters of the brain: mitochondrial dysfunction and oxidative status in epilepsy. *Journal of bioenergetics and biomembranes*, *42*(6), 457-459.
[29] Aguiar, C. C. T., Almeida, A. B., Araújo, P. V. P., Vasconcelos, G. S., Chaves, E. M. C., do Vale, O. C., & Vasconcelos, S. M. M. (2012). Anticonvulsant effects of agomelatine in mice. *Epilepsy & Behavior*, *24*(3), 324-328.
[30] Betting, L. E., & Guerreiro, C. A. (2008). Tratamento das epilepsias parciais. *Journal of Epilepsy and Clinical Neurophysiology*, *14* (Suppl 2), 25-31. [Treatment of partial epilepsies. *Journal of Epilepsy and Clinical Neurophysiology*, *14* (Suppl 2), 25-31.]
[31] Betting, L. E., Kobayashi, E., Montenegro, M. A., Min, L. L., Cendes, F., Guerreiro, M. M., & Guerreiro, C. A. (2003). Treatment of epilepsy: consensus of the Brazilian specialists. *Arquivos de neuro-psiquiatria*, *61*(4), 1045-1070.
[32] French, J. A., Kanner, A. M., Bautista, J., Abou-Khalil, B., Browne, T., Harden, C. L., & Bergen, D. (2004). Efficacy and tolerability of the new antiepileptic drugs I: Treatment of new onset epilepsy Report of the Therapeutics and Technology Assessment Subcommittee and Quality Standards Subcommittee of the American

Academy of Neurology and the American Epilepsy Society. *Neurology*, *62*(8), 1252-1260.

[33] Kwan, P., & Brodie, M. J. (2000). Early identification of refractory epilepsy. *New England Journal of Medicine*, *342*(5), 314-319.

[34] Reiter, R. J., TAN, D. X., Poeggeler, B., MENENDEZ- PELAEZ, A. R. M. A. N. D. O., CHEN, L. D., & Saarela, S. (1994). Melatonin as a Free Radical Scavenger: Implications for Aging and Age- Related Diseases. *Annals of the New York Academy of Sciences*, *719*(1), 1-12.

[35] Carlberg, C. (2000). Gene regulation by melatonin. *Annals of the New York Academy of Sciences*, *917*(1), 387-396.

[36] León, J., Macías, M., Escames, G., Camacho, E., Khaldy, H., Martín, M., & Acuña-Castroviejo, D. (2000). Structure-related inhibition of calmodulin-dependent neuronal nitric-oxide synthase activity by melatonin and synthetic kynurenines. *Molecular pharmacology*, *58*(5), 967-975.

[37] Wan, Q. I., Man, H. Y., Liu, F., Braunton, J., Niznik, H. B., Pang, S. F., & Wang, Y. T. (1999). Differential modulation of GABAA receptor function by Mel1a and Mel1b receptors. *Nature neuroscience*, *2*(5), 401-403.

[38] Cheng, X. P., Sun, H., Ye, Z. Y., & Zhou, J. N. (2012). Melatonin modulates the GABAergic response in cultured rat hippocampal neurons. *Journal of pharmacological sciences*, *119*(2), 177-185.

[39] Jou, M. J., Peng, T. I., Reiter, R. J., Jou, S. B., Wu, H. Y., & Wen, S. T. (2004). Visualization of the antioxidative effects of melatonin at the mitochondrial level during oxidative stress- induced apoptosis of rat brain astrocytes. *Journal of pineal research*, *37*(1), 55-70.

[40] Baydas, G., Reiter, R. J., Nedzvetskii, V. S., Nerush, P. A., & Kirichenko, S. V. (2002). Altered glial fibrillary acidic protein content and its degradation in the hippocampus, cortex and cerebellum of rats exposed to constant light: reversal by melatonin. *Journal of pineal research*, *33*(3), 134-139.

[41] Rennie, K., De Butte, M., & Pappas, B. A. (2009). Melatonin promotes neurogenesis in dentate gyrus in the pinealectomized rat. *Journal of pineal research*, *47*(4), 313-317.

[42] Ramírez-Rodríguez, G., Klempin, F., Babu, H., Benítez-King, G., & Kempermann, G. (2009). Melatonin modulates cell survival of new neurons in the hippocampus of adult mice. *Neuropsychopharmacology*, *34*(9), 2180-2191.

[43] Ding, K., Wang, H., Xu, J., Li, T., Zhang, L., Ding, Y., & Zhou, M. (2014). Melatonin stimulates antioxidant enzymes and reduces oxidative stress in experimental traumatic brain injury: the Nrf2–ARE signaling pathway as a potential mechanism. *Free Radical Biology and Medicine*, *73*, 1-11.

[44] Deng, Y., Jiao, C., Mi, C., Xu, B., Li, Y., Wang, F., & Xu, Z. (2015). Melatonin inhibits manganese-induced motor dysfunction and neuronal loss in mice:

involvement of oxidative stress and dopaminergic neurodegeneration. *Molecular neurobiology*, *51*(1), 68-88.

[45] Das, A., Belagodu, A., Reiter, R. J., Ray, S. K., & Banik, N. L. (2008). Cytoprotective effects of melatonin on C6 astroglial cells exposed to glutamate excitotoxicity and oxidative stress. *Journal of pineal research*, *45*(2), 117-124.

[46] Assaf, N., Shalby, A. B., Khalil, W. K., & Ahmed, H. H. (2012). Biochemical and genetic alterations of oxidant/antioxidant status of the brain in rats treated with dexamethasone: protective roles of melatonin and acetyl-L-carnitine. *Journal of physiology and biochemistry*, *68*(1), 77-90.

[47] Costantino, G., Cuzzocrea, S., Mazzon, E., & Caputi, A. P. (1998). Protective effects of melatonin in zymosan-activated plasma-induced paw inflammation. *European journal of pharmacology*, *363*(1), 57-63.

[48] Cuzzocrea, S., Zingarelli, B., Gilad, E., Hake, P., Salzman, A. L., & Szabó, C. (1997). Protective effect of melatonin in carrageenan- induced models of local inflammation: relationship to its inhibitory effect on nitric oxide production and its peroxynitrite scavenging activity. *Journal of pineal research*, *23*(2), 106-116.

[49] Hung, M. W., Tipoe, G. L., Poon, A. M. S., Reiter, R. J., & Fung, M. L. (2008). Protective effect of melatonin against hippocampal injury of rats with intermittent hypoxia. *Journal of pineal research*, *44*(2), 214-221.

[50] Carrillo- Vico, A., Lardone, P. J., Naji, L., Fernández- Santos, J. M., Martín-Lacave, I., Guerrero, J. M., & Calvo, J. R. (2005B). Beneficial pleiotropic actions of melatonin in an experimental model of septic shock in mice: regulation of pro-/anti- inflammatory cytokine network, protection against oxidative damage and anti- apoptotic effects. *Journal of pineal research*, *39*(4), 400-408.

[51] Gitto, E., Reiter, R. J., Sabatino, G., Buonocore, G., Romeo, C., Gitto, P., & Barberi, I. (2005). Correlation among cytokines, bronchopulmonary dysplasia and modality of ventilation in preterm newborns: improvement with melatonin treatment. *Journal of pineal research*, *39*(3), 287-293.

[52] Liu, F., Ng, T. B., & Fung, M. C. (2001). Pineal indoles stimulate the gene expression of immunomodulating cytokines. *Journal of neural transmission*, *108*(4), 397-405.

[53] Ijff, D. M., & Aldenkamp, A. P. (2013). Cognitive side-effects of antiepileptic drugs in children. *Handbook of clinical neurology*, *111*, 707-718.

[54] Coppola, G., Iervolino, G., Mastrosimone, M., La Torre, G., Ruiu, F., & Pascotto, A. (2004). Melatonin in wake–sleep disorders in children, adolescents and young adults with mental retardation with or without epilepsy: a double-blind, cross-over, placebo-controlled trial. *Brain and Development*, *26*(6), 373-376.

[55] Peled, N., Shorer, Z., Peled, E., & Pillar, G. (2001). Melatonin effect on seizures in children with severe neurologic deficit disorders. *Epilepsia*, *42*(9), 1208-1210.

[56] Elkhayat, H. A., Hassanein, S. M., Tomoum, H. Y., Abd-Elhamid, I. A., Asaad, T., & Elwakkad, A. S. (2010). Melatonin and sleep-related problems in children with intractable epilepsy. *Pediatric neurology*, *42*(4), 249-254.

[57] Gupta, M., Aneja, S., & Kohli, K. (2005). Add-on melatonin improves sleep behavior in children with epilepsy: randomized, double-blind, placebo-controlled trial. *Journal of child neurology*, *20*(2), 112-115.

[58] Goldberg-Stern, H., Oren, H., Peled, N., & Garty, B. Z. (2012). Effect of melatonin on seizure frequency in intractable epilepsy: a pilot study. *Journal of child neurology*, *27*(12), 1524-1528.

[59] Bazil, C. W. (2003). Effects of Antiepileptic Drugs on Sleep Structure. *CNS drugs*, *17*(10), 719-728.

[60] Tchekalarova, J., Moyanova, S., De Fusco, A., & Ngomba, R. T. (2015). The role of the melatoninergic system in epilepsy and comorbid psychiatric disorders. *Brain research bulletin*, *119*, 80-92.

[61] Uberos, J., Augustin- Morales, M. C., Molina Carballo, A., Florido, J., Narbona, E., & Muñoz- Hoyos, A. (2011). Normalization of the sleep–wake pattern and melatonin and 6- sulphatoxy- melatonin levels after a therapeutic trial with melatonin in children with severe epilepsy. *Journal of pineal research*, *50*(2), 192-196.

[62] Manni, R., De Icco, R., Cremascoli, R., Ferrera, G., Furia, F., Zambrelli, E., & Terzaghi, M. (2016). Circadian phase typing in idiopathic generalized epilepsy: Dim light melatonin onset and patterns of melatonin secretion—Semicurve findings in adult patients. *Epilepsy & Behavior*, *61*, 132-137.

[63] Molina-Carballo, A., Munoz-Hoyos, A., Sánchez-Forte, M., Uberos-Fernández, J., Moreno-Madrid, F., & Acuna-Castroviejo, D. (2007). Melatonin increases following convulsive seizures may be related to its anticonvulsant properties at physiological concentrations. *Neuropediatrics*, *38*(03), 122-125.

[64] Dabak, O., Altun, D., Arslan, M., Yaman, H., Vurucu, S., Yesilkaya, E., & Unay, B. (2016). Evaluation of plasma melatonin levels in children with afebrile and febrile seizures. *Pediatric neurology*, *57*, 51-55.

[65] Reiter, R. J. (1996). Antioxidant actions of melatonin. *Advances in pharmacology*, *38*, 103-117.

[66] Gilad, E., Cuzzocrea, S., Zingarelli, B., Salzman, A. L., & Szabó, C. (1997). Melatonin is a scavenger of peroxynitrite. *Life sciences*, *60*(10), PL169-PL174.

[67] Gupta, M., Aneja, S., & Kohli, K. (2004). Add-on melatonin improves quality of life in epileptic children on valproate monotherapy: a randomized, double-blind, placebo-controlled trial. *Epilepsy & Behavior*, *5*(3), 316-321.

[68] Fauteck, J. D., Schmidt, H., Lerchl, A., Kurlemann, G., & Wittkowski, W. (1999). Melatonin in epilepsy: first results of replacement therapy and first clinical results. *Neurosignals*, *8*(1-2), 105-110.

[69] Borowicz, K. K., Kamiński, R., Gasior, M., Kleinrok, Z., & Czuczwar, S. J. (1999). Influence of melatonin upon the protective action of conventional anti-epileptic drugs against maximal electroshock in mice. *European neuropsychopharmacology*, *9*(3), 185-190.

[70] Forcelli, P. A., Soper, C., Duckles, A., Gale, K., & Kondratyev, A. (2013). Melatonin potentiates the anticonvulsant action of phenobarbital in neonatal rats. *Epilepsy research*, *107*(3), 217-223.

[71] Uz, T., Giusti, P., Franceschini, D., Kharlamov, A., & Manev, H. (1996). Protective effect of melatonin against hippocampal DNA damage induced by intraperitoneal administration of kainate to rats. *Neuroscience*, *73*(3), 631-636.

[72] Mohanan, P. V., & Yamamoto, H. A. (2002). Preventive effect of melatonin against brain mitochondria DNA damage, lipid peroxidation and seizures induced by kainic acid. *Toxicology letters*, *129*(1), 99-105.

[73] Aydin, L., Gundogan, N. U., & Yazici, C. (2015). Anticonvulsant efficacy of melatonin in an experimental model of hyperthermic febrile seizures. *Epilepsy research*, *118*, 49-54.

[74] Moezi, L., Shafaroodi, H., Hojati, A., & Dehpour, A. R. (2011). The interaction of melatonin and agmatine on pentylenetetrazole-induced seizure threshold in mice. *Epilepsy & Behavior*, *22*(2), 200-206.

[75] Yildirim, M., & Marangoz, C. (2006). Anticonvulsant effects of melatonin on penicillin-induced epileptiform activity in rats. *Brain research*, *1099*(1), 183-188.

[76] Chang, C. F., Huang, H. J., Lee, H. C., Hung, K. C., Wu, R. T., & Lin, A. M. Y. (2012). Melatonin attenuates kainic acid- induced neurotoxicity in mouse hippocampus via inhibition of autophagy and α- synuclein aggregation. *Journal of pineal research*, *52*(3), 312-321.

[77] Tchekalarova, J., Petkova, Z., Pechlivanova, D., Moyanova, S., Kortenska, L., Mitreva, R., & Stoynev, A. (2013). Prophylactic treatment with melatonin after status epilepticus: effects on epileptogenesis, neuronal damage, and behavioral changes in a kainate model of temporal lobe epilepsy. *Epilepsy & Behavior*, *27*(1), 174-187.

[78] Janjoppi, L., de Lacerda, A. F. S., Scorza, F. A., Amado, D., Cavalheiro, E. A., & Arida, R. M. (2006). Influence of pinealectomy on the amygdala kindling development in rats. *Neuroscience letters*, *392*(1), 150-153.

[79] Mevissen, M., & Ebert, U. (1998). Anticonvulsant effects of melatonin in amygdala-kindled rats. *Neuroscience letters*, *257*(1), 13-16.

[80] de Lima, E., Soares, J. M., Garrido, Y. D. C. S., Valente, S. G., Priel, M. R., Baracat, E. C., & Amado, D. (2005). Effects of pinealectomy and the treatment with melatonin on the temporal lobe epilepsy in rats. *Brain research*, *1043*(1), 24-31.

[81] Lima, E., Cabral, F. R., Cavalheiro, E. A., da Graça Naffah-Mazzacoratti, M., & Amado, D. (2011). Melatonin administration after pilocarpine-induced status epilepticus: a new way to prevent or attenuate postlesion epilepsy?. *Epilepsy & Behavior, 20*(4), 607-612.

[82] Ma, Y., Sun, X., Li, J., Jia, R., Yuan, F., Wei, D., & Jiang, W. (2017). Melatonin Alleviates the Epilepsy-Associated Impairments in Hippocampal LTP and Spatial Learning Through Rescue of Surface GluR2 Expression at Hippocampal CA1 Synapses. *Neurochemical research, 42*(5), 1438-1448.

[83] Anton-Tay, F. (1974). Melatonin: effects on brain function. *Advances in biochemical psychopharmacology, 11*, 315.

[84] Reiter, R. J., Tan, D. X., Manchester, L. C., & Qi, W. (2001). Biochemical reactivity of melatonin with reactive oxygen and nitrogen species. *Cell biochemistry and biophysics, 34*(2), 237-256.

[85] Niles, L. (1991). Melatonin interaction with the benzodiazepine-GABA receptor complex in the CNS. In *Kynurenine and Serotonin Pathways* (pp. 267-277). Springer New York.

[86] Muñoz-Hoyos, A., Sánchez-Forte, M., Molina-Carballo, A., Escames, G., Martin-Medina, E., Reiter, R. J., & Acuña-Castroviejo, D. (1998). Melatonin's role as an anticonvulsant and neuronal protector: experimental and clinical evidence. *Journal of Child Neurology, 13*(10), 501-509.

[87] Seabra, M. D. L. V., Bignotto, M., Pinto Jr, L. R., & Tufik, S. (2000). Randomized, double- blind clinical trial, controlled with placebo, of the toxicology of chronic melatonin treatment. *Journal of pineal research, 29*(4), 193-200.

[88] Vimala, P. V., Bhutada, P. S., & Patel, F. R. (2014). Therapeutic potential of agomelatine in epilepsy and epileptic complications. *Medical hypotheses, 82*(1), 105-110.

[89] Dastgheib, M., & Moezi, L. (2014). Acute and chronic effects of agomelatine on intravenous penthylenetetrazol-induced seizure in mice and the probable role of nitric oxide. *European journal of pharmacology, 736*, 10-15.

[90] Azim, M. S., Agarwal, N. B., & Vohora, D. (2017). Effects of agomelatine on pentylenetetrazole-induced kindling, kindling-associated oxidative stress, and behavioral despair in mice and modulation of its actions by luzindole and 1-(m-chlorophenyl) piperazine. *Epilepsy & Behavior, 72*, 140-144.

[91] Aygün, H., Aydin, D., Inanir, S., Ekici, F., Ayyildiz, M., & AĞAR, E. (2015). The effects of agomelatine and melatonin on ECoG activity of absenceepilepsy model in WAG/Rij rats. *Turkish Journal of Biology, 39*(6), 904-910.

[92] Fenoglio-Simeone, K., Mazarati, A., Sefidvash-Hockley, S., Shin, D., Wilke, J., Milligan, H., & Maganti, R. (2009). Anticonvulsant effects of the selective melatonin receptor agonist ramelteon. *Epilepsy & Behavior, 16*(1), 52-57.

[93] de Alencar Rocha, A. K. A., de Lima, E., Amaral, F., Peres, R., Cipolla-Neto, J., & Amado, D. (2017). Altered MT1 and MT2 melatonin receptors expression in the hippocampus of pilocarpine-induced epileptic rats. *Epilepsy & Behavior*, *71*, 23-34.

In: Melatonin: Medical Uses and Role in Health and Disease ISBN: 978-1-53612-987-8
Editors: Lore Correia and Germaine Mayers © 2018 Nova Science Publishers, Inc.

Chapter 9

MELATONIN AS A TREATMENT FOR ANTIPSYCHOTIC INDUCED WEIGHT GAIN: A REVIEW OF THE EVIDENCE

Trevor R. Norman[*]*, PhD*

Department of Psychiatry, University of Melbourne,
Austin Hospital, Heidelberg, Victoria, Australia

ABSTRACT

A concerning aspect of the use of antipsychotic medication has been the issue of weight gain and the consequent increase of visceral adiposity leading to the development of the metabolic syndrome. The issue has reached prominence in recent years with the introduction of so-called second generation (SGAs) or 'atypical' antipsychotics. Because of their perceived benefits, with respect to neurological side effects, these medications have achieved first line status for the treatment of psychotic states in adults, adolescents and children. While behavioural strategies and life style modifications remain the corner stone for management of antipsychotic-induced weight gain, where such techniques fail adjunctive pharmacological treatments are used. Melatonin as a potential adjunctive treatment relies on the association between the development of diabetes and abnormal circadian rhythmicity. Limited pre-clinical studies suggest an effect of melatonin on visceral fat and other parameters associated with the metabolic syndrome but the effects on weight gain are equivocal. Clinical trials in antipsychotic medicated patients also suggest some positive effects for melatonin but the studies have significant shortcomings. The mechanism by which melatonin exerts any clinical effects remains largely un-investigated. The strength of the clinical evidence for an effect of melatonin, like that of the pre-clinical studies, is equivocal.

Keywords: melatonin, antipsychotic, weight gain, metabolic syndrome, rodents, atypical antipsychotic, clinical trials, humans

---

[*] Corresponding Author Email: trevorrn@unimelb.edu.au.

INTRODUCTION

Antipsychotic medications introduced in the 1950's were exemplified by chlorpromazine and related compounds. While these medications were effective in various psychotic states, they produced a plethora of side effects, which made them less than ideal treatments. Principally, the development of extrapyramidal or Parkinson-like symptoms and tardive dyskinesia was the cause of significant iatrogenic functional impairment for patients. Recent clinical practice has been to replace these so-called 'first generation' or 'typical' antipsychotics with newer, 'second generation' or 'atypical agents.' In contrast to the 'first generation' of antipsychotics, these more recent agents are less likely to be associated with neurological side effects (Coccurello and Moles, 2010). Pharmacological actions of the newer drugs, most notably the modulatory effects of 5HT2A antagonism on dopamine neurotransmission in the nigro-striatal system, is associated with this reduced neurological burden (Meltzer, 1992). Nevertheless, while lower extrapyramidal effects may offer some advantages in terms of patient compliance with medication, the second-generation agents are not without certain disadvantages. Of most concern is the propensity for weight gain, albeit with some variance in susceptibility between the various agents in clinical use (Newcomer, 2005).

Weight gain assessed as body weight change, change in body mass index or clinically relevant ($\geq 7\%$) weight change from baseline is an established side effect of atypical antipsychotics during both acute and maintenance treatment (De Hert et al., 2011a). Estimates vary, but 15 to 72% of patients with schizophrenia treated with these agents experience weight gain. There are marked inter-individual variations in weight gain, irrespective of prescribed agent with weight loss, as well as maintenance of weight and weight gain reported with the same agent (Bak et al., 2014). Meta-analyses and other studies show that the propensity for weight gain differs between antipsychotics (Correll et al., 2015). Among second-generation agents, clozapine and olanzapine are associated with higher weight gain for patients than for those receiving quetiapine, risperidone and paliperidone, who are at intermediate risk. Aripiprazole, amisulpride, ziprasidone, sertindole, asenapine and lurasidone have a lower risk of increased body weight (Bak et al., 2014). Prior exposure to treatment may affect the degree of weight gain (De Hert et al., 2011b). Weight gain with antipsychotics is often rapid following the initiation of treatment then slows reaching a plateau within about a year (Hasnain et al., 2012). Of itself, weight gain may not be problematic, for example, in a patient who is already below ideal weight for height norms. On the other hand, in addition to an increase in body mass, administration of atypical antipsychotics has been associated with hyperlipidaemia, glucose dysregulation and insulin resistance leading to the possible development of the metabolic syndrome and Type II diabetes (De Hert et al., 2011a). Changes in metabolic factors may occur in the absence of weight gain. Furthermore, such factors are associated with an increased overall mortality (Tiihonen et al., 2009). For example, in patients with schizophrenia a reduced life expectancy of 20 years or more is recognised, a finding confirmed by studies in different geographic regions (Olfson et al., 2015). The excess of mortality is due, in no small part, to physical illness with cardiovascular disease (CVD) a major contributor. Death from CVD is twice as likely in patients with schizophrenia compared to the general population (Osborn et al., 2007, 2015). The prevalence of cardiovascular risk factors in schizophrenia is well recognised (Casey et al., 2004; De Hert et al., 2011a).

The mechanism(s) of weight gain remain obscure but both clinical and pre-clinical data suggest increased appetite and food intake, as well as delayed satiety signalling are involved (Bak et al., 2014; Deng, 2013; Ballon et al., 2015). A frequently cited reason for weight gain is shared pharmacological actions, particularly antagonist effects at 5-HT2C and H1 receptors. Antipsychotics with the greatest affinities for 5-HT2C and H1 receptors are associated with the highest weight gain (Reynolds, Kirk, 2010). Genetic factors play a role in weight gain as investigations in twin pairs and sibling studies estimate up to 60-80% contributions for antipsychotic-related weight gain (Reynolds, 2012). Aside from schizophrenia, administration of atypical antipsychotic medications may occur in other conditions such as bipolar disorder and in disorders of childhood and adolescence, for example, autism spectrum disorder and conduct disorders. As with schizophrenia, bipolar disorder has a diminished life expectancy associated with a higher prevalence of Type II diabetes and CVD related to weight gain (Correll et al., 2015).

ANTIPSYCHOTIC INDUCED WEIGHT GAIN

Pre-Clinical Models

The most commonly reported side effect of atypical antipsychotic medications in humans is weight gain. This clinical finding, coupled with other metabolic effects of the drugs in patients, stimulated attempts to replicate the consequences in animal models with a view to understanding the mechanisms involved. While the pre-clinical models provide considerable insight they have not always proven to be consistent. Effects of antipsychotic medications on various parameters of the metabolic syndrome, including weight gain, using *in vivo* animal models, as well as using *in vitro* studies have been comprehensively reviewed (Boyda et al., 2010). Studies in rats show glucose dysregulation, characterised by increased fasting blood glucose levels and glucose intolerance. Acute or chronic treatment with atypical and some typical antipsychotics results in elevated fasting plasma glucose concentrations. Both fasting and non-fasting levels of insulin, a hormone involved in the regulation of glucose metabolism, show either increases or no change, but rarely decreases. Rats receiving atypical antipsychotic drugs showed insulin resistance in some studies, with the greatest effects observed for clozapine and olanzapine. Risperidone or chlorpromazine administration also caused insulin resistance in rats. The effects of antipsychotics on plasma lipid concentrations are notably inconsistent. While chronic olanzapine administration had no effect on free lipids or cholesterol levels, there were changes in adipose tissue content. Subcutaneous and visceral adipose deposits proliferated after chronic drug administration. Changes in leptin and ghrelin levels are inconsistent, but an increase in adiponectin after olanzapine and clozapine is noted in female and male rats (Boyda et al., 2010).

While weight gain may be a secondary consequence of the primary disturbances in glucose and lipid regulation, it is a frequently reported clinical side effect (Newcomer, 2005). Thus, studies directed toward the effects of melatonin on the increase in weight are more obviously pertinent to the topic of this analysis. Although weight gain, at first sight, would appear to be a relatively straightforward effect to model in rodents, this has not proven to be the case in practice. Typical and atypical antipsychotic agents produced conflicting findings on weight when administered to rodents. For example self-regulated, chronic administration of either clozapine or olanzapine in rodents (Sprague-Dawley or Wistar rats; C57Bl/6J or A/J mice) increased body weight in female but not male rats treated with olanzapine (Albaugh et al., 2006). Weight gain was detectable within 2 to 3 days of drug administration and was associated with a hyperphagic effect. Chronic administration (up to 29 days) led to adiposity. Similarly, a dose response study examined the ability of clozapine to induce weight gain in female rats (Cooper et al., 2007). High doses of clozapine (6 and 12 mg/kg i.p., b.i.d.) and an intermediate dose (1-4 mg/kg, i.p., b.i.d.) showed no evidence for clozapine-induced weight gain, indeed weight loss was noted. This contrasted with previous results by the same authors (Cooper et al., 2005), which noted olanzapine-induced weight gain, hyperphagia, enhanced adiposity and metabolic changes at similar doses. Low doses of clozapine (0.25-0.5 mg/kg, i.p, b.i.d.) were also found to induce weight loss. By contrast, the administration of olanzapine (2.0 or 7.5 mg/kg, via osmotic mini-pump for 4 weeks) to female Sprague-Dawley rats did not change body weight or food intake (Chintoh et al., 2008). There was a significant accumulation of visceral fat accompanied by decreases in locomotor activity. The findings in rodents are in contrast to the clinical situation where both olanzapine and clozapine have been associated with the greatest weight gain in men and women (Newcomer 2005).

Enhanced adiposity is more readily demonstrated in rats than is weight gain. Animal data clearly do not always mimic the clinical situation, which suggests that findings from such models need a nuanced interpretation. The enhanced adiposity observed in the absence of antipsychotic-induced weight gain and hyperphagia might represent a direct drug effect on adipocyte function independent of drug-induced hyperphagia (Minet-Ringuet et al., 2006).

Contrasting effects of atypical antipsychotic medications on weight gain in male and female rats mirrored earlier findings with typical agents (Baptista et al., 1987, 1988).

Both adult females and pre-pubertal males demonstrated weight gain, but not adult male rats, which often tended to display a non-significant body weight loss. This gender difference challenges both the face and predictive validity of rat models of drug-induced weight gain. Within this limitation, gender specific animal models are relevant for providing insights into treatment modalities for weight gain prevention or reversal.

Other limitations of rodent models of antipsychotic induced weight gain are also apparent (Boyda et al., 2010). Drug pharmacokinetic differences exist between rodents and humans, which suggest that dosing schedules used in animals may not achieve comparable steady state exposure (Kapur et al., 2003). For example, olanzapine is more rapidly metabolised in rodent than human; the plasma elimination half-life in rats for example is 2–3 h whereas in humans it is typically >24 h (Aravagiri et al., 1999; Callaghan et al., 1999). Thus, under-dosing in rodents may conceivably affect weight outcomes. Attempts to overcome this pharmacokinetic difference have utilised extended release systems, such as osmotic mini-pumps. The comparative effects of osmotic mini pump versus daily subcutaneous (s.c.) or intraperitoneal (i.p.) injections on body weight, food intake and body composition for olanzapine (7.5 mg/kg/day) versus placebo (n = 8/group) in female Sprague–Dawley rats was examined over 14 days of treatment (Mann et al., 2013). Significantly greater weight gain was observed in the olanzapine treated rats when the drug was delivered by mini pump compared to either s.c. or i.p. injections. Bodyweight gain of ≥7% from baseline occurred in twice as many animals (75%) after mini pump administration than with daily s.c. or i.p. injections. The olanzapine treated mini pump group consumed more kilocalories than vehicle, and consumed more than the s.c. or i.p. groups. Visceral fat in olanzapine treated animals versus vehicle treated animals was greatest in the mini pump sample. On the other hand, despite this result, further studies using extended delivery methods have failed to produce consistent effects on weight (Choi et al., 2007).

While humans treated with atypical agents report cravings for carbohydrate rich foods, the inconsistent hyperphagic effects in rodents might be due to their less palatable chow (Snigdha et al., 2008). Nevertheless, food preference may only be one aspect of weight gain, as increased consumption of high fat / high carbohydrate diet does not reliably translate to greater weight gain in antipsychotic treated animals (Smith et al., 2009). Factors that may also influence outcomes in rodent studies are the age of the animals (juvenile versus adult) and drug dose (Boyda et al., 2010).

Clearly further detailed exploration of factors influencing antipsychotic induced weight gain in rodents is necessary to develop a reliable model mimicking the effects noted in humans. A distinct possibility, that rats might not be an optimal species for modelling this side effect, remains.

Studies in Healthy Volunteers

As with pre-clinical studies, investigations in healthy volunteer subjects allow for the control of a number of confounding variables, which might otherwise contribute to antipsychotic induced weight gain. Patients with psychiatric disorders commonly have comorbid conditions, while lifestyle factors may account for some of the effects on weight and metabolic parameters seen in patients with schizophrenia and other disorders.

A double blind, randomized, placebo-controlled crossover trial in normal weight (BMI 18.5–25) healthy volunteers who received placebo or olanzapine (10 mg/day) for three days, reported no effect on body weight or BMI (Albaugh et al., 2011). On the other hand, this short administration of drug was associated with alterations in parameters involved with obesity and diabetes e.g., altered oral glucose tolerance, increased leptin and triglycerides. Clearly, even after short administration, olanzapine at least is associated with important metabolic alterations, which may lead to longer-term weight gain. Similar findings occurred in a parallel group study comparing olanzapine and aripiprazole with placebo over a 9-day drug administration period (Teff et al., 2013). While there was no change in weight in either drug group, olanzapine, but not aripiprazole, induced insulin resistance. No change in body weight or BMI occurred following treatment with olanzapine (10 mg/d) or haloperidol (3 mg/d) for 8 days in healthy normal-weight men (Vidarsdottir et al., 2010). Olanzapine reduced fasting plasma free fatty acid concentrations and hampered insulin action on glucose disposal, whereas the effects of haloperidol were not as clear-cut. Moreover, olanzapine, but not haloperidol, blunted the insulin-induced decline of free fatty acids and triglycerides. The effects occurred without a measurable change in body fat mass.

Using a crossover design Fountaine et al., (2010) examined the effect of olanzapine or placebo on body weight gain in male volunteers. Subjects received olanzapine 5 mg/day for 3 days followed by 10 mg/day for 12 days or a matching placebo with a minimum 12-day washout between treatments. Olanzapine caused 2.62 kg increase in body weight compared to a minimal effect of placebo. The authors attributed the increase in weight to an increased food intake. They did not find any evidence of decreased energy expenditure, decreased activity level, or short-term alterations in insulin sensitivity. Similarly, olanzapine (5 mg/day for 7 days, then 10 mg/day for 7 days) produced a significant increase in weight compared to placebo in healthy males (Daurignac et al., 2014). Metabolic parameters (triglycerides, insulin and leptin) increased in the olanzapine group. Olanzapine administration (10 mg/day) for 10 days produced a small but statistically significant increase from baseline in BMI in healthy male volunteers (Sacher et al., 2008). By contrast, oral intake of ziprasidone (80 mg/day) for 10 days did not alter BMI. Using a hyper-insulinemic euglycemic challenge in both groups only olanzapine treated subjects exhibited acute insulin resistance. Together these data suggest, but do not prove conclusively, that the increase in body weight and adiposity underlies the development of the metabolic syndrome in response to the administration of atypical antipsychotic medications.

Body weight increases between 1.95 to 2.8 kg followed olanzapine administration to healthy volunteers over periods ranging from one to 3 weeks (Sowell et al., 2002, 2003; Roerig et al., 2005; Literáti-Nagyy et al., 2010). Doses of medication employed in the studies coupled with the initial weight of the subjects may be important determinants of the effects observed.

Studies in healthy volunteers in general reflect the findings of the rodent investigations. Weight gain, while observed in some studies, is somewhat less reproducible than alterations in metabolic parameters implicated in the development of diabetes. Methodologically healthy volunteer studies are subject to certain limitations such as the dose of antipsychotic medication and the period of drug administration. Within these limitations, the results suggest that antipsychotic medications induce metabolic changes independent of weight gain and that weight itself is probably the result of hyperphagia, rather than decreased energy consumption (Ballon et al, 2014).

MELATONIN EFFECTS IN ANTIPSYCHOTIC INDUCED WEIGHT GAIN

Antipsychotic induced weight gain, the associated changes in metabolic and cardiovascular parameters, coupled with diminished life expectancy for patients receiving these agents long term suggests that counteractive measures are in the best interests of patients. In the absence of a clear mechanism of action, definitive recommendations are lacking. Life style advice or the co-prescription of other medications, are among the options employed, but only metformin has a robust data base to support its use when life style advice alone is not effective (Mizuno et al., 2014).

Melatonin, the circadian hormone whose secretion is synchronised to the light-dark cycle, is involved in important biological functions in the body: sleep regulation, circadian rhythm, immune modulation, reproduction, anti-inflammatory responses, antioxidant and energy metabolism (Bartness et al., 2002; Pandi-Perumal et al., 2006; Wyatt et al., 2006; Koziróg et al., 2011). Exogenous administration of the hormone is linked to the suppression of body weight gain and visceral adiposity in rodents maintained on a high fat, high cholesterol diet (Rasmussen et al., 1999; Wolden-Hanson et al., 2000; Prunet-Marcassus et al., 2003; Raskind et al., 2007; Terrón et al., 2013). Furthermore, exogenous administration improves various aspects of the metabolic syndrome in rodent models (Hoyos et al., 2000; Puchalski et al., 2003). In humans too there is evidence that melatonin attenuates metabolic dysfunctions (Cardinali et al., 2011). For example, supplementation with melatonin in patients with the metabolic syndrome ameliorated most components of the disorder (Goyal et al., 2014).

These effects may be due to the potent antioxidant activity of melatonin (Korkmaz et al. 2009a) as adiposity has been associated with increased oxidative activity (Savini et al., 2013). A chronobiologic mechanism cannot be overlooked as convincing evidence exists for the association between chrono-disruption, sleep deprivation and melatonin suppression in the metabolic syndrome and obesity (Reiter et al., 2012). Consequently the use of supplementary melatonin administration as a potential treatment for weight gain (and the associated metabolic alterations accompanying it) from antipsychotic medication administration would appear to be a viable therapeutic option.

Pre-Clinical Models

In what is essentially a prevention trial, the effect of melatonin on olanzapine induced weight gain employed a parallel group design to investigate the effect in female Sprague Dawley rats (Raskind et al., 2007). Four groups of animals (n = 11 per group) were allocated to 8 weeks treatment with olanzapine, melatonin, olanzapine-melatonin combination, or placebo in their drinking water. During the course of the experiment, the maintenance dose of olanzapine was 2mg/kg/day, while the dose of melatonin added to the drinking water was a final concentration of 0.4 µg/ml. After 8 weeks of treatment, weight differed significantly among the four groups of rats with olanzapine treated animals exhibiting the highest body weight. There were no statistically significant differences between the other three groups of animals. Weight gain over the 8 weeks was ~20.0% of baseline for the olanzapine treated rats compared to 10% for olanzapine + melatonin, 5% melatonin alone, and 7% for vehicle alone animals. Nocturnal plasma melatonin concentrations were determined by RIA from tail blood samples collected at the midpoint of the dark period in week 7. Olanzapine suppressed melatonin concentrations by ~55% compared to vehicle treated animals. Melatonin co-administered with olanzapine reversed the nocturnal suppression. As with some other studies, a hyperphagic effect was evident in the initial 2-weeks of treatment in the olanzapine-treated rats, but addition of melatonin treatment did not prevent this. Thereafter food consumption was not significantly different between treatment groups. Visceral fat pad weight increased most in rats treated with olanzapine and was not different from controls in rats treated with olanzapine + melatonin. Rats treated with melatonin alone did not differ from controls with regard to visceral fat.

The data provide some evidence for a preventative effect of melatonin co-administration on metabolic alterations, including weight gain, induced by olanzapine in female rodents. However, the relatively short duration of treatment begs the question of the sustainability of the effect over treatment periods more akin to those used in clinical settings. Despite this, the olanzapine-metabolic effects might be, at least in part, due to olanzapine-induced changes in melatonin secretion. Olanzapine diminished locomotor activity but melatonin did not restore this, suggesting that changes in activity are probably not responsible for the metabolic effects.

The role of gonadal hormones in protecting from weight gain is unknown. Sexual dimorphism of olanzapine on weight gain is a feature observed in other studies. Might this be a protective effect of testosterone?

Clinical Studies

Pre-clinical studies provide a rationale for the use of melatonin, and perhaps melatonin agonists, as a treatment for the prevention of weight gain in antipsychotic treated patients with psychotic disorders. The effects of melatonin administration on antipsychotic induced weight gain are summarised in Table 1. An additional study (also summarised in Table 1) examined the effect of ramelteon, a melatonin agonist, on weight gain in patients with schizophrenia.

Table 1. Clinical Studies of Melatonin and Antipsychotic Induced Weight Gain

Diagnosis (N)	Study Design	Dose of Melatonin	Weight Gain Effect	Other Metabolic Effects	Reference
Bipolar (38) 11-17yrs	DBPC 12 weeks	3mg/day	NR	TG, SBP	Mostafavi et al., 2014
Schizophrenia (36) 18-65yrs	DBPC 8weeks	3mg/day	Mel>PBO	Waist circumference	Modabbernia et al., 2014
Schizophrenia or Bipolar (44) 18-45yrs	DBPC 8 weeks	5mg/day	Mel>PBO	DBP, fat mass; Bipolar only	Romo-Nava et al., 2014
Schizophrenia (25) 18-65yrs	DBPC 8 weeks	8mg/day*	No Effect	TC, TC/HDL ratio	Borba et al., 2011

* This study used the melatonin MT1 / MT2 agonist ramelteon.

Melatonin was associated with significantly less weight gain than placebo in first episode schizophrenia patients treated with olanzapine (Modabbernia et al., 2014). In this randomised, double blind, placebo-controlled study, patients with DSM-IV-TR schizophrenia were assigned to either melatonin 3mg/day or placebo on a 1:1 basis for an 8-week treatment period with melatonin given as a single tablet at night. As the patients were in the first episode of illness, it appeared that commencement of melatonin treatment was concurrent with the commencement of olanzapine treatment, which was titrated to final mean dose of ~20mg/day in both groups. Anthropometric and biochemical parameters were assessed at baseline, weeks 4 and 8. A mixed effects model was used to analyse the changes in parameters over time. The final sample size of 18 patients per treatment arm was, according to the authors *a priori* power calculation, sufficient to give a power of 80% at the 0.05 level based on an observed mean weight difference of 3kg between groups. Groups were matched at baseline for weight, other anthropometric and biochemical parameters as well as severity of illness. At week 8, melatonin treated patients gained significantly less weight (3.3% of the baseline) than the placebo group (8.5% of the baseline). Mean absolute difference between groups was 3.2kg. The mixed model analysis showed a significant time by treatment interaction for both week 4 and week 8. Based on the reported changes in weight from baseline to week 8 in the two groups, a moderate to large effect size (Cohen's d = 0.74) could be deduced. The effect of melatonin is therefore relatively robust. Changes in other metabolic parameters (cholesterol, insulin, and blood sugar concentrations) were not significantly different between the two groups, while triglyceride concentration approached significance. Despite the small sample size and an inability to control for factors such as diet and exercise, 3mg/day melatonin has some beneficial effects as a preventative agent in this group of first episode patients over a limited time span. A weakness of the study is that *a priori* the number of patients likely to experience weight gain cannot be determined. Therefore matching on the propensity to gain weight could not occur.

Subsequent studies have addressed the efficacy of melatonin in weight gain and metabolic parameter amelioration in older and different diagnostic patient groups. In bipolar disorder, melatonin reduced the metabolic effects of olanzapine in a short-term study in adolescent patients (Mostafavi et al., 2014). The study was a 12-week, parallel group, randomized, double blind, placebo-controlled trial in which patients aged 11-17years were allocated to either placebo or melatonin (3mg/day) added to their treatment regimen. All patients received lithium combined with olanzapine as their bipolar disorder treatment. Patients were within the normal BMI range at the commencement of treatment. While 48 patients were allocated to treatment, only 38 patients were analysed for effects on metabolic parameters. Total cholesterol, triglycerides and fasting blood glucose along with blood pressure were measured at baseline, week 6 and week 12. The data were analysed using repeated measures ANOVA. While the repeated measures analysis showed a statistically significant difference between melatonin and placebo on total cholesterol levels, inspection of the data suggests that both groups had similar week 12 concentrations. The difference over time was accentuated by the fact that the melatonin group had a higher baseline cholesterol than the placebo group. The failure to test for a significant difference at baseline and to include this as a covariate in the analysis may have confounded the findings, such that melatonin offered no advantages over placebo. The rise in systolic blood pressure was slower in the melatonin than the placebo group but baseline differences may again have confounded the findings. The authors reported a gender difference in effects on triglyceride concentrations (melatonin had a larger effect in males to reduce triglycerides) but a separate analysis was not presented. Given the small sample size, (11 males in the melatonin group and 9 in the placebo group) this effect might be a statistical artefact. This study provides little if any convincing evidence for an effect of melatonin to ameliorate key biochemical parameters of the metabolic syndrome. Such effects as claimed by the authors are possibly due to the failure to account for important baseline differences in the statistical analysis, while the small sample size suggests the study lacks sufficient power to demonstrate differences should they have existed. The authors failed to report dispersions for the change scores from baseline to week 12 for any of the parameters reported so an effect size for the findings could not be calculated.

Using similar methodology, the effects of added melatonin on weight gain and other metabolic parameters in a mixed group of schizophrenia and bipolar disorder patients treated with atypical antipsychotic medications was examined (Romo-Nava et al., 2014). The study was conducted over eight-weeks using a double blind, randomised, placebo controlled, parallel group design. The metabolic effect of melatonin in atypical antipsychotic treated patients was evaluated in terms of weight, blood pressure, lipids, glucose, body composition, and anthropometric measures. The study enrolled 44 patients (20 bipolar disorder; 24 schizophrenia) who received placebo (n = 24) or a prolonged release formulation of melatonin 5 mg/day (n = 20). Mean changes in parameters from baseline to end-point were analysed using a t-test. According to this analysis, there was a significant difference in the mean change of diastolic blood pressure between the placebo and melatonin groups (1.1 versus 5.1 mmHg, respectively). When accounting for baseline differences in weight, as a covariate in the analysis, the difference in mean weight increase was significant between the placebo and melatonin groups (2.2 versus 1.5kg, respectively). Melatonin reportedly produced beneficial metabolic effects on fat mass and diastolic blood pressure in the bipolar disorder but not in the schizophrenia group. This trial differs from the previous studies in that it can be regarded as a treatment rather than a preventative one i.e., patients were already receiving medications before the addition of melatonin rather than at the commencement of treatment. Although the duration since the first diagnosis of illness was long, the length of medication usage was not reported. This might conceivably bias the study with respect to different propensities to metabolic abnormalities between groups e.g., sustained treatment durations in schizophrenia may be associated with more weight gain. Furthermore, in addition to a mix of atypical antipsychotic agents (quetiapine, olanzapine, risperidone and clozapine) some patients received mood stabilisers, antidepressants or benzodiazepines. While the authors could statistically separate relevant metabolic effects in 'high' versus 'medium' risk (for weight gain) antipsychotic medications, there was no analysis to account for the effects of these other agents. Given the sample size, it seems unlikely that there would have been sufficient power to account adequately for this in a statistical analysis. Melatonin was shown to reduce weight gain in patients receiving 'medium' risk drugs (risperidone, quetiapine) an effect that was not apparent in the 'high' risk (olanzapine, clozapine) treated patients, where melatonin seemed to be associated with greater weight gain. This result would appear to be paradoxical for a hypothetical weight reducing substance. However, there was an imbalance between groups with only five melatonin treated patients receiving 'high' risk drugs compared to 11 patients receiving 'high' risk drugs in the placebo group. The lack of effect may again be due to statistical power issues due to a small sample size.

Ramelteon is a melatonin agonist with high selectivity for MT1 and MT2 receptors and little affinity for quinone reductase-2 binding (Kato et al., 2005). The selectivity of ramelteon for MT1 is found to be >1000-fold over that of MT2 receptors. It has negligible affinity for other neuronal receptors. A number of studies have established the efficacy of ramelteon in treating patients with chronic insomnia (Kuriyama et al., 2014). An 8-week double blind, placebo controlled study evaluated its effect on weight gain, some metabolic and inflammatory markers in 25 subjects with DSM-IV diagnosed schizophrenia or schizoaffective disorder (Borba et al., 2011). Patients in the study received a variety of atypical antipsychotic medications: clozapine, olanzapine, risperidone or quetiapine. In addition to the psychiatric diagnosis subjects were required to have a body mass index (BMI) >27kg/m^2 and evidence of a component of the metabolic syndrome (see Table 2). Randomization of subjects to either ramelteon 8 mg/day or placebo occurred in a 2:1 fashion. From baseline to week 8 assessments, body weight did not change significantly in either group nor were there significant differences between ramelteon and placebo treated patients when adjusted for the baseline weight differences. Total cholesterol and cholesterol to HDL ratio decreased significantly in ramelteon treated patients but other metabolic parameters were unaltered. Calculated effect sizes for the changes were generally small to medium (Cohen's d = 0.24 for total cholesterol and d = 0.16 for cholesterol to HDL ratio). The major methodological issue with this study is the small sample size (only 20 patients: 6 placebo, 14 ramelteon completed the full course of treatment). Clearly, replication of these findings in larger samples is required before ramelteon is recommended for this indication in broader clinical use. The dose of ramelteon used in this study (8mg/day) is that usually recommended for sleep disturbance. In clinical trials as a hypnotic higher doses (up to 32mg/day) have been evaluated (Kuriyama et al., 2014). Dose response data for the medication needs evaluation as a treatment for antipsychotic induced weight gain. Higher, or indeed lower, doses may be more effective for ameliorating the metabolic and weight gain effects of atypical antipsychotics.

None of the studies described here have attempted to establish the mechanism of action of melatonin or rameleteon in patients. Antioxidant actions and normalisation of circadian rhythms are obvious possibilities. If the former case, then other agents also suggest themselves but if the latter, then melatonin would prove to be a doubly useful therapeutic agent.

Table 2. Criteria for Metabolic Syndrome*

Metabolic syndrome occurs, when a person has three or more of the following measurements:

Abdominal obesity†
Triglyceride level of ≥150 mg/dL (1.7 mmol/L)
HDL cholesterol ≤40 mg/dL (1.03 mmol/L)in men or ≤ 50 mg/dL (1.29 mmol/L) in women
Systolic blood pressure of 130 mm Hg or greater, or diastolic blood pressure of 85 mm Hg or greater
Fasting glucose of 100 mg/dL (5.6 mmol/L)or greater

*American Heart Association criteria.
https://www.heart.org/HEARTORG/Conditions/More/MetabolicSyndrome/About-Metabolic-Syndrome_UCM_301920_Article.jsp accessed 26th July 2017.
† Depends on ethnicity: European/North American Men ≥102cm Women ≥88cm; Asian Men ≥90 cm Women ≥80 cm.

CONCLUSION

In the use of antipsychotic medication, weight gain is a significant ongoing concern. The concomitant effects of dyslipidaemia, insulin resistance and cardiovascular effects no doubt contribute to a shortened life expectancy for patients receiving these agents. Strategies directed at attenuation of weight gain hold the prospect of an improvement in the quality of life for patients receiving the drugs. While there is a general consensus that life style changes such as an increase in daily exercise, healthy dietary choices, cessation of smoking, is preferable to additional medical treatment, such approaches are not always successful. Where such life style modifications are not successful, pharmacological management may be appropriate. However, such adjunctive strategies require careful mechanism based assessment.

Animal models offer an approach to evaluating the usefulness of medications for weight control. The practical value of such models however is fraught with difficulties. Choice of strain, as well as species, produce inconsistent results even when the medication investigated is one with a well-recognised weight gain issue in clinical usage (olanzapine). Sexual dimorphic effects are also apparent in some species with female rats more likely to experience weight gain than males, again a departure from clinical experience. Furthermore, pharmacokinetic differences between rodent and man mean that total systemic exposure to medication across a dosing interval in rodents is not comparable to the clinical situation. This can be overcome, to some extent, using a continuous release drug delivery system (osmotic mini-pumps for example) or by different routes of drug administration (oral versus intramuscular versus intraperitoneal). Furthermore, different animal strains may show wide variations in the susceptibility to weight gain and metabolic changes. Nevertheless, in animal strains where weight gain is reproducible, the effectiveness of pharmacological interventions is testable. The antioxidant, anti-inflammatory and circadian restoring effects of melatonin provide a rationale for its evaluation in pre-clinical and clinical studies as an adjunctive treatment for the prevention of weight gain.

Thus, melatonin administration prevented weight gain in female rats when co-administered with olanzapine from the beginning of treatment. While encouraging, exploration of the effect in male rats as well as dose response relationships for melatonin are lacking. Further, it is not clear if melatonin can reverse the weight gain seen with antipsychotic medication. In the clinical situation, not every patient is equally susceptible to the weight-gaining effect of antipsychotic medications. In any preventative trial therefore, the efficacy of the preventative agent may appear to be poor because some patients would not have gained weight in any event, making the study apparently null. Examining melatonin effects in patients with weight gain and the early signs of the metabolic syndrome would address the issue of treatment as opposed to prophylaxis. No clinical trials to date have addressed this issue. While the methodology used in the studies to date is *de rigueur,* durations of treatment are short and the sample sizes small. Thus, these studies are pilot evaluations rather than definitive efficacy trials. Furthermore, the studies did not evaluate a dose-response effect of melatonin or the melatonin agonist. In the absence of such investigations the possibility that doses chosen for the studies were ineffective remains.

Is melatonin an effective treatment for the weight gain noted in patients receiving antipsychotic medications? Present evidence would suggest that there is too little data to conclude definitively and that in-depth evaluations are required along the lines suggested here. Comparative efficacy of melatonin versus agents such as the widely used diabetes agent metformin are also lacking.

REFERENCES

Albaugh VL, Henry CR, Bello NT, Hajnal A, Lynch SL, Halle B, Lynch CJ. (2006) Hormonal and Metabolic Effects of Olanzapine and Clozapine Related to Body Weight in Rodents. *Obesity (Silver Spring)* 14 (1), 36-51.

Albaugh VL, Singareddy R, Mauger D, Lynch CJ (2011) A Double Blind, Placebo-Controlled, Randomized Crossover Study of the Acute Metabolic Effects of Olanzapine in Healthy Volunteers *PLoS ONE*, 6: e22662.

Aravagiri M, Teper Y, Marder SR (1999) Pharmacokinetics and tissue distribution of olanzapine in rats. *Biopharm Drug Dispos* 20:369–77.

Bak M, Fransen A, Janssen J (2014). Almost all antipsychotics result in weight gain: a meta-analysis. *PLoS One,* 9: e94112.

Ballon JS, Pajvani U, Freyberg Z (2014). Molecular pathophysiology of metabolic effects of antipsychotic medications. *Trends Endocrinol Metab.* 25:593-600.

Baptista, T, Herna´ndez, L., Parada, MA (1987) Long term administration of some antipsychotic drugs increases body weight and feeding in rats: are dopamine D2 receptors involved? *Pharmacol. Biochem. Behav.* 27, 399–405.

Baptista, T, Parada, MA, Murzi, E (1988) Puberty modifies sulpiride effects on body weight in rats. *Neurosci. Lett.* 92, 161–164.

Bartness TJ, Demas GE, Song CK (2002). Seasonal changes in adiposity: the roles of the photoperiod, melatonin and other hormones, and sympathetic nervous system. *Exp Biol Med* 227:363–376.

Borba CPC, Fan X, Copeland PM, Paiva A, Freudenreich O, Henderson DC (2011) Placebo-controlled pilot study of ramelteon for adiposity and lipids in patients with schizophrenia. *J Clin Psychopharmacol.*, 31: 653–658.

Boyda HN, Tse L, Procyshyn R, Honer WG, Barr AM (2010) Preclinical models of antipsychotic drug-induced metabolic side effects. *Trends in Pharmacological Sciences.* 31: 484-497.

Callaghan JT, Bergstrom RF, Ptak LR, Beasley CM (1999) Olanzapine. Pharmacokinetic and pharmacodynamic profile. *Clin Pharmacokinet* 37:177–93.

Cardinali DP, Cano P, Jiménez-Ortega V, Esquifino AI (2011). Melatonin and the metabolic syndrome: physiopathologic and therapeutical implications. *Neuroendocrinology* 93:133–142.

Casey DE, Haupt DW, Newcomer JW, (2004) Antipsychotic induced weight gain and metabolic abnormalities: implications for increased mortality in patients with schizophrenia. *J Clin Psychiatry* 65 (Suppl 7): 4–18.

Chintoh A, Mann SW, Lam TKT, Giacca A, Remington G (2008) Insulin resistance following continuous, chronic olanzapine treatment: An animal model, *Schizophrenia Research* 104: 23-30.

Coccurello R, Moles A. (2010) Potential mechanisms of atypical antipsychotic-induced metabolic derangement: Clues for understanding obesity and novel drug design. *Pharmacology & Therapeutics* 127: 210–251.

Cooper, GD Pickavance, LC, Wilding, JPH, Halford, JCG, Goudie, AJ. (2005) A parametric analysis of olanzapine-induced weight gain in female rats. *Psychopharmacology (Berl)* 181, 80–89.

Cooper, GD Pickavance, LC, Harrold, JA, Wilding, JPH, Halford, JCG, Goudie, AJ. (2007) Effects of olanzapine in male rats: enhanced adiposity in the absence of hyperphagia, weight gain or metabolic abnormalities. *J. Psychopharmacol.* 21, 405–413.

Correll CU, Detraaux J, LePeleire J, Dehert M. (2015) Effects of antipsychotics, antidepressants and mood stabilizers on risk for physical diseases in people with schizophrenia, depression and bipolar disorder. *World Psychiatry* 14:119–136.

Daurignac E, Leonard KE, Dubovsky SL (2015) Increased lean body mass as an early indicator of olanzapine-induced weight gain in healthy men *International Clinical Psychopharmacology* 30:23–28.

De Hert M, Detraux J, van Winkel R (2011a). Metabolic and cardiovascular adverse effects associated with antipsychotic drugs. *Nat Rev Endocrinol* 8:114-26.

De Hert M, Correll CU, Bobes J (2011b). Physical illness in patients with severe mental disorders. I. Prevalence, impact of medications and disparities in health care. *World Psychiatry* 10: 52-77.

Deng C. (2013). Effects of antipsychotic medications on appetite, weight, and insulin resistance. *Endocrinol Metab Clin North Am* 42:545-63.

Eder U, Mangweth B, Ebenbichler C, Weiss E, Hofer A, Hummer M, Kemmler G, Lechleitner M, Fleischhacker WW (2001) Association of Olanzapine-Induced Weight Gain With an Increase in *Body Fat Am J Psychiatry* 158:1719–1722.

Fell MJ, Anjum N, Dickinson K, Marshall KM, Peltola LM, Vickers S, Cheetham S, Neill JC. (2007) The distinct effects of subchronic antipsychotic drug treatment on macronutrient selection, body weight, adiposity, and metabolism in female rats. *Psychopharmacology* (Berl) 194, 221–231

Fountaine RJ, Taylor AE, MancusoJP, Greenway FL, Byerley LO, Smith SR, Most MM, Fryburg DA. (2010). *Increased Food Intake and Energy Expenditure Following Administration of Olanzapine to Healthy Men Obesity* 18, 1646–1651.

Goyal A, Terry PD, Superak HM, Nell-Dybdahl CL, Chowdhury R, Phillips LS, Kutner MH (2014). Melatonin supplementation to treat the metabolic syndrome: a randomized controlled trial. *Diabetol Metab Syndr* 6:124.

Hasnain M, Vieweg WV, Hollett B. (2012) Weight gain and glucose dysregulation with second-generation antipsychotics and antidepressants: a review for primary care physicians. *Postgrad Med* 124:154-67.

Hoyos M, Guerrero JM, Perez-Cano R, Olivan J, Fabiani F, Garcia-Pergañeda A, Osuna C (2000). Serum cholesterol and lipid peroxidation are decreased by melatonin in diet-induced hypercholesterolemic rats. *J Pineal Res* 28:150–155.

Kapur S, VanderSpek SC, Brownlee BA, Nobrega JN. (2003) Antipsychotic dosing in preclinical models is often unrepresentative of the clinical condition: a suggested solution based on in vivo occupancy. *J. Pharmacol. Exp. Ther.* 305, 625–631

Kato K, Hirai K, Nishiyama K, et al., "Neurochemical properties of ramelteon (TAK-375), a selective MT1/MT2 receptor agonist," *Neuropharmacology,* 48: 301–310 (2005).

Korkmaz A, Topal T, Tan DX, Reiter RJ. (2009). Role of melatonin in metabolic regulation. *Rev Endocr Metab Disord.* 10:261 – 70.

Koziróg M, Poliwczak AR, Duchnowicz P, Koter-Michalak M, Sikora J, Broncel M (2011). Melatonin treatment improves blood pressure, lipid profile, and parameters of oxidative stress in patients with metabolic syndrome. *J Pineal Res* 50:261–266.

Kuriyama A, Honda M, Hayashino Y, Ramelteon for the treatment of insomnia in adults: a systematic review and meta-analysis, *Sleep Medicine*, 15 (2014) 385–392.

Literáti-Nagy B, Péterfai E, Kulcsár E, Literáti-Nagy Z, Buday B, Tory K, (2010). Beneficial effect of the insulin sensitizer (HSP inducer) BGP-15 on olanzapine-induced metabolic disorders. *Brain Res Bull* 83:340–344.

Mann S, Chintoh A, Giacca A, Fletcher P, Nobrega J, Hahn M, Remington G (2013) Chronic olanzapine administration in rats: Effect of route of administration on weight, food intake and body composition Pharmacology, *Biochemistry and Behavior* 103 717–722.

Meltzer, H. Y. (1992). The importance of serotonin–dopamine interactions in the action of clozapine. *Br J Psychiatry Suppl* 17, 22−29.

Minet-Ringuet J, Even PC, Guesdon B, Tome D, de Beaurepaire R (2005) Effects of chronic neuroleptic treatments on nutrient selection, body weight, and body composition in the male rat under dietary self-selection. *Brain Res. Behav. Brain. Res.* 163, 204–211.

Minet-Ringuet, J. Even PC, Goubern M, Tome D, de Beaurepaire R (2006) Long term treatment with olanzapine mixed with the food in male rats induces body fat deposition with no increase in body weight and no thermogenic alteration. *Appetite* 46, 254–262.

Mizuno Y, Suzuki T, Nakagawa A, Yoshida K, Mimura M, Fleischhacker WW, Uchida H. (2014). Pharmacological Strategies to Counteract Antipsychotic-Induced Weight Gain and Metabolic Adverse Effects in Schizophrenia: *A Systematic Review and Meta-analysis Schizophrenia Bulletin* 40, 1385–1403.

Modabbernia A, Heidari P, Soleimani R, Sobhani A, Roshan ZA, Taslimi S, (2014). Melatonin for prevention of metabolic side-effects of olanzapine in patients with first-episode schizophrenia: randomized double-blind placebo controlled study. *J Psychiatr Res* 53:133–140.

Mostafavi A, Solhi M, Mohammadi MR, Hamedi M, Keshavarzi M, Akhondzadeh S (2014). Melatonin decreases olanzapine induced metabolic side-effects in adolescents with bipolar disorder: a randomized double-blind placebo controlled trial. *Acta Med Iran* 52:734–739.

Newcomer, J. W. (2005). Second-generation (atypical) antipsychotics and metabolic effects: a comprehensive literature review. *CNS Drugs* 19, 1−93.

Olfson M, Gerhard T and Huang C (2015) Pre-mature mortality among adults with schizophrenia in the United States. *JAMA Psychiatry* 72: 1172–1181.

Osborn DPJ, Levy G, Nazareth I, (2007) Relative risk of cardiovascular and cancer mortality in people with severe mental illness from the United Kingdom's general practice research database. *Arch Gen Psychiatry* 64: 242–249.

Osborn D, Hardoon S, Omar RZ (2015) Cardiovascular risk prediction models for people with severe mental illness. Results from the Prediction and Management of

Cardiovascular Risk in People with Severe Mental Illnesses (PRIMROSE) Research Program. *JAMA Psychiatry* 72: 143–151.

Pandi-Perumal SR, Srinivasan V, Maestroni GJ, Cardinali DP, Poeggeler B, Hardeland R (2006). Melatonin: nature's most versatile biological signal? *FEBS J* 273:2813–2838.

Prunet-Marcassus B, Desbazeille M, Bros A, Louche K, Delagrange P, Renard P (2003). Melatonin reduces body weight gain in Sprague Dawley rats with diet-induced obesity. *Endocrinology* 144:5347–5352.

Puchalski SS, Green JN, Rasmussen DD (2003). Melatonin effect on rat body weight regulation in response to high-fat diet at middle age. *Endocrine* 21:163–167.

Raskind MA, Burke BL, Crites NJ, Tapp AM, Rasmussen DD (2007). Olanzapine induced weight gain and increased visceral adiposity is blocked by melatonin replacement therapy in rats. *Neuropsychopharmacology* 32:284–288.

Rasmussen DD, Boldt BM, Wilkinson CW, Yellon SM, Matsumoto AM (1999). Daily melatonin administration at middle age suppresses male rat visceral fat, plasma leptin, and plasma insulin to youthful levels. *Endocrinology* 140:1009–1012.

Reiter RJ, Tan D-X, Korkmaz A, Ma S (2012) Obesity and metabolic syndrome: Association with chronodisruption, sleep deprivation, and melatonin suppression *Annals of Medicine* 44: 564–577.

Reynolds GP. (2012). Pharmacogenetic aspects of antipsychotic drug induced weight gain – a critical review. *Clin Psychopharmacol Neurosci.* 10:71-7.

Reynolds GP, Kirk SL. (2010). Metabolic side effects of antipsychotic drug treatment – pharmacological mechanisms. *Pharmacol Ther* 125:169-79.

Roerig JL, Mitchell JE, de Zwaan M (2005) A comparison of the effects of olanzapine and risperidone versus placebo on eating behaviors. *J Clin Psychopharmacol* 25:413–418.

Romo-Nava F, Alvarez-Icaza González D, Fresán-Orellana A, Saracco Alvarez R, Becerra-Palars C, Moreno J (2014). Melatonin attenuates antipsychotic metabolic effects: an eight-week randomized, double-blind, parallel-group, placebo-controlled clinical trial. *Bipolar Disord* 16:410–421.

Sateia MJ, Kirby-Long P, Taylor JL. Efficacy and clinical safety of ramelteon: an evidence-based review. *Sleep Med Rev.* 2008; 12:319–332.

Sacher J, Mossaheb N, Spindelegger C, Klein N, Geiss-Granadia T, Sauermann R, Lackner E, Joukhadar C, Muller M, Kasper S (2008). Effects of Olanzapine and Ziprasidone on Glucose Tolerance in Healthy Volunteers. *Neuro psycho pharmacology.* 33, 1633–1641.

Savini I, Catani MV, Evangelista D, Gasperi V, Avigliano L. (2013). Obesity-Associated Oxidative Stress: Strategies Finalized to Improve Redox State. *Int. J. Mol. Sci.* 14: 10497-10538.

Smith GC, Vickers MH, Cognard E, Shepherd PR (2009) Clozapine and quetiapine acutely reduce glucagon-like peptide-1 production and increase glucagon release in obese rats: implications for glucose metabolism and food choice behaviour. *Schizophr. Res.* 115, 30–40.

Snigdha S, Thumbi C, Reynolds GP, Neill JC (2008) Ziprasidone and aripiprazole attenuate olanzapine-induced hyperphagia in rats. *J. Psychopharmacol.* 22, 567–571.

Sowell M, Mukhopadhyay N, Cavazzoni P (2003) Evaluation of insulin sensitivity in healthy volunteers treated with olanzapine, risperidone, or placebo: a prospective, randomized study using the two-step hyperinsulinemic, euglycemic clamp. *J Clin Endocrinol Metab* 88:5875–5880.

Sowell M, Mukhopadhyay N, Cavazzoni P (2002) Hyperglycemic clamp assessment of insulin secretory responses in normal subjects treated with olanzapine, risperidone, or placebo. *J Clin Endocrinol Metab* 87: 2918–2923.

Teff KL, Rickels MR, Grudziak J, Fuller C, Nguyen HL, Rickels K (2013). Antipsychotic-induced insulin resistance and postprandial hormonal dysregulation independent of weight gain or psychiatric disease. *Diabetes* 62:3232–3240.

Terrón MP, Delgado-Adámez J, Pariente JA, Barriga C, Paredes SD, Rodríguez AB (2013). Melatonin reduces body weight gain and increases nocturnal activity in male Wistar rats. *Physiol Behav* 118:8–13.

Tiihonen J, Lonnqvist J, Wahlbeck K. (2009) 11-Year follow-up of mortality in patients with schizophrenia: a population-based cohort study (FIN11 study). *Lancet* 374: 620–627.

Vidarsdottir S, de Leeuw van Weenen JE, Frö"lich M, Roelfsema F, Romijn JA, Pijl H (2010). *Effects of Olanzapine and Haloperidol on the Metabolic Status of Healthy Men J Clin Endocrinol Metab.* 95:118–125

Wolden-Hanson T, Mitton DR, McCants RL, Yellon SM, Wilkinson CW, Matsumoto AM, Rasmussen DD (2000). Daily melatonin administration to middle-aged male rats suppresses body weight, intraabdominal adiposity, and plasma leptin and insulin independent of food intake and total body fat. *Endocrinology* 141:487–497.

Wyatt JK, Dijk DJ, Ritz-de Cecco A, Ronda JM, Czeisler CA (2006). Sleep facilitating effect of exogenous melatonin in healthy young men and women is circadian-phase dependent. *Sleep* 29:609–618.

BIOGRAPHICAL SKETCH

Trevor Ronald Norman

Address: Department of Psychiatry, University of Melbourne, Austin Hospital, Heidelberg, Victoria 3084, Australia. Tel: (03) 9496-5680, FAX (03) 9459-0821, Email: trevorrn@unimelb.edu.au.

Present Position: Associate Professor, Honorary Principal Fellow, Department of Psychiatry, University of Melbourne.

Qualifications: BSc (University of Adelaide) 1969
BSc (Hons) University of Adelaide) 1970
PhD (University of Adelaide) 1974

Publications (Last Five Years)

Bochsler, L., Olver, J. S. and Norman, T. R. A review of duloxetine in the acute and continuation treatment of major depressive disorder. *Expert Reviews on Neurotherapeutics*, 11: 1525-1539 (2011).

Norman, T. R. The Decline and fall of the psychopharmacology empire? Australian and *New Zealand Journal of Psychiatry*, 45: 1005 (2011).

Norman, T. R., Cranston, I., Irons, J. A., Gabriel, C., Dekeyne, A., Millan, M. J. and Mocaër, E. Agomelatine suppresses locomotor hyperactivity in olfactory bulbectomised rats: A comparison to melatonin and to the $5-HT_{2C}$ antagonist, S32006, *European Journal of Pharmacology*, 674: 310-312 (2012).

Norman, T. R. The effect of agomelatine on 5HT2C receptors in humans: a clinically relevant mechanism? *Psychopharmacology* 221: 177-178, (2012)

Norman, T. R. Agomelatine, melatonin and depressive disorder, *Expert Opinion on Investigational Drugs*, 22: 407-410, (2013).

Hamer, J., Norman, T. R. and Kanaan, R. K. One year treatment continuation in patients switched to paliperidone palmitate: A retrospective study, *Journal of Clinical Psychopharmacology*, 75: 1267-1269 (2014).

Olver, J. S., Pinney, M., Maruff, P. and Norman, T. R. Impairments of spatial working memory and attention following acute psychosocial stress, *Stress and Health*, 31: 115–123, (2015).

Bryson, A., Carter, O., Norman, T. R. and Kanaan, R. $5HT_{2A}$ agonists: A novel therapy for functional neurological disorders? *International Journal of Neuropsychopharmacology*, 20: 422-427, (2017).

Norman, T. R. Problematic medical marijuana, but not all cannabinoids? Australian and *New Zealand Journal of Psychiatry*, Accepted for publication (2017).

Norman, T. R., Activity of fluoxetine in animal models of depression and anxiety, In: *Fluoxetine/SSRI: Pharmacology, Mechanisms of Action and Potential Side Effects*, ed. G. Pinna, Nova Science Publishers, Hauppauge, NY, USA, 2014, pp

Norman, T. R., Prospects for the development of animal models of bipolar disorder, In: *Bipolar Disorders: Basic Mechanisms and Therapeutic Implications,* 3rd edition. Ed. J. Soares and A. Young, Cambridge University Press, Cambridge, UK, 2016. pp 8-20.

Norman, T. R., Agomelatine, melatonin and depression, In: *Melatonin, Neuroprotective Agents and Antidepressant Therapy*, Eds F. López-Muñoz, V. Srinivasan, D. de Berardis, C. Álamo and T. A. Kato, Springer, India, 2016, pp 229-247.

Norman, T. R., Multi-modal pharmacological treatments for major depressive disorder: Testing the hypothesis, In: *Frontiers in Clinical Drug Research – Central Nervous System*, Volume 2, Ed Atta-ur-Rahman, Bentham Science Publishers, United Arab Emirates, 2016, pp 203-231.

Norman, T. R., Pharmacotherapy for mood and anxiety disorders, In: *Mental Health and Illness Worldwide. Mental Health and Illness of the Elderly*, Eds H. Chiu, K. Shulman, D. Ames. Springer, 2017, pp 1-27.

In: Melatonin: Medical Uses and Role in Health and Disease   ISBN: 978-1-53612-987-8
Editors: Lore Correia and Germaine Mayers   © 2018 Nova Science Publishers, Inc.

Chapter 10

MELATONIN: BENEFICIAL ASPECTS AND UNDERLYING MECHANISMS

Stephen C. Bondy[*]

Environmental Toxicology Program,
Center for Occupational and Environmental Health,
Department of Medicine, University of California,
Irvine, CA, US

ABSTRACT

Melatonin has a reportedly beneficial effect in the treatment of numerous disorders. While it is not certain what the most important primary mechanism of melatonin's actions is, its application can lead to a variety of favorable outcomes in a range of pathological conditions. In addition to the prevention or amelioration of disease states, melatonin has attributes that can be of benefit to the normal healthy organism. The most notable of these, is the potential of melatonin to slow down the progression of several indices characterizing the aging process. The incidence of many secondary undesirable states associated with aging, may also be reduced simply by deceleration of normal aging. The utility of melatonin is further enhanced by its very low level of toxicity and absence of any capacity to induce or promote tumor development. As a natural product, melatonin cannot be patented and is thus inexpensive and readily available. Melatonin can be regarded either as a dietary supplement or as a remedy. A range of clinical reports supports the validation of the therapeutic use of melatonin for many disorders. These are buttressed by parallel descriptions of studies on experimental animal models. This review is intended firstly as a short summary of the range of disorders and conditions, where there is evidence that they may be improved by melatonin administration. This is followed by consideration of the processes that may underlie these advantageous changes. It is proposed that melatonin's ability to retard aging events may be the source

---

[*] Corresponding Author: Stephen C. Bondy, Ph. D. Center for Occupational & Environmental Health, Department of Medicine, University of California, Irvine, 100 Theory, Suite 100, Irvine, CA 92617-1830, US, Email: scbondy@uci.edu; Tel.: 949 824 8077; Fax: 949 824 2070.

of many of its positive attributes. Finally, the conclusion looks forward to future work focusing on the possibility of studies investing the possible advantages conferred by melatonin on the epigenetic profile.

Keywords: melatonin, clinical, aging, inflammation, oxidative stress

1. INTRODUCTION

Melatonin was originally recognized for its role in biological rhythms, its daily variations displaying high nocturnal levels changing seasonally according to photoperiod act as a clock and a calendar. This circadian role remains an important feature of melatonin. However, an increasing number of properties and functions have been identified for this neurohormone in the last 25 years. This issue summarizes the current status of melatonin as a possible treatment for a wide range of diseases. This specific article will briefly outline some major areas where melatonin treatment has been described as having utility in a broad range of clinical areas. These reports are based to a large extent on promising results obtained in animal models and in isolated cell systems and only to a lesser extent by unambiguous, well-controlled clinical accounts. This listing is incomplete, since many of the areas are described in greater detail in other reviews in this special issue more specifically focused on a defined disorder. The main thrust of this article is to examine what molecular targets underlie the ability of melatonin to ameliorate a large number of adverse health issues, which are not always very closely related from a medical perspective. It is likely that the many beneficial properties of melatonin reside in a much smaller number of primary sites of action.

2. DISORDERS WHERE MELATONIN ADMINISTRATION HAS BEEN REPORTED AS BENEFICIAL

2.1. Diseases Where Immune Function Is Compromised

2.1.1. Cancer

Melatonin treatment of aged experimental animals has been shown to lead to a major reduction of overall incidence of tumors [1]. In this study, male mice were given dietary melatonin (40 ppm) in a series of studies performed over nine years. Animals were sacrificed at 26 months age and examined for tumors, which are commonly found in many strains of mice during senescence. Such aged mice previously supplemented for 3 months with dietary melatonin, had 60% fewer tumors than untreated controls. Moreover, the size of any tumors found in melatonin treated mice was markedly smaller than those present in control mice. Since this study was conducted over an extended period of time, with many groups of animals, potentially confounding factors such as seasonal or dietary variations are unlikely to have interfered with the findings. The use of several mouse strains strengthens the likelihood that the study may have general relevance. The overall spontaneous death rate was significantly reduced in mice receiving melatonin, so that the total mortality in the most aged mice was halved, from 18% to 9%. There is also evidence of both reduced incidence of tumors and increased latency of their development, in animals exposed to carcinogens [2-4]. These findings are likely to be relevant to human populations, as age-related changes in gene expression in mice, have close parallels in humans [5]. In humans, the inclusion of melatonin improves the efficacy of standard chemotherapeutic agents and radiation procedures, and reduces the incidence of their undesirable side effects [6]. Therefore, melatonin offers the possibility of dramatic reductions in cancer incidence among human populations at little expense and with minimal risk of toxic side effects.

2.1.2. Multiple Sclerosis

Multiple sclerosis is a neurodegenerative disease causing demyelination. The disease characteristically has a waxing and waning nature and relapses often have a seasonal relation. Circulating melatonin levels, whose production is modulated by seasonal variations in night length, negatively correlate with multiple sclerosis activity in humans. Since melatonin ameliorates an experimental analogue of multiple sclerosis, experimental autoimmune encephalitis [7], this has implications for its use as an adjunct therapy for multiple sclerosis.

2.1.3. Wound Healing and Repair

Wound healing is reported as delayed in pinealectomized animals. This effect is largely reversed by administration of melatonin to these animals [8]. There are also reports of increased regenerative effects in several other tissues form animal studies, including crushed muscle [9]. A few clinical reports in this area indicate that melatonin may be a valuable adjunct in humans in speeding up tissue repair such as healing of duodenal
ulcers [10].

2.2. Diseases Associated with Persistent Inflammation

2.2.1. Sepsis

Sepsis is a common and severe complication of many disorders and has a high mortality rate. However, there is a large body of evidence derived both from animal studies and from a few limited human studies, which strongly suggests that melatonin may have significant utility in the treatment of sepsis [11, 12]. More work in this area is likely to be very rewarding.

2.2.2. Edema

A broadly protective effect of melatonin upon several types of induced edema has been reported. This includes hypoxic-induced systemic and cerebral edema [13]. Acute damage to the blood brain barrier, accompanied by vasogenic edema, resulting from following traumatic brain injury, is attenuated by melatonin treatment [14]. The only clinical report of melatonin use in the context of edema concerns central serous chorioretinopathy, in which when fluid builds up under the retina, which can lead to retinal detachment. Melatonin is effective in the treatment of this disorder [15].

2.2.3. Hypertension

Melatonin has generally been reported to depress blood pressure but this is not always the case. Melatonin may depress nocturnal blood pressure while it increases it during the day [16]. The basis for somewhat inconsistent findings in this area is perhaps that, in humans exogenous melatonin differentially alters vascular blood flow [17]. Fluctuations in daily expression of melatonin receptors may also account for variability. Although melatonin may be beneficial for certain types of nocturnal hypertension [18], it is unlikely to have widespread utility in the treatment of hypertension.

2.2.4. Cerebral Ischemia and Injury

In experimental animals' models, melatonin has been reported to be neuroprotective after hypoxic-ischemic brain injury [19]. Compared to animals subjected to transient hypoxia in the absence of melatonin, animals treated with this agent just after surgery exhibited much less TUNEL staining, evidence of cell loss and astroglial activation (as revealed by GFAP staining). Melatonin treatment may also constitute a promising approach to traumatic brain injury [20]. There are very few clinical reports in this area but melatonin has been shown to have value in effecting improved outcomes following neonatal hypoxic-ischemic encephalopathy [21]. This is potentially valuable approach to neuroprotection following hypoxic-ischemic brain injury, and other clinical studies are currently in progress [22]. Melatonin appears to promote neurogenesis following induction of various forms of ischemia [23-25]. A similar induction of neurogenesis has been reported following cranial irradiation [26].

The protective effect of melatonin against cardiac ischemia-reperfusion injury has been well documented in experimental animals and in isolated cell systems [27]. There are also a limited number of clinical reports of positive results in treatment of cardiovascular disease [28, 29].

2.2.5. Alzheimer's Disease

Melatonin is as major regulator of circadian rhythms but also displays antioxidant and anti-inflammatory properties that may have implications in Alzheimer's disease. For this reason, it has often proved useful in restoring an aberrant sleep cycle. In some reports, melatonin treatment has failed to either improve sleep efficacy or "sundowning" agitation in Alzheimer's (AD) patients [30], and in yet other accounts, melatonin has been found supportive of both sleep maintenance and cognitive functioning [31]. In a randomized meta-analysis of 7 well-controlled trials, this quality has been reported as beneficial in improving sleep in various dementias but with no beneficial effects on cognitive function [32]. These conflicting reports are in contrast to several reports using transgenic animal models of AD, where melatonin has consistently been found protective, preventing appearance of both morphological markers of AD and behavioral indices of neurological deficits [33-35]. Failure to convert beneficial pharmacological effects found using such models to the human condition is unfortunately rather common with many other drugs intended to ameliorate AD. The use of doses of melatonin several orders of magnitude greater than those generally used to induce sleep, or very potent melatonin analogs [36], represents a rather desperate means of resolving this problem.

2.2.6. Organ Transplantation

Melatonin has been posited as a means of reducing the extent of graft rejection following organ [37]. This is based on successful use of melatonin in a wide range of animal models of transplantation but very few clinical reports exist in the area as yet. This is a promising approach but there remain two significant caveats. Firstly, as in the case of AD described above, success in animal models can often lead to disappointment when applied to the human condition where more chronic factors can play a critical role. Secondly, the use of very high levels of melatonin in several of these trials such as 200-500 mg/kg daily [38], which corresponds to a human dose of up to 35 g per adult, may lead to unforeseen side effects. This is in contrast to the widespread usage of melatonin at levels of 1-3 mg daily. Melatonin cannot be assumed to be non-toxic at such levels many orders of magnitude higher than physiological. In fact the response of inflammatory cytokines to various high levels of administered melatonin, is biphasic, leading to heighted inflammation at some levels [39]. The sites upon which very high levels of melatonin may act, cannot be predicted by what is known about its normal physiology.

2.3. Diseases thought to Involve Pro-Oxidant Conditions

2.3.1. Parkinson's Disease

The use of melatonin in treatment of Parkinson's disease (PD) was proposed by Cotzias as long ago as 1973 [40]. The beneficial effects of melatonin on the symptoms of

PD in both chemical and genetic animal models of this disorder are marked [41-43]. However, the effect of melatonin in PD models is complex and multifactorial. Both melatonin agonists and antagonists have been described as of utility in the treatment of PD. The blockade of MT2 receptors may exert antidepressant effects while MT2 agonists can increase striatal dopamine levels [44]. As with other neurodegenerative disorders, translation of findings from model systems to the human disease has been equivocal. While the quality of sleep in PD patients is significantly improved, other aspects of PD remain unaffected [45].

2.3.2. Epilepsy

There are several reports of seizure control by melatonin and its agonists, in experimental animals [46, 47]. While there are also clinical descriptions of seizure reduction by melatonin [48], a recent meta-analysis found few reports with adequate methodological quality, systematic evaluation of seizure frequency and adverse events, and so was unable to draw a conclusion [49]. While not possessing strong anti-seizure activity directly, melatonin may be a useful adjunct therapy when combined with classical anti-convulsants such a barbiturate [50].

2.4. Disorders Involving Disrupted of Circadian Rhythms and Sleep Disturbance, Where Excessive Inflammation Contribute to Pathology

2.4.1. Health-associated Issues Relating to Nocturnal Shift Work

Prolonged exposure to light can reduce life span, and induce tumorigenesis in both rats and mice. This can be ameliorated by administration of melatonin [51]. This is paralleled by reports of the likely carcinogenic effects of circadian disruption by night shift work among humans, especially with respect to breast cancer [52]. It is relevant that blind women appear to have a reduced risk of breast cancer [53]. There is evidence that night shift work can lead to elevation several markers of inflammation including C-reactive protein [54].

2.4.2. Autistic Spectrum Disorder

An abnormal circadian sleep–wake cycle has been repeatedly reported in autism. This complex disease involving impaired social communication has been associated with asynchrony of other biological rhythms [55]. Melatonin given at appropriate times may not merely, improve sleep quality but may also improve communication, reduce social withdrawal, and decrease stereotyped behaviors [55]. Large-scale studies are needed to place these preliminary suggestions on a firmer basis. Autism has been found to involve to excessive basal levels of inflammation [56].

2.4.3. Alzheimer's Disease

The trials of melatonin in the treatment of 'sundowning' are described above in the section on Alzheimer's disease.

The preceding list of disorders is not intended to be exhaustive. For example derangement of immune function and sleep disorders are both known to be present in post-traumatic stress syndrome (PTSD) and melatonin efficacy has been discussed in this context [57]. However, an underlying feature of this incomplete list is the ubiquity of deranged immune responses and hazard of excess production of oxidant radicals in disorders where there are suggestions of the utility of melatonin therapy.

3. MECHANISMS UNDERLYING MELATONIN PHYSIOLOGY

Most of the conditions reported to benefit from melatonin treatment express both evidence of excess inflammation conditions and signs of oxidative stress. Many of these disorders have been reported to improve after administration of either antioxidant or anti-inflammatory agents. It is very likely that melatonin's properties combine both of these attributes. To discern whether primary the basis of melatonin actions is antioxidant or immunomodulatory is very difficult due to the close inter-relation between these mechanisms. Oxidative and nitrosative pathways are induced by inflammatory responses, but oxidative stress may initiate events by upregulation of the expression of inflammatory genes via activation of redox-responsive transcription factors. The very low tissue concentration of free melatonin, around 1pM [58] is many orders of magnitude less than that of the primary water soluble antioxidant within the cell, glutathione, (present in mM amounts) or the lipophilic α-tocopherol. Interactions between hydrophilic and lipophilic antioxidants and lipid/water interfaces, allow the transfer of the ability to quench reactive oxygen species throughout the cell. While amphiphilic is unlikely to have a significant direct anti-oxidant effect, in view of the major circadian flux of melatonin levels, any such effect would be strongest by night [59]. Melatonin more probably acts by way of its receptors, whose activation may enable the transfer of key transcription factors to the nucleus, thereby activating a suite of an-inflammatory and anti-oxidant genes. Indeed, melatonin induction of Nrf-2 followed by consequent activation of the antioxidant response element (ARE) has often been described [60,61]. Regulation of these genes, generally accompanied by down-regulation of the inflammatory NF-κB transcription factor [62], broadly expands defensive responses. Their activation can subdue inflammatory and pro-oxidant events, while their inhibition can lead to oxidative stress and persistent inflammatory events. Many cardioprotective effects of melatonin are completely blocked in the presence of inhibitors of MT1 and MT2 receptors [63], and genetic deletion of these receptors alters cognitive and motor behavior [64]. However, some neuroprotective events triggered by melatonin, involve pathways other than by known melatonin receptors [65, 66].

A general mechanistic trajectory may be by melatonin binding to a receptor thus leading to activation of a second messenger (e.g., calcium entry into the cell of altered levels of cyclic nucleotides). This is followed by kinase-effected or proteolytic activation or de-ubiquitination of transcription factors, which leads to their nuclear translocation. Once within the nucleus, active transcription factors such as Nrf-2, can effect derepression of a key suite of genes coding for antioxidant and immune modulation enzymes. Other factors may reduce the expression of less well-defined sets of inflammation-related genes.

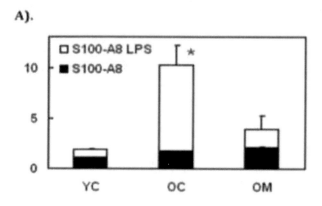

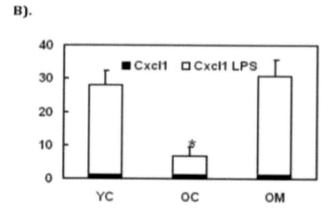

Figure 1. Effect of age and dietary melatonin on cortical expression of mRNAs for 2 chemokines, following an inflammatory challenge. A). mRNA for chemokine S100-A8, B). mRNA forCxcl1 YC = 6 month old mice, OC = 27 month-old mice, OM = old mice receiving 40 ppm dietary melatonin for 15 weeks. LPS = single i.p. injection of 30 µg lipopolysaccharide 4 hours prior to sacrifice. Scale value of untreated control = 1. *: P<0.05 that value differs from corresponding control value [adapted from 70].

Melatonin cannot be considered to be a simple non-selective anti-oxidant or anti-inflammatory agent. The sophisticated nature of melatonin regulation is illustrated by the fact that, in normal cells, melatonin increases levels of the anti-apoptic gene, Bcl-2 [67] while in malignant cells; Bcl-2 levels are decreased [68]. Abnormal forms of this gene are oncogenic, involved in many tumor types [69]. Opposing effects such as this may account for the pro-apoptotic effect of melatonin on tumor cells and the stimulation of cell division and enhanced cell survival effected in normal tissues. A similar apparently contradictory effect is seen when considering the immune response of tissues to melatonin. The response of mice to the inflammogen, lipopolysaccharide (LPS), is exacerbated with aging in the case of expression of chemokine S100-A8, but attenuated for chemokine Cxcl1. In either case however, pretreatment with melatonin caused the response profile of older animals to revert to that of the younger [70]. Consequently melatonin can either inhibit or augment the inflammatory response of aged animals but always in a direction toward more closely resembling the reaction of young animals (Figure 1). Melatonin should thus be considered as a modulator rather than an indiscriminate repressor of immune and inflammatory events.

4. Hypothesis: Much of the Protective Effect of Melatonin May Be by Way of Reversal of Age-Related Changes in Gene Expression

In recent years it has been widely recognized that extended inflammation forms part of the aging process [71]. This inflammation is especially pronounced in nervous tissue, which seems to retain memory of previous inflammatory events long after the original stimulus has dissipated [72, 73]. This may account for the finding that during aging, the expression of many genes relating to immune function and inflammation, is selectively elevated. It has recently been reported that immune stimulation leads to impaired and more heterogeneous responses in cells derived from aged animals. Immune stimulation up-regulated a core activation program in young animals but with age, this was disrupted, leading to increased expression heterogeneity [74]. The administration of dietary melatonin to aged animals, can reverse this trend in the brain, and lead to a less diffuse overall gene expression profile [75], which then closely resembles that of much younger animals [76, 77]. Although gene expression and protein levels are not always fully consonant [78], the restoration of basal gene expression to a more youthful state is likely to result in a diminution of basal levels of inflammatory cytokines. Melatonin is a good candidate for attenuation of inflammatory conditions associated with aging [79]. The subduing of inappropriate inflammation in the absence of a provocative stimulus, also allows a more effective immune response to be mounted in the presence of an inflammogen [80]. Such adjustment of expression of genes relating to immune function may be the cause of many of the advantageous properties of melatonin.

CONCLUSION

Melatonin administration, as an adjuvant therapy, can be applied to a wide range of conditions. In fact, its use has been advocated as part of a spectrum of nutritional agents that may be useful in the management of several age-related neurodegenerative disease such as Parkinson's disease [81]. Melatonin is not essential for life but its deficiency in pinealectomized animals, results in a less healthy organism, not as able to respond to physiological challenges [82]. Melatonin treatment is not a panacea and in fact generally does not establish a 'cure' for most disorders. However, what it almost invariably accomplishes, is to lead to a moderate but consistent metabolic 'push' in a beneficial direction. This positive pressure allows melatonin to be a valuable adjunct in the treatment of very many disorders. Perhaps its widespread utility reflects a commonality between many disease states, such as the presence of excessive and futile inflammation, and heightened pro-oxidant conditions. Mitigation of these adverse factors can initiate a major step toward correction of more specific pathological events. Melatonin is non-toxic at most levels, and thus it is a safe and inexpensive pharmaceutical agent. It follows that its utility may be extended by its usage by healthy older individuals whose endogenous production of this neurohormone has declined. This would tend alleviate the undesirable effects of senescence and could delay the onset of many chronic age-related disorders.

Future research concerning processes that explain the favorable properties of melatonin are likely to emphasize epigenetic changes triggered by melatonin [83, 84] including the activation of some sirtuins [85, 86]. As is discussed in a recent review, melatonin has the opposite effect on tumor cells where SIRT1 is inhibited [87]. This again illustrates the subtle regulatory effects of melatonin. Alteration of expression of miRNAs by melatonin is also going to be an area of increasing interest. This level of regulation promises to a key means of accounting for some of melatonin's anticancer effects [88, 89]. Melatonin also blocks the effects of LPS on several miRNAs, and this epigenetic control may be another means by which it is protective [90]. Other melatonin-based epigenetic promotions may play a role in retardation of senescence [91].

The paucity of reliable clinical data, despite very strong suggestive evidence that more widespread usage of melatonin could have major positive effects, is a serious shortcoming. Perhaps this is in part due to the reluctance of pharmaceutical companies to support studies, which promise very limited financial returns. However, from a societal health perspective, better well-controlled studies with a large number of subjects are urgently needed. Placing information concerning the value of melatonin therapy on a firmer and more extensive clinical footing could dramatically change the overall profile of population health.

REFERENCES

[1] Sharman, E. H.; Sharman, K. G.; Bondy, S. C. Extended exposure to dietary melatonin reduces tumor number and size in aged male mice. *Exp. Gerontol.* 2011, *46*, 18-22.

[2] Cos, S.; Sanchez-Barcelo, E. J. Melatonin and mammary pathological growth. *Front. Neuroendocrinol.* 2000, *21*, 133-170.

[3] Lenoir, V.; de Jonage-Canonico, M. B.; Perrin, M. H.; Martin, A.; Scholler, R.; Kerdelhué, B. Preventive and curative effect of melatonin on mammary carcinogenesis induced by dimethylbenz[a]anthracene in the female Sprague-Dawley rat. *Breast Cancer Res.* 2005, *7*, R470-476.

[4] Anisimov, V. N.; Egormin, P. A.; Piskunova, T. S. Metformin extends life span of HER-2/neu transgenic mice and in combination with melatonin inhibits growth of transplantable tumors in vivo. *Cell Cycle* 2010, *9*, 188-197.

[5] Sharman, E. H.; Sharman, K. G.; Bondy, S. C. Parallel changes in gene expression in aged human and mouse cortex. *Neurosci. Lett.* 2005, *390*, 4-8.

[6] Sanchez-Barcelo, E. J.; Mediavilla, M. D.; Alonso-Gonzalez, C.; Reiter, R. J. *Expert Opin. Investig. Drugs* 2012, *21*, 819-831.

[7] Farez, M. F.; Mascanfroni, I. D.; Méndez-Huergo, S. P.; Yeste, A.; Murugaiyan, G.; Garo, L P.; Balbuena Aguirre, M. E.; Patel, B.; Ysrraelit, M. C.; Zhu, C.; Kuchroo, V. K.; Rabinovich, G. A.; Quintana, F. J.; Correale, J. Melatonin contributes to the seasonality of multiple sclerosis relapses. *Cell* 2015, *162*, 1338-1352.

[8] Ozler, M.; Simsek, K.; Ozkan, C.; Akgul, E. O.; Topal, T.; Oter, S.; Korkmaz, A. Comparison of the effect of topical and systemic melatonin administration on delayed wound healing in rats that underwent pinealectomy. *Scand. J. Clin. Lab. Invest.* 2010, *70*, 447-452.

[9] Stratos, I.; Richter, N.; Rotter, R.; Li, Z.; Zechner, D.; Mittlmeier, T.; Vollmar, B. Melatonin restores muscle regeneration and enhances muscle function after crush injury in rats. *J. Pineal Res.* 2012, *52*, 62-70.

[10] Celinski, K.; Konturek, P. C.; Konturek, S. J.; Slomka, M.; Cichoz-Lach, H.; Brzozowski.; T.; Bielanski, W. Effects of melatonin and tryptophan on healing of gastric and duodenal ulcers with Helicobacter pylori infection in humans. *J. Physiol. Pharmacol.* 2011, *62*, 521-526.

[11] Sharman, E. H. and Bondy, S. C. Melatonin as a therapeutic agent for sepsis. In *Melatonin, Therapeutic Value and Neuroprotection,* Srinivasan, V; Gobbi, G. V.; Shillcutt, S. D.; Suzen S. Eds.; CRC Press, Boca Raton, Florida, USA, 2014; pp. 71-88.

[12] Hu, W.; Deng, C.; Ma, Z.; Wang, D.; Fan, C.; Li, T.; Di, S.; Gong, B.; Reiter, R J.; Yang, Y. *Br. J. Pharmacol.* 2017, *174*, 754-768.

[13] Xu, L. X.; Lv, Y.; Li, Y. H.; Ding, X.; Wang, Y.; Han, X.; Liu, M. H.; Sun, B.; Feng, X. Melatonin alleviates brain and peripheral tissue edema in a neonatal rat model of hypoxic-ischemic brain damage: the involvement of edema related proteins. *BMC Pediatr.* 2017, *17*, 90.

[14] Alluri, H.; Wilson, R. L.; Anasooya Shaji, C.; Wiggins-Dohlvik, K.; Patel, S.; Liu, Y.; Peng, X.; Beeram, M. R.; Davis, M. L.; Huang, J. H.; Tharakan, B. Melatonin preserves blood-brain barrier integrity and permeability via matrix metalloproteinase-9 inhibition. *PLoS One* 2016, *11*, e0154427.

[15] Gramajo, A. L.; Marquez, G. E.; Torres, V. E.; Juárez, C. P.; Rosenstein, R. E.; Luna, J. D. Therapeutic benefit of melatonin in refractory central serous chorioretinopathy. *Medscape. Eye* (Lond). 2015, *29*, 1036-1045.

[16] Rechciński, T.; Trzos, E.; Wierzbowska-Drabik, K.; Krzemińska-Pakulea, M.; Kurpesa, M. Melatonin for nondippers with coronary artery disease: assessment of blood pressure profile and heart rate variability. *Hypertension Res.* 2010, *33*, 56–61.

[17] Cook, J. S.; Sauder, C. L.; Ray, C. A. Melatonin differentially affects vascular blood flow in humans. *Am. J. Physiol. Heart Circ. Physiol.* 2011, *300*, H670–H674.

[18] Grossman, E.; Laudon, M.; Yalcin, R.; Zengil, H.; Peleg, E.; Sharabi, Y.; Kamari, Y.; Shen-Orr, Z.; Zisapel, N. Melatonin reduces night blood pressure in patients with nocturnal hypertension. *Am. J. Med.* 2006, *119*, 898-902.

[19] Alonso-Alconada, D.; Alvarez, A.; Lacalle, J.; Hilario, E. Histological study of the protective effect of melatonin on neural cells after neonatal hypoxia-ischemia. *Histol. Histopathol.* 2012, *27*, 771–783.

[20] Naseem, M.; Parvez, S. Role of melatonin in traumatic brain injury and spinal cord injury. *Scientific World J.* 2014, *2014*, 586270.

[21] Aly, H.; Elmahdy H.; El-Dib M.; Rowisha, M., Awny, M.; El-Gohary, T.; Elbatch, M.; Hamisa, M.; El-Mashad, A. R. Melatonin use for neuroprotection in perinatal asphyxia: a randomized controlled pilot study. *J. Perinatol.* 2015, *35*, 186-191.

[22] Juul, S. E.; Ferriero, D. M. Pharmacological neuroprotective strategies in neonatal brain injury. *Clin. Perinatol.* 2014, *41*, 119–131.

[23] Kilic, E.; Kilic, U.; Bacigaluppi, M.; Guo, Z.; Abdallah, N. B.; Wolfer, D. P.; Reiter, R. J.; Hermann, D. M.; Bassetti, C. L. Delayed melatonin administration promotes neuronal survival, neurogenesis and motor recovery, and attenuates hyperactivity and anxiety after mild focal cerebral ischemia in mice. *J. Pineal Res.* 2008, *45*, 142-148.

[24] Ajao, M. S.; Olaleye, O.; Ihunwo, A. O. Melatonin potentiates cells proliferation in the dentate gyrus following ischemic brain injury in adult rats. *J. Anim. Vet. Adv.* 2010, *9*, 1633–1638.

[25] Chern, C. M.; Liao, J. F.; Wang, Y. H.;Shen. Y. C. Melatonin ameliorates neural function by promoting endogenous neurogenesis through the MT2 melatonin receptor in ischemic-stroke mice. *Free Radic. Biol. Med.* 2012, *52*,1634–1647.

[26] Manda, K.; Ueno, M.; Anzai, K. Cranial irradiation-induced inhibition of neurogenesis in hippocampal dentate gyrus of adult mice: attenuation by melatonin pretreatment. *J. Pineal Res.* 2009, *46*, 71–78.

[27] Yang, Y.; Sun, Y.; Yi, W.; Li, Y.; Fan, C.; Xin, Z.; Jiang, S.; Di, S.; Qu, Y.; Reiter RJ.; Yi, D. A review of melatonin as a suitable antioxidant against myocardial ischemia-reperfusion injury and clinical heart diseases. *J. Pineal Res.* 2014, *57*, 357-366.

[28] Zaslavskaia, R. M.; Shcherban, E. A.; Lilitsa, G. V.; Logvinenko, S. I. Melatonin in the combined treatment of cardiovascular diseases. *Klin. Med. (Mosk).* 2010, *88*, 26-30.

[29] Gögenur, I.; Kücükakin, B.; Panduro Jensen, L.; Reiter, R. J.; Rosenberg, J. Melatonin reduces cardiac morbidity and markers of myocardial ischemia after elective abdominal aortic aneurism repair: a randomized, placebo-controlled, clinical trial. *J. Pineal Res.* 2014, *57*, 10-5.

[30] Gehrman, P. R.; Connor, D. J.; Martin, J. L.; Shochat, T.; Corey-Bloom, J.; Ancoli-Israel, S. Melatonin fails to improve sleep or agitation in double-blind randomized placebo-controlled trial of institutionalized patients with Alzheimer disease. *Am. J. Geriatr. Psychiatry.* 2009, 17, 166-169.

[31] Wade, A. G.; Farmer, M.; Harari, G.; Fund, N.; Laudon, M.; Nir, T.; Frydman-Marom, A.; Zisapel, N. Add-on prolonged-release melatonin for cognitive function and sleep in mild to moderate Alzheimer's disease: a 6-month, randomized, placebo-controlled, multicenter trial. *Clin. Interv. Aging* 2014,*9*, 947-961.

[32] Xu, J.; Wang, L. L.; Dammer, E. B.; Li, C. B.; Xu, G.; Chen, S. D.; Wang, G. Melatonin for sleep disorders and cognition in dementia: a meta-analysis of randomized controlled trials. *Am. J. Alzheimers. Dis. Other Demen.* 2015, *30*,439-447.

[33] García-Mesa, Y.; Giménez-Llort, L.; López, L. C.; Venegas, C.; Cristòfol, R.; Escames, G.; Acuña-Castroviejo, D.; Sanfeliu C. Melatonin plus physical exercise are highly neuroprotective in the 3xTg-AD mouse. *Neurobiol. Aging.* 2012, *33*, 1124.e13-29.

[34] Olcese, J. M.; Cao, C.; Mori, T.; Mamcarz, M. B.; Maxwell, A.; Runfeldt, M. J.; Wang, L.; Zhang, C.; Lin, X.; Zhang, G.; Arendash, G. W. Protection against cognitive deficits and markers of neurodegeneration by long-term oral administration of melatonin in a transgenic model of Alzheimer disease. *J. Pineal Res.* 2009, *47*, 82-96.

[35] Rudnitskaya, E. A.; Muraleva, N. A.; Maksimova, K. Y.; Kiseleva, E.; Kolosova, N. G.; Stefanova, N. A. Melatonin attenuates memory impairment, amyloid-β accumulation, and neurodegeneration in a rat model of sporadic Alzheimer's disease. *J. Alzheimers Dis.* 2015, *47*,103-116.

[36] Cardinali, D. P.; Pagano, E. S.; Scacchi Bernasconi, P. A.; Reynoso, R.; Scacchi, P. Melatonin and mitochondrial dysfunction in the central nervous system. *Horm. Behav.* 2013, *63*,322-330.

[37] Esteban-Zubero, E.; García-Gil, F. A.; López-Pingarrón, L.; Alatorre-Jiménez, M. A.; Iñigo-Gil, P.; Tan, D. X.; García, J. J.; Reiter, R. J. Potential benefits of melatonin in organ transplantation: a review. *J. Endocrinol.* 2016, *229*, R129-R146.

[38] Lim, A. A.; Wall, M. P.; Greinwald, J. H. Effects of dimethylthiourea, melatonin, and hyperbaric oxygen therapy on the survival of reimplanted rabbit auricular composite grafts. *Otolaryngol.-Head Neck Surg.* 1999, *121*, 231–237.

[39] Hemadi, M.; Shokri, S.; Moramezi, F.; Nikbakht, R.; Sobhani, A. Potential use of melatonin supplementation to protect vitrified testicular grafts from hypoxic-ischaemic damage. *Andrologia* 2014, *46*, 513–521.

[40] Papavasiliou, P. S.; Cotzias, G. C.; Düby, S. E.; Steck, A. J.; Bell, M.; Lawrence, W. H. Melatonin and parkinsonism. *JAMA*. 1972, *221*, 88-89.

[41] Sharma, R.; McMillan, C .R.; Tenn, C. C.; Niles L. P. Physiological neuroprotection by melatonin in a 6-hydroxydopamine model of Parkinson's disease. *Brain Res.* 2006, *1068*, 230-236.

[42] Su, L. Y.; Li, H.; Lu, L.; Feng, Y. M.; Li, G. D.; Luo, R.; Zhou, H. J.; Lei, X. G.; Ma, L.; Li, J. L.; Xu, L.; Hu, X. T.; Yao, Y. G. Melatonin attenuates MPTP-induced neurotoxicity via preventing CDK5-mediated autophagy and SNCA/α-synuclein aggregation. *Autophagy* 2015, *11*, 1745-1759.

[43] Sun, X.; Ran, D.; Zhao, X.; Huang, Y.; Long, S.; Liang, F.; Guo, W.; Nucifora, F. C., Gu, H.; Lu, X.; Chen, L.; Zeng, J.; Ross, C. A, Pei Z.; *Mol. Med. Rep.* 2016, *13*,3936-3944.

[44] Noseda, A. C.; Rodrigues, L. S.; Targa, A. D.; Aurich, M. F.; Vital, M. A.; Da Cunha, C.; Lima. M. M. Putative role of monoamines in the antidepressant-like mechanism induced by striatal MT2 blockade. *Behav. Brain Res.* 2014, *275*,136-145.

[45] Paus, S.; Schmitz-Hübsch, T.; Wüllner, U.; Vogel, A.; Klockgether, T.; Abele, M. Bright light therapy in Parkinson's disease: a pilot study. *Mov. Disord.* 2007, *22*, 1495-1498.

[46] Fenoglio-Simeone, K.; Mazarati, A.; Sefidvash-Hockley, S.; Shin, D.; Wilke, J.; Milligan, H.; Sankar, R.; Rho, J. M.; Maganti, R. Anticonvulsant effects of the selective melatonin receptor agonist ramelteon. *Epilepsy Behav.* 2009, *16*, 52-57.

[47] Aydin, L.; Gundogan, N. U.; Yazici, C. Anticonvulsant efficacy of melatonin in an experimental model of hyperthermic febrile seizures. *Epilepsy Res.* 2015, *118*, 49-54.

[48] Goldberg-Stern, H.; Oren, H.; Peled, N.; Garty, B. Z. Effect of melatonin on seizure frequency in intractable epilepsy: a pilot study. *J. Child Neurol.* 2012, *27*, 1524-1528.

[49] Brigo, F.; Igwe S. C.; Del Felice, A. Melatonin as add-on treatment for epilepsy. *Cochrane Database Syst. Rev.* 2016, *8*, CD006967.

[50] Forcelli, P. A.; Soper, C.; Duckles, A.; Gale, K.; Kondratyev, A. Melatonin potentiates the anticonvulsant action of phenobarbital in neonatal rats. *Epilepsy Res.* 2013, *107*, 217-223.

[51] Anisimov, V. N.; Vinogradova, I. A.; Panchenko, A. V.; Popovich, I. G.; Zabezhinski, M. A. Light-at-night-induced circadian disruption, cancer and aging. *Curr. Aging Sci.* 2012, 5,170-177.

[52] Stevens, R,G.; Brainard, G. C.; Blask, D. E.; Lockley, S. W.; Motta, M. E. CA Breast cancer and circadian disruption from electric lighting in the modern world. *Cancer J. Clin.* 2014, *64*,207-218.

[53] Flynn-Evans, E. E.; Stevens, R. G.; Tabandeh, H.; Schernhammer, E. S.; Lockley, S. W. Total visual blindness is protective against breast cancer. *Cancer Causes Control* 2009, *20*,1753–1756.

[54] Kim, S.; Jang, E.; Kwon, S.; Han, W.; Kang, M.; Nam, Y.; Lee, Y. Night shift work and inflammatory markers in male workers aged 20–39 in a display manufacturing company. *Ann. Occup. Environ. Med.* 2016, *28*, 48.

[55] Tordjman, S.; Davlantis, K. S.; Georgieff, N.; Geoffray, M. M.; Speranza, M.; Anderson, G. M.; Xavier, J.; Botbol, M.; Oriol C, Bellissant E, Vernay-Leconte J, Fougerou C, Hespel A, Tavenard A, Cohen D, Kermarrec S, Coulon N, Bonnot, O.; Dawson, G. Autism as a disorder of biological and behavioral rhythms: toward new therapeutic perspectives. *Front. Pediatr.* 2015,3,1.

[56] Bjorklund, G.; Saad, K.; Chirumbol,o S.; Kern, J. K.; Geier, D. A.; Geier, M. R.; Urbina, M. A. Immune dysfunction and neuroinflammation in autism spectrum disorder. *Acta Neurobiol. Exp. (Wars).* 2016,76,257-268.

[57] Wang, Z.; Young, M. R. PTSD, a disorder with an immunological component. *Front. Immunol.* 2016,*7*,219.

[58] Lahiri, D. K., Ge, Y-W, Sharman, E. H., Bondy, S. C. Age-related changes in serum melatonin in mice, higher levels of combined melatonin and melatonin sulfate in the brain cortex than serum, heart, liver and kidney tissues. *J. Pineal Res.* 2004, *36,* 217-223.

[59] de Almeida, E. A.; Di Mascio, P.; Harumi, T. Spence, D. W.; Moscovitch, A.; Hardeland, R.; Cardinali, D. P.; Brown, G. M.; Pandi-Perumal, S. R. Measurement of melatonin in body fluids: standards, protocols and procedures. *Childs Nerv. Syst.* 2011, *27*, 879-891.

[60] Ding, K.; Wang, H.; Xu, J.; Li, T.; Zhang, L.; Ding, Y.; Zhu, L.; He, J.; Zhou M. Melatonin stimulates antioxidant enzymes and reduces oxidative stress in experimental traumatic brain injury: the Nrf2-ARE signaling pathway as a potential mechanism. *Free Radic. Biol. Med.* 2014, *73*, 1-11.

[61] Vriend, J.; Reiter, R. J. The Keap1-Nrf2-antioxidant response element pathway: a review of its regulation by melatonin and the proteasome. *Mol. Cell. Endocrinol.* 2015, *401*, 213-220.

[62] Hao, J.; Li, Z.; Zhang, C.; Yu, W.; Tang, Z.; Li, Y.; Feng, X.; Gao, Y.; Liu, Q.; Huang, W.; Guo, W.; Deng W. Targeting NF-κB/AP-2β signaling to enhance antitumor activity of cisplatin by melatonin in hepatocellular carcinoma cells. *Am. J. Cancer Res.* 2017, *7*, 13-27.

[63] Grossini, E.; Molinari, C.; Uberti, F.; Mary, D. A.; Vacca, G.; Caimmi, P. P. Intracoronary melatonin increases coronary blood flow and cardiac function through β-adrenoreceptors, MT1/MT2 receptors, and nitric oxide in anesthetized pigs. *J. Pineal Res.* 2011, *51*, 246-257.

[64] O'Neal-Moffitt, G.; Pilli, J.; Kumar, S. S.; Olcese, J. Genetic deletion of MT_1/MT_2 melatonin receptors enhances murine cognitive and motor performance. *Neuroscience* 2014, *277*, 506-521.

[65] Kilic, U.; Yilmaz, B.; Ugur, M.;Yüksel, A.; Reiter, R. J.; Hermann, D. M.; Kilic, E. Evidence that membrane-bound G protein-coupled melatonin receptors MT1 and MT2 are not involved in the neuroprotective effects of melatonin in focal cerebral ischemia. *J. Pineal Res.* 2012, *52*, 228-235.

[66] Radogna, F.; Nuccitelli, S.; Mengoni, F.; Ghibelli, L. Neuroprotection by melatonin on astrocytoma cell death. *Ann. N. Y. Acad. Sci.* 2009, *1171*, 509-513.

[67] Guo, X. H.; Li, Y. H.; Zhao, Y. S.; Zhai, Y. Z.; Zhang, L. C. Anti-aging effects of melatonin on the myocardial mitochondria of rats and associated mechanisms. *Mol. Med. Rep.* 2017, *15*, 403-410.

[68] Chuffa, L. G.; Alves, M. S.; Martinez, M.; Camargo, I. C.; Pinheiro, P. F.; Domeniconi, R. F.; Júnior, L. A.; Martinez, F. E. Apoptosis is triggered by melatonin in an in vivo model of ovarian carcinoma. *Endocr. Relat. Cancer* 2016, *23*, 65-76.

[69] Youle, R. J.; Strasser, A. The BCL-2 protein family: opposing activities that mediate cell death. *Nat. Rev. Mol. Cell Biol.* 2008, *9*, 47-59.

[70] Sharman, E. H.; Sharman, K. G.; Bondy, S. C. Melatonin causes gene expression in aged animals to respond to inflammatory stimuli in a manner differing from that of young animals. *Curr. Aging Sci.* 2008, *1*, 152-158.

[71] Shaw, A. C.; Goldstein, D. R.; Montgomery, R. R. Age-dependent dysregulation of innate immunity. *Nat. Rev. Immunol.* 2013, *13*, 875-887.

[72] Qin, L.; Wu, X.; Block, M. L.; Liu, Y.; Breese, G. R.; Hong, J. S.; Knapp, D. J.; Crews, F. T. Systemic LPS causes chronic neuroinflammation and progressive neurodegeneration. *Glia* 2007, *55*, 453-462.

[73] Bondy, S. C.; Sharman, E. H. Melatonin, oxidative stress and the aging brain. In: *Aging and Age-Related Disorders*; Bondy, S. C. and Maiese, K. Eds.; Humana Press, Totowa NJ, USA, 2010; pp. 339-357.

[74] Martinez-Jimenez, C. P., Eling, N et al. Ageing increases cell-to-cell transcriptional variability upon immune stimulation. *Science* 2017, *355*, 1433-1436.

[75] Sharman, E. H.; Sharman, K. Z.; Bondy, S. C. Dietary melatonin alters age-related diffusion of cerebral gene expression and restores a more youthful degree of individual variation. *Int. J. Neuroprot. Neuroregen.* 2006, *2*, 190-193.

[76] Sharman, E.; Sharman, K. Z.; Lahiri, D. K.; Bondy, S. C. Age-related changes in murine CNS mRNA gene expression are modulated by dietary melatonin. *J. Pineal Res.* 2004, *36*, 165-170.

[77] Sharman, E. H.; Bondy, S. C.; Sharman, K. Z.; Lahiri, D.; Cotman, C. W.; Perreau, V. M. Effects of melatonin and age on gene expression in mouse CNS using microarray analysis. *Neurochem. Internat.* 2007,*50*, 336-344.

[78] Campbell, A.; Sharman, E. H.; Bondy, S. C. Age-related differences in the response of the brain to chronic dietary melatonin. *Age* 2013, *36*, 49-55.

[79] Hardeland, R.; Cardinali, D. P.; Brown, G. M.; Pandi-Perumal, S. R. Melatonin and brain inflammaging. *Prog. Neurobiol.* 2015, *127-128*, 46-63.

[80] Perreau, V. M.; Bondy, S. C.; Cotman, C. W.; Sharman, K. Z.; Sharman, E. H. Melatonin treatment in old mice enables a more youthful response to LPS in the brain. *J. Neuroimmunol.* 2007, *182*, 22-31.

[81] Phillipson, O. T. Management of the aging risk factor for Parkinson's disease. *Neurobiol. Aging* 2014, *35*, 847-857.

[82] Reiter, R. J.; Tan, D.; Kim, S. J.; Manchester, L. C.; Qi, W.; Garcia, J. J.; Cabrera, J. C.; El-Sokkary, G.; Rouvier-Garay, V. Augmentation of indices of oxidative damage in life-long melatonin-deficient rats. *Mech. Ageing Dev.* 1999, *110*, 157-173.

[83] Sharma, R.; Ottenhof, T.; Rzeczkowska, P. A.; Niles, L. P. Epigenetic targets for melatonin: induction of histone H3 hyperacetylation and gene expression in C17.2 neural stem cells. *J. Pineal Res.* 2008, *45*, 277-284.

[84] Lee, S. E.; Kim, S. J.; Yoon, H. J.; Yu, S. Y; Yang, H.; Jeong, S. I.; Hwang, S. Y.; Park, C. S,.;Park, Y. S. Genome-wide profiling in melatonin-exposed human breast cancer cell lines identifies differentially methylated genes involved in the anticancer effect of melatonin. *J. Pineal Res.* 2013, *54*, 80-88.

[85] Tajes, M.; Gutierrez-Cuesta, J.; Ortuno-Sahagun, D.; Camins, A.; Pallas, M. Anti-aging properties of melatonin in an in vitro murine senescence model: involvement of the sirtuin 1 pathway. *J. Pineal Res.* 2009, *47*,228–237.

[86] Bai, X. Z.; He, T.; Gao, J. X.; Liu, Y.; Liu, J. Q.; Han, S. C.; Li, Y.; Shi, J. H.; Han, J. T.; Tao, K.; Xie, S. T.; Wang, H. T.; Hu, D. H.; Melatonin prevents acute kidney injury in severely burned rats via the activation of SIRT1. *Sci. Rep.* 2016, *6*, 32199.

[87] Mayo, J. C.; Sainz, R. M.; González Menéndez, P.; Cepas, V. Tan, D-X.; Reiter, R. J. Melatonin and sirtuins: A "not-so unexpected" relationship. *J. Pineal Res.* 2017; *62*:e12391.

[88] Sohn EJ, Won G, Lee J, Lee S, Kim SH. Upregulation of miRNA3195 and miRNA374b mediates the anti-angiogenic properties of melatonin in hypoxic PC-3 prostate cancer cells. *J. Cancer* 2015, 6, 19-28.

[89] Mori, F.; Ferraiuolo, M.; Santoro, R.; Sacconi, A,; Goeman, F.; Pallocca, M.; Pulito, C.; Korita., E.; Fanciulli, M.; Muti, P,; Blandino, G.; Strano S. Multitargeting activity of miR-24 inhibits long-term melatonin anticancer effects. *Oncotarget* 2016, 20532-20548.

[90] Carloni, S.; Favrais, G.; Saliba, E.; Albertini, M. C.; Chalon, S.; Longini, M.; Gressens, P.; Buonocore, G.; Balduini, W. Melatonin modulates neonatal brain inflammation through endoplasmic reticulum stress, autophagy, and miR-34a/silent information regulator 1 pathway. *J. Pineal Res.* 2016, *61*, 370-380.

[91] Cai, B.; Ma, W.; Bi, C.; Yang, F.; Zhang, L.; Han, Z.; Huang, Q.; Ding, F.; Li, Y.; Yan, G.; Pan, Z.; Yang, B.; Lu, Y. Long noncoding RNA H19 mediates melatonin inhibition of premature senescence of c-kit(+) cardiac progenitor cells by promoting miR-675. *J. Pineal Res.* 2016, *61*, 82-95.

In: Melatonin: Medical Uses and Role in Health and Disease   ISBN: 978-1-53612-987-8
Editors: Lore Correia and Germaine Mayers   © 2018 Nova Science Publishers, Inc.

Chapter 11

MELATONIN AND ORAL DISORDERS: POTENTIAL THERAPEUTIC APPLICATIONS

Ana Capote-Moreno[1],, MD, PhD, Paloma Patiño[2], and Alejandro Romero[3], PhD*

[1]Department of Oral and Maxillofacial Surgery,
University Hospital La Princesa,
Autonomous University of Madrid, Madrid, Spain
[2]Paediatric Unit, La Paz University Hospital, Madrid, Spain
[3]Department of Toxicology and Pharmacology,
Faculty of Veterinary Medicine,
Complutense University of Madrid, Madrid, Spain

ABSTRACT

In the last decade, melatonin has emerged as an important therapeutic alternative against numerous pathologies. Its capacity as free radical scavenger and antioxidant joined to its immunomodulatory action, oncostatic activity and anti-inflammatory properties, has been widely reported. Each of these actions are tightly controlled by the regulation of different signaling pathways. So, the purpose of this chapter is briefly summarize the data documenting the use of melatonin and its beneficial effects in oral related-diseases, such as bisphosphonate-related osteonecrosis of the jaw, periodontitis, mucositis, cancer, and/or oral infections. Furthermore, taking into account the low cost of melatonin and its reduced toxicity, we hypothesize on novel therapeutic strategies for the treatment of diseases of the oral cavity. Nevertheless, the molecular and cellular mechanisms underlying the actions of melatonin need further exploration and accordingly, new clinical studies should be conducted in human patients to assess its efficacy.

---

* Corresponding Author: Ana Capote-Moreno, MD, Departmentof Oral and Maxillofacial Surgery, University Hospital La Princesa, Autonomous University of Madrid, Calle Diego de León 62, 28006-Madrid, Spain, Phone: +34-915202200, Email: anacapotemoreno@gmail.com.

Keywords: melatonin, oral cavity, oral squamous cell carcinoma, bisphosphonates, osteonecrosis, periodontal disease

INTRODUCTION

Melatonin (*N*-acetyl-5-methoxy-tryptamine) is an indoleamine synthesized and secreted mainly by the pineal gland and it is released into blood and cerebrospinal fluid (CSF) to exert modulatory actions that regulate circadian rhythms under suprachiasmatic nucleus control [1]. In this sense, numerous reports have evidenced the ubiquitous melatonin-presence in various extra-pineal organs including immune system cells, brain, airway epithelium, bone marrow, gut, ovary, testes, skin and likely other tissues [2-4]. Melatonin secretion is synchronized to the light/dark cycle, with a nocturnal maximum and low diurnal baseline levels. The natural age-related decline of serum and CSF melatonin levels takes place progressively after 40-45 years, which may contribute to numerous pathophysiological changes [5, 6]. Due to its lipophilic nature, most of the melatonin secreted to blood circulates bounded to plasma proteins. However, free melatonin in blood (around 30% of blood melatonin not albumin-bound) is able to diffuse to surrounding tissues, such as oral cavity and mucosa through salivary gland secretion. The proportion of salivary melatonin ranges from 24% to 33% and seems to be stable [7]. Some authors have suggested the measurement of the melatonin levels in saliva as a reliable non-invasive technique for monitoring melatonin secretion and circadian rhythm [8]. The measurement of urine excretion of the main metabolite of melatonin, 6-sulfatoxymelatonin, which is mostly excreted during the night, is another useful biomarker to measure circadian rhythmicity. At physiological concentrations, melatonin acts in the oral mucosa cells by receptor-mediated actions. The two-major membrane-associated melatonin receptors, MT1 and MT2, belong to the G-protein-coupled receptor superfamily. A third melatonin-binding site (MT3 receptor) has been purified and characterized as the enzyme quinone reductase 2 (NQO2), whose melatonin-dependent stimulation may contribute to the antioxidant potential of this neurohormone. Furthermore, melatonin may mediate its actions through the nuclear receptor ROR/RZR (retinoid orphan receptors/retinoid Z receptors) [9]. All of them, could define the heterogeneity of the multiple mechanisms of melatonin observed in the oral cavity. It is known that melatonin crosses all morphophysiological barriers and passively enters in the oral mucosa through salivary secretion. Recent novel studies have demonstrated that this neurohormone is also synthesized in the oral cavity and salivary glands cells, where it may be involved in autocrine/paracrine signaling actions binding to the MT1 receptor [10, 11]. In this sense, Shimozuma and co-workers (2011), in cell subpopulations of salivary glands, evidenced and identified the expression of two well characterized enzymes which participate in the synthesis of melatonin: alkylamine-*N*-acetyltransferase

(AANAT), which converts serotonin to *N*-acetylserotonin (NAS), and hydroxyindole-O-methyltransferase (HIOMT), which converts NAS to melatonin.

Melatonin exhibits anti-inflammatory and oncostatic effects, as well as properties as an indirect antioxidant and also as a direct free radical scavenger. Both antioxidant and anti-inflammatory activities play an important role in bone formation and reduction of bone resorption, as well as in periodontal diseases. Added to these actions, melatonin exerts an oncostatic and antiproliferative activity through the modulation of multiple signaling pathways, including stimulating anticancer immunity, oncogene expression regulation, induced cancer cell apoptosis and antiangiogenic effects in different tumors [12]. Nevertheless, the molecular and cellular mechanisms underlying melatonin´s actions in oral squamous cell carcinoma (OSCC) requires further exploration.

In this chapter, we summarize the potential actions of melatonin in different oral pathologies and its therapeutic applications including periodontal disease, mucositis, oral infections, oral implantology and bone pathologies as bisphosphonate-related osteonecrosis of the jaws and OSCC.

ANTIOXIDANT AND ANTI-INFLAMMATORY ACTIVITIES OF MELATONIN IN THE ORAL CAVITY

The role of melatonin as a versatile antioxidant and scavenger of reactive oxygen and nitrogen species (RONS) has been widely documented [12, 13]. Under pathologic conditions the oxidative stress (OS), which is an important event of early injury in oral cavity, is developed due to an excess of production of free radicals. These oxidative responses may be generated in oral cavity by several conditions, as well as inflammation of periodontal tissues caused by different microorganisms, osteoclasts activity in bone resorption, dental trauma and tooth extractions, oral infections and oral cytotoxicity by different agents (drugs or dental materials) [13].

Recently, it has been evidenced that melatonin reduces bone resorption process mediated by osteoclasts and blocks the activity of prooxidant enzymes as superoxide dismutase (SOD). Melatonin downregulates osteoclast formation and activation by the modulation of different nuclear signaling pathways. The upregulation of antioxidant and downregulation of prooxidant enzymes induced by melatonin, has been studied in periodontal tissues by several authors [14]. In this context, melatonin may act as a pleiotropic molecule, mitigating periodontal disease and periodontitis caused by the oxidative damage of increased RONS, through direct free-radical scavenging mechanisms or by increasing the activity and expression of antioxidant enzymes [14].

Inflammation plays a crucial role in a wide variety of diseases that involve the oral cavity. Protective anti-inflammatory profile of melatonin is achieved in early stages of periodontitis and in other inflammatory disorders that affect the oral cavity. Melatonin is effective inhibiting proinflammatory factors such as nitric oxide (NO), interleukin-6 (IL-6), tumor necrosis factor-alpha (TNF-α) and C-reactive protein [14, 15]. So, in addition to its known anti-oxidant and anti-inflammatory activities, melatonin exerts a significant immunomodulatory effect in several processes as it stimulates the immune system [16, 17]. Melatonin can upregulate the production of IL-2, IL-12, and interferon gamma (IFN-γ). Moreover, melatonin induces the response of the monocyte to granulocyte-macrophage colony-stimulating factor (GM-CSF), IL-3, IL-4 and IL-6. An increase in natural killer (NK) cells activity and in the production of granulocytes, macrophages, neutrophils, and erythrocytes is also observed after treatment with melatonin [17-19].

MELATONIN AND BONE METABOLISM

Bone metabolism is an active process of bone resorption and formation that takes place continuously in all body's bones including the maxilla and the jaw. The first stage consists in the activation of osteoclasts precursors that differentiated into mature osteoclasts, which are responsible of bone resorption by acidification and proteolytic digestion. Afterwards, bone formation begins with osteoblast precursor recruitment to the resorption area that differentiate to mature osteoblasts and form new bone by osteoid secretion, and final matrix mineralization [20]. This active process is mediated by several hormones and cytokines, growth factors and also by melatonin. Melatonin directly interacts with osteoblasts and osteoclasts, promotes type I collagen expression, the main component of the extracellular bone matrix, stimulates secretion of other bone proteins as alkaline phosphatase, osteocalcin, osteopontin and bone sialoprotein and stimulates matrix mineralization by hydroxyapatite crystal deposition. Indirectly, melatonin regulates bone metabolism through the interaction with systemic hormones such as Parathyroid hormone (PTH), calcitonin and estradiol. The antioxidant and free-radical scavenger actions of melatonin also protect bone cells from OS generated by osteoclasts activity [21].

Another signalling mechanisms where melatonin may be implicated are bone repair after fractures and bone defects after tooth extraction. Bone repair implies initial inflammation, proliferation of bone cells, neoangiogenesis, collagen deposition and mineralization. In this complex scenario, Halici et al. [22] evidenced in a rat model that melatonin acts controlling the first inflammatory phase, through regulation of antioxidant enzymes as superoxide dismutase and myeloperoxidase and acting as a free radical scavenger Regarding bone maturation, melatonin may be considered an important player in the proliferative and collagen deposition phases. In reference to this consideration, experimental studies have revealed that melatonin may stimulate angiogenesis in bone repair through modulating the expression of growth factors such as vascular endothelial growth factor (VEGF) and other pro-angiogenic cytokines [14]. Tooth extraction consist in a traumatic process where RONS are generated and contribute to postsurgical inflammation and risk of local infection. Some authors have suggested that the use of melatonin in teeth socks may accelerate bone healing and reduce inflammation and infection [13, 23].

A different bone pathology that could affect the jaws and that has become more relevant in the last decade is Bisphosphonate-related Osteonecrosis of the jaw (BRONJ). Bisphosponates (BP) are pyrophosphate composite analogues with high affinity to hydroxyapatite that reduce bone resorption in several pathological conditions as bone metastases, multiple myeloma, osteoporosis and bone primary disorders as Paget´s disease or osteogenesis imperfecta. BRONJ is an adverse side-effect described in the last fifteen years that affects the jaws with bone exposition, inflammation, secondary infection, oral fistula, local pain and bone necrosis [24]. This complication may appear secondary to bone injury, such as dental extractions or other bone surgeries (dental implant placement), but may also occur spontaneously, mainly with intravenous BP. Zoledronic acid (ZA) is a nitrogenous BP endowed with potent antiresorptive actions when it is administered intravenously. These drugs induce osteoclast apoptosis although they can also reduce osteoblast viability at high doses [25]. The recent introduction of other antiresorptive drugs instead of BP with similar adverse effects, as antiangiogenic therapy with Denobsumad, have changed the nomenclature to Medication-related Osteonecrosis of the jaws (MROJ) [26].

Recent experimental studies in culture cells have evaluated the possible protective effect of melatonin in BROJ [25, 27]. In a recent report, Rodriguez-Lozano et al. [27] have evidenced the cytoprotective effect of melatonin in cell proliferation and for promoting osteogenic differentiation in bone marrow culture cells, using ZA-treated mesenchymal stem cells of periodontal ligament and bone marrow [27]. Otherwise, the study of Camacho-Alonso et al. [25] carried out in ZA-treated human osteoblast also has evidenced a beneficial effect of melatonin in reducing the cytotoxic action of ZA on osteoblast cell viability. Therefore, these beneficial properties make melatonin suitable as a potential treatment against BROJ. Anyway, further *in vivo* and clinical studies should be conducted in advancing melatonin as a therapeutic strategy to reduce the effects resulting from other antiresorptive drug as Denobsumad.

MELATONIN AND PERIODONTAL DISEASE

Periodontal disease is an inflammatory chronic disorder that affects periodontal tissues (periodontal ligament, cementum and alveolar bone) with a high global prevalence although different treatment strategies. Local inflammation mediated by neutrophils and macrophages and the exacerbated immune response against bacteria, generates an oxidative response and free radical´s production that facilitate gingival bleeding, periodontal pocket formation and destruction of connective tissue attachment.

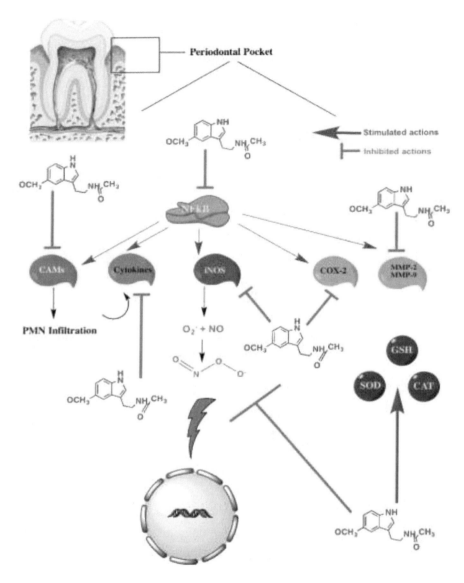

Figure 1. Schematic diagram of the main pathological events subsequent to periodontitis and critical points where melatonin exerts its neuroprotective effects. After periodontitis onset, proinflammatory cytokines, free radicals (FR), Matrix metalloproteinases (MMPs) and infiltration of polymorphonuclear cells (PMNs) are activated. Melatonin inhibits the cell adhesion molecules (CAMs) and impairs the recruitment of PMNs. Furthermore, this indoleamine downregulates NFκB, a proinflammatory transcription factor, inhibiting its eventual translocation to the nucleus and ulterior binding to DNA, thereby reducing inflammation. Periodontitis also involves an excess of FR, and melatonin is thus a highly effective antioxidant. Melatonin directly scavenges RONS ($O_2^{•-}$; NO) and indirectly upregulates the expression of antioxidant enzymes; CAT (catalase); GSH (glutathione) and SOD (superoxide dismutase) preventing DNA damage. Consequently, melatonin exerts effective protective modulating multiple points in the cascade of events in periodontal disease.

Advanced periodontitis results in a progressive destruction of dental tissue support with the subsequent loss of teeth. Melatonin has been proposed as a mediator in periodontitis [13-15, 28]. As previously mentioned, melatonin has a wide variety of well-known actions such as a direct free radical scavenger activity, indirect antioxidant properties, anti-inflammatory, and immunomodulatory effects. All the mentioned beneficial actions play an important role in periodontal disease progression (Figure 1). Down-regulation of expression pro-inflammatory factors (IL-6, TNF-alpha, C-reactive protein) may occur at the same time of up-regulation of osteoblast product mediators (acid phosphatase, alkaline phosphatase, osteocalcin, osteopontin) that correlates with reduction of inflammation of the periodontium and with alveolar bone resorption. The direct antimicrobial action of melatonin synergistically acts in these processes [13]. Cutando et al. [28] correlated low levels of salivary melatonin with a high degree of periodontal destruction. They suggested that monitoring melatonin levels could be used as a biomarker of periodontal severity. These authors also suggested the potential usefulness of melatonin to reduce periodontitis and bone destruction. In a different study of the same research group, it was demonstrated the beneficial effect of the topic application of melatonin in diabetic patients with periodontal disease, who are more susceptible for periodontitis and with more severe forms, reducing gingivitis, the pocket depth and preventing progression of the disease [29].

MELATONIN AND ORAL CARCINOMA

OSCC is the most prevalent malignant tumor of the oral cavity that shows a global annual incidence around 300000 new cases [30]. Although advances in new therapeutic approaches, surgical techniques and cytotoxic drugs, 5-year global mortality rates for oral cancer remain high around 50%. OSCC is a locally aggressive tumor that could metastasize to regional lymph nodes and other organs by lymphatic and haematological dissemination. The survival rates drop drastically when lymph node metastases occur, so diagnosis of OSCC in the early stages and prevention of distant progression is critical. Most of these tumors are sporadic and several risk factors have been identified as tobacco and alcohol use, poor oral and dental condition, immunosuppressive status, or premalignant lesions as, for example, leucoplakia, erythroplakia, hypertrophic candidiasis or lichen planus. In the last decade, diverse molecular and genetic alterations have been correlated to OSCC and tumor progression in different studies. Hypoxia, is a common characteristic of the microenvironment found in these tumors [31]. This hypoxic condition regulates transcription of hypoxia inducible factor 1-alpha (HIF1-α) and promotes overexpression pro-angiogenic factors such as vascular endothelial growth factor (VEGF). Overexpression of these factors has been correlated with tumor aggressiveness and progression. Polymorphisms of HIF 1-α have also been related to tumor relapse and poor prognosis [29]. Other mechanisms that may play an important role in tumor development are OS and immunosuppressive status. RONS species generated under the oxidative microenvironment may damage nitrogen bases of DNA and proteins in normal oral cells and contribute to mutations and tumoral transformation [30]. So, when the antioxidant systems fail, the increase of free radicals cause DNA damage and cells accumulate genetic alterations that facilitate transformation to tumoral phenotype. The immune system that is involved in tumoral development, plays an important role in controlling neoplastic growth. Cell immunity activity as NK, T-lymphocytes and monocytes/macrophages detects and removes cancerous cells by several mechanisms mediated by cytokines as IL-2, IL-6, IL-12 or TNF-α. It is for this reason that immunosuppressive status leads to tumor growth and dissemination.

Oncostatic actions of melatonin have been proposed in a wide variety of tumors as prostate, colorectal, breast, neural, ovarian and cervix, sarcomas, melanoma, larynx carcinomas, hepatocarcinoma and skin cancers [12, 33]. However, the knowledge about its relationship with OSCC is preliminary. An oncostatic effect has been observed at pharmacological doses that are higher than physiological concentrations [30]. Anti-oxidant properties and free-radical scavenger activity of melatonin may play an important role in the hypoxic condition that is generated in local advanced oral tumors. Melatonin may also exert a direct pro-apoptotic action in tumoral keratinocytes and an oncogene regulator function, as well as antiangiogenic effects by inhibiting VEGF and other growth factors as the epidermal growth factor (EGF) or the insulin growth factor-1 (IGF-1) [12]. In several studies carried out in OSCC cell lines, the addition of melatonin in the culture medium has demonstrated subexpression of both HIF 1-α and VEGF factors and a reduced cell viability, supporting the antiangiogenic effect of melatonin in oral cancer previously seen in other tumors [31]. Added to this, melatonin may also exert an immunoenhancer action in OSCC by activating cell-mediated mechanisms and secretion of cytokines with regulatory immune function [3]. All of these promising results need to be confirmed by new studies *in vivo,* as well as to the via of administration and oncostatic doses of melatonin that should be used in oncologic patients.

Radiotherapy and chemotherapy are frequently used in OSCC, as an adjuvant treatment to surgery and also as the main therapy in locally advanced or disseminated cases. One of the most frequent local adverse side effects of radiotherapy, enhanced by cytotoxic drugs, is mucositis. Mucositis can affect the oral cavity and the entire gastrointestinal tract. Multiple factors contribute to mucositis development [34]. Initially, oral mucosa cells are exposed to cytotoxic agents that cause an oxidative reaction and RONS formation with the subsequent DNA damage and cell death. OS induce both a direct tissue injury and also the activation of inflammatory pathways that contributes to the progression phase of mucositis. In the initial phase of mucositis, oral mucosa erythema and desquamation of the submucosa and basal cell layers occur, with preservation of the mucosa integrity. When the injury progresses, this integrity is breached and ulceration can be observed in prolonged severe mucositis. A bacterial colonization of the ulcers may contribute to inflammation and may increases morbidity and pain. This sobreinfection prolongs symptoms of dysphagia and odynophagia and decreases quality of life of these patients, retarding also the last healing phase of mucositis where the epithelium integrity is restored and symptoms begin to abate [35].

Multiple treatments have been proved to alleviate symptoms and to short disease course, as well as to prevent the development of gastrointestinal mucositis. Melatonin has also been employed in this disorder due to both antioxidant and anti-inflammatory properties [35]. This indoleamine enhances activity of endogenous antioxidant enzymes as superoxide dismutase, catalase, glutathione peroxidase, glutathione reductase and glutamyl-cystein synthase, inhibits pro-inflammatory cytokines and can maintain homeostasis of mitochondria, whose function is altered because of cytotoxicity of chemo-radiotherapy. Several reports have evidenced the pleiotropy of melatonin to counteract mucositis, as it may be more effective that other alternatives (*N*-acetyl-cysteine, amifostine and others). Recently, Ortiz et al. [10] have demonstrated the effectiveness of the melatonin gel in preventing epithelium disruption in mucositis and ulcers formation, lowering the loss of proliferative progenitor stems cells and enhancing the healing process. Melatonin has emerged as a potential therapeutic tool to ameliorate the molecular and organ/tissue damage associated with radio and/or chemotherapy-induced mucositis, which may improve patients' quality of life [35].

MELATONIN AND IMPLANTOLOGY

Several studies have demonstrated the beneficial effect of melatonin in oral dental implants. Nowadays, dentition missing situation is mostly restored with osseointegrated titanium implants. The process of osseointegration consists on direct apposition of new bone in contact with the implant surface, as well as remodelling of the adjacent pre-existing bone [14]. Different biological mechanisms have been implicated in this process including cell maturation, apposition, neoangiogenesis and bone mineralization. Primary stability is critical for avoiding implant failure. Several local growth factors may accelerate this process of integration, and in this sense, melatonin may be useful. The effect of melatonin in promoting bone regeneration in dental implants has been assessed by diverse authors [36] in different experimental animal models, in which melatonin alone or in combination with other bone promoters have been used for this aim [37-40]. Cutando et al. [37] analyzed the effect of topical application of lyophilized powder melatonin in Beagle dogs at 2 weeks of the implant placement and revealed that bone-implant contact (BIC), interthread bone density and bone neoformation were increased in treated implants in comparison with the control group. Guardia et al. [39] found similar results at 5- and 8-weeks of treatment. Comparable results were obtained by Yamazaki et al. [40], using a rabbit model after daily intraperitoneal injection of melatonin. Altogether, these findings have documented that melatonin may act as a biomimetic agent in endo-osseous dental implants, as it promotes osteogenesis, increases BIC density and improves peri-implant osteogenesis.

After implant placement, adjacent bone necrosis around the implant often takes place as well as activation of an inflammatory response to the surgical procedure itself [41]. Inflammatory cells migrate to the peri-implant tissues and increase free-radical production, promote an OS and activate bone resorption by osteoclasts. Anti-inflammatory and antioxidant properties of melatonin may play again an important role in postsurgical dental implant healing and in later osseointegration [14, 41].

MELATONIN AND OTHER ORAL DISEASES

Melatonin displays other recognized potential actions in oral cavity diseases. This neurohormone has antimicrobial properties against several microorganisms that may affect the oral cavity, such as candidiasis, herpes-virus infection and cariogenic bacteria as *Lactobacillus* and *Staphylococcus mutans* [13]. In herpes-virus infection, the beneficial effects of melatonin have been compared with conventional antiviral treatment with acyclovir, showing a reduction in severity of the infection as effective as antiviral therapy [42]. This effect may be related to the immunomodulatory action of melatonin, by the stimulation of both cell and humoral immunity, with the activation of NK and CD4 lymphocytes and with enhanced expression of cytokines like IL-1β [15]. Similar immunoenhancer effect have been proposed in oral candidiasis and candida sepsis, which supports the possible useful treatment with topic or systemic melatonin in oral and systemic mycosis [43].

The protective effect of melatonin against cytotoxicity and genotoxicity of dental materials may be another application of this indoleamine. Dental composite materials commonly used in dental restoration are a mixture of different inorganic polymersin situ formed from methacrylate-acid (MAA-) monomer as: 2-hidroxymethyl methacrylate (HEMA) and methylmethacrylate (MMA), urethane methacrylate (UDMA), bisphenol A-diglycidyl dimethacrylate (bis-GMA) and others. Usually, a considerable fraction of these materials may not be polymerized and it is releases to oral cavity due to mechanical stress, chewing or due to the action of salivary enzymes or oral bacteria [15]. Product degradation of these polymers may migrate through dentinal tubules and cause direct damage in pulp cells or pass to the bloodstream. These materials have shown a cytotoxic effect both *in vivo* and *in vitro*experimental studies, by oxidative mechanisms and DNA bases damage [44, 45]. Due to antioxidant, antiinflammatory and oncostatic actions of melatonin, it may be considered as a cytoprotective agent against adverse effects of dental material.

Different systemic conditions can impair some oral disorders such as periodontitis. Diabetic patients may be more susceptible to severe periodontal disease and it may also contribute to hyperglycemic decompensation [41]. Melatonin may act as a protective agent of periodontal status in these patients and it may exert an important role in glucose metabolism [29, 46]. Psychiatric disorders as depression, schizophrenia and bipolar disorder, can contribute to poor oral health and periodontal status and also to teeth lost. Melatonin has been used in these patients with promising results, as it improves the circadian rhythm alteration frequently seen in these pathologies, attenuates weight gain associated to antipsychotic drugs and improves other metabolic disorders as lipid profile alteration or blood pressure decompensation [41]. This issue, associated to the direct effect of melatonin on oral health, makes it a reliable therapeutic strategy in psychiatric disorders.

POTENTIAL THERAPEUTIC APPLICATIONS OF MELATONIN IN ORAL DISORDERS

As mentioned above, melatonin is a neurohormone with multiple beneficial effects in oral pathologies. To date, therapeutic utility of melatonin has been limited to *in vitro* and *in vivo* experimental models. Melatonin gel may be useful in certain oral applications as the treatment and prevention of periodontitis and mucositis, the prevention of tooth extraction complications, or the improvement of osseointegration and bone regeneration after dental implant placement. However, the systemic use of melatonin may be more effective in other oral disorders for instance, OSSC, oral infections as candidiasis and herpes-virus, and also in MROJ patients. The dosage is substantially different depending on the indication, thus, clinical studies have demonstrated that the doses of melatonin to reach a measurable anti-inflammatory/scavenger effect in oral diseases are higher than those applied as chronobiotic on sleep disturbances. In this context, the treatment with melatonin would require further clinical research [30]. In patients treated with bisphosphonates or Denobsumad, melatonin may prevent MROJ in those cases that are submitted for teeth extractions and other risk surgeries for the jaws. It may also be useful in patients with established osteonecrosis, to improve bone and mucosa healing and regeneration and to avoid progression of the necrotic osseous tissue. According to *in vitro* studies, diary high oral doses may be needed for this purpose [23, 25]. In OSSC, actual knowledge is limited to few studies with cultured cells and review reports. Conversely, melatonin´s oncostatic effect in other tumors is well-known. A high percentage of patients with oral carcinomas may present other oral secondary tumors due to a "field cancerization" of the oral mucosa and also secondary neoplasms in other locations [47]. These patients may be susceptible for treatment with oral melatonin in order to prevent secondary neoplasm and also local recurrences that frequently occur in OSSC. Some authors have speculated the possible utility of this therapy in preventing the development of an invasive carcinoma in patients with premalignant lesions such as oral leucoplakia, lichen planus or erythroplakia [13, 48]. It has also been suggested that restoration of MT1 expression may inhibit the growth of OSCC, showing promising results for new oral oncologic strategies [48]. In this scenario, the administration of adjuvant melatonin with current chemotherapy protocols in OSSC may significantly benefit oncologic patients. Consequently, the possible synergistic effects of the combined therapy require further evaluation in humans.

CONCLUSION

Melatonin has been widely reported as a potent antioxidant, anti-inflammatory, free radical scavenger, immunomodulatory and oncostatic agent that makes this molecule an excellent therapeutic "magic bullet" in oral cavity diseases. To date, no significant negative effects after melatonin administration had been reported [49]. Based on the diverse of therapeutic useful pharmacological properties of melatonin, it may be beneficial in multiple oral conditions such as periodontal disease, bone regeneration and bone disorders as MROJ, oral infections, dental implant surgery, mucositis and oral carcinomas. Since it has an amphiphilic nature, which facilitates passage across cell compartments, and it shows an excellent safety profile, therapeutical application of melatonin is a reliable and versatile option in the oral cavity. At this point, timing and dosage of melatonin administration seems to be critical to reach therapeutic success against conventional treatments. However, additional clinical research is needed to determine and evaluate the beneficial impact of melatonin in oral diseases.

CONFLICT OF INTEREST

The authors report no conflicts of interest.

REFERENCES

[1] Reiter, R.J., (1991). Pineal melatonin: cell biology of its synthesis and of its physiological interactions. *Endocr. Rev.* 12, 151-180.

[2] Menendez-Pelaez, A., Reiter, R.J., (1993). Distribution of melatonin in mammalian tissues: the relative importance of nuclear versus cytosolic localization. *J. Pineal. Res.* 15:59-69.

[3] Venegas, C., García, J.A., Escames, G., Ortiz, F., Lopez, A., Doerrier, C., Garcia-Corzo, L.,Lopez, L.C.,Reiter, R.J., Acuña-Castroviejo, D., (2012). Extrapineal melatonin: analysis of its subcellular distribution and daily fluctuations. *J. Pineal. Res.* 52**:**217-227.

[4] Acuña-Castroviejo, D., Escames, G., Venegas, C., Diaz-Casado, M.E., Lima-Cabello, E., López, L.C., Rosales-Corral, S., Tan, D.X., Reiter, R.J., (2014). Extrapineal melatonin: sources, regulation, and potential functions. *J. Cell. Mol. Life. Sci.* 71**:**2997-3025.

[5] Reiter, R.J., Richardson, B.A., Johnson, L.Y., Ferguson, B.N., Dinh, D.T., (1980). Pineal melatonin rhythm: reduction in aging Syrian hamsters. *Science.* 210:1372-1373.

[6] Reiter, R.J., Craft, C.M., Johnson, J.E.,Jr., King, T.S., Richardson, B.A., Vaughan, G.M., Vaughan, M.K., (1981). Age-associated reduction in nocturnal pineal melatonin levels in female rats. *Endocrinology.* 109:1295-1297.

[7] Laakso, M.L., Porkka-Heiskanen, T., Alila, A., Stenberg, D., Johansson, G., (1990). Correlation between salivary and serum melatonin: dependence on serum melatonin levels. *J. Pineal. Res.* 9, 39-50.

[8] Nowak, R., McMillen, I.C., Redman, J., Short, R.V., (1987). The correlation between serum and salivary melatonin concentrations and urinary 6-hydroxymelatonin sulphate excretion rates: two non-invasive techniques for monitoring human circadian rhythmicity. *Clin. Endocrinol.* 27, 445-452.

[9] Carrillo-Vico, A., García-Pergañeda, A., Naji, L., Calvo, J.R., Romero, M.P., Guerrero, J.M., (2003). Expression of membrane and nuclear melatonin receptor mRNA and protein in the mouse immune system. *Cell. Mol. Life. Sci.* 60, 2272-2278.

[10] Ortiz, F., Acuña-Castroviejo, D., Doerrier, C., Dayoub, J.C., López,L.C., Venegas, C., García, J.A., López, A., Volt, H., Luna-Sanchez, M., Escames, G., (2015). Melatonin blunts the mitochondrial/NLRP3 connection and protects against radiation-induced oralmucositis. *J. Pineal. Res.* 58, 34-49.

[11] Shimozuma, M., Tpkuyama, R., Tatehara,S., Umeki, H., Ide, S., Mishima, K., Saito, I., Satomura, K., (2011). Expression and cellular localization of melatonin-synthesizing enzymes in rat and human salivary glands. *Histochem. Cell. Biol.* 135, 389-396.

[12] Cutando, A., López-Valverde, A., De Vicente, J., López-Giménez, J., Alias-García, I., Gómez-De Diego, R.,(2014). Action of melatonin on squamous cell carcinoma and other tumors of the oral cavity (review). *Oncol. Lett.* 7, 923-926.

[13] Najeeb, S., Khurshid, Z., Zohaib, S., Zafar, M.S., (2016). Therapeutic potential of melatonin in oral medicine and periodontology. *Kaohsiung. J. Med. Sci.* 32, 391-396.

[14] Permuy, M., López-Peña, M., González-Cantalapiedra, A., Muñoz, F., (2017). Melatonin: a review of its potential functions and effects on dental diseases. *Int. J. Mol. Sci.* 18, 865-877.

[15] Cengiz, M.I., Cengiz, S., Wang, H.L., (2012). Melatonin and oral cavity. *Int. J. Dent.* 2012, 491872:1-491872:9.

[16] Szczepanik, M., (2007). Melatonin and its influence on immune system. *J. Physiol. Pharmacol.* 58 Suppl 6, 115-124.

[17] Carrillo-Vico, A., Reiter, R.J., Lardone, P.J., Herrera, J.L., Fernandez-Montesinos, R., Guerrero, J.M., Pozo, D., (2006). The modulatory role of melatonin on immune responsiveness. *Curr. Opin. Investig. Drugs.* 7, 423-431.

[18] Currier, N.L., Sun, L-Y., Miller, S.C., (2000). Exogenous melatonin: quantitative enhancement in vivo of cells mediating non-specific immunity. *J. Neuroimmunol.* 104, 101–108.

[19] Kaur, C., Ling, E.A., (1999). Effects of melatonin on macrophages/microglia in postnatal rat brain. *J. Pineal. Res.* 26, 158-168.

[20] Liu, J., Huang, F., He, H.W., (2013). Melatonin effects on hard tissues: bone and tooth. *Int. J. Mol. Sci.* 14, 10063-10074.

[21] López-Martínez, F., Olivares-Ponce, P.N., Guerra-Rodríguez, M., Martínez-Pedraza, R., (2012). Melatonin: bone metabolism in oral cavity. *Int. J. Dent.* 2012, 628406:1-628406:5.

[22] Halici, M., Öner, M., Güney, A., Canöz, O., Narin, F., Haici, C., (2010). Melatonin promotes fracture healing in the rat model. *Eklem. Hastalik. Cerrahisi.* 21, 172-177.

[23] Cobo-Vázquez, C., Fernández-Tresguerres, I. Ortega-Aranegui, R., López-Quiles, J., (2014). Effects of local melatonin applications on postextraction sockets after third molar surgery. A pilot study.*Med. Oral. Patol.Oral. Cir. Bucal.* 19, e628-633.

[24] Ruggiero, S.L., Dodson, T.B., Assael, L.A., Landersberg, R., Marx, R.E., Mehrotra, B., (2009). American Association of Oral and Maxillofacial Surgeons position paper on bisphosphonated-related osteonecrosis of the jaw- 2009 update. *J. Oral. Maxillofac. Surg.* 67, 2-12.

[25] Camacho-Alonso, F., Urrutia-Rodríguez, I., Oñate-Cabrerizo, D., Oñate-Sánchez, R.E., Rodríguez-Lozano, F.J., (2017). Cytoprotective effects of melatonin on zoledronic acid-treated human osteblasts. *J. Craniomaxillofac. Surg.* 45, 1251-1257.

[26] Ruggiero, S.L., American Association of Oral and Maxillofacial Surgeons position paper on medication-related osteonecrosis of the jaw- 2014 update. *J. Oral. Maxillofac. Surg.* 72, 1938-1956.

[27] Rodríguez-Lozano, F.J., García-Bernal, D., Ros-Roca M.A., Algueró, M.C., Oñate-Sánchez, R.E., Camacho-Alonso, F., Moraleda, J.M., (2015). Cytoprotective effects of melatonin on zoledronic acid-treated human mesenchymalstemcells in vitro. *J. Craniomaxillofac. Surg.* 43, 855-862.

[28] Cutando, A., Galindo, P., Gómez-Moreno, G., Arana, C., Bolaños, J., Acuña-Castroviejo, D., Wang, H.L., (2006). Relationship between salivary melatonin and severity of periodontal disease. *J. Periodontol.* 77, 1533-1538.

[29] Cutando, A., López-Valverde, A., Gómez-de-Diego, R., Arias-Santiago, S., Vicente-Jiménez, J., (2013). Effect of gingival application of melatonin on alkaline and acid phosphatase, osteopontin and osteocalcin in patients with diabetes and periodontal disease.*Med. Oral. Patol.Oral. Cir. Bucal.* 18, e657-663.

[30] Mehta, A., Kaur, G., (2014). Potential role of melatonin in prevention and treatment of oral carcinoma. *Indian. J. Dentistry.* 5, 86-91.

[31] Do Nascimento-Gonçalves, N., Ventura-Rodrigues, R., Jardim-Perassi B.V., Gobbe-Moschetta, M., Ramos-Lopes, J., Colombo, J., Pires de Campos-Zuccari, D.A., (2014). Molecular markers of angiogenesis and metastasis in lines of oral carcinoma after treatment with melatonin. *Anticancer. Agents. Med. Chem.* 14, 1304-1311.

[32] Muñoz-Guerra, M.F., Fernández-Contreras, M.E., Moreno, A.L., Martin, I.D., Herraez, B., Gamallo, C., (2009). Polymorphism in the hypoxia inducible factor 1-alpha and the impact in the prognosis of early stages of oral cancer. *Ann. Surg. Oncol.* 16, 2351-2358.

[33] Li, Y., Li, S., Zhou, Y., Meng, X., Zhang, J.J., Xu, D.P., Li, H.B., (2017) Melatonin for the prevention and treatment of cancer. *Oncotarget.* 8, 39896-39921.

[34] Sonis, S.T., (2004). Oral mucositis in cancer therapy. *J. Support. Oncol.* 2, 3-8.

[35] Abdel-Moneim, A.E., Guerra-Librero, A., Florido,J., Shen, Y-Q., Fernández-Gil, B., Acuña-Castroviejo, D., Escames, G., (2017). Oral mucositis: melatonin gel an effective new treatment. *Int. J. Mol. Sci.* 18, 1003-1013.

[36] Arora, H., Ivanovski, S., (2017). Melatonin as a pro-osteogenic agent in oral implantology: a systematic review of histomorphometric outcomes in animals and quality evaluation using ARRIVE guidelines. *J. Periodont. Res.* 52, 151-161.

[37] Cutando, A., Gómez-Moreno, G., Arana, C., Muñoz, F., López-Peña, M., Stephenson, J., Reiter, R.J., (2008). Melatonin stimulates osteointegration of dental implants. *J. Pineal. Res.* 45, 174-179.

[38] Calvo-Guirado, J.L., Gómez-Moreno, G., Arana, C., Cutando, A., Alcaraz-Baños, M., Chiva, F., López-Marí, L., Guardia, J., (2009). Melatonin plus porcine bone on discrete calcium deposit implant surface stimulates osteointegration in dental implants. *J. Pineal. Res.* 47, 164-172.

[39] Guardia, J., Gómez-Moreno, G.,Ferrera, M.J., Cutando, A., (2011). Evaluation of effects of topic melatonin on implant surface at 5 and 8 weeks in Beagle dogs. *Clin Implant. Dent. Relat. Res.* 13, 262-268.

[40] Yamazaki, S-I., Ochi, M., Hirose, Y., Nakanishi, Y., Nakade, O., (2008). Melatonin enhances peri-implant osteogenesis in the femur of rabbits. *J. Oromax. Biomech.* 14, 34-38.

[41] Carpentieri, A.R., Peralta-López, M.E., Aguilar, J., Solá, V.M., (2017). Melatonin and periodontal tissues: molecular and clinical perspectives. *Pharmacol. Res.* in press.

[42] Nunes, O.D.S., Pereira, R,.D.S., (2008). Regression of herpes viral infection symptoms using melatonin and SB-73: comparison with acyclovir. *J. Pineal. Res.* 44, 373-378.

[43] Yavuz, T., Kaya, D., Behçet, M., Ozturk, E., Yavuz, O., (2007). Effects of melatonin on Candida sepsis in an experimental rat model. *Adv. Ther.* 24, 91-100.

[44] Schweikl, H., Spagnuolo, G., Schmalz G., (2006). Genetic and cellular toxicology of dental resin monomers. *J. Dental. Res.* 85, 870-877.

[45] Blasiak, J., Kasznicki, J., Drzewoski, E., Pawlowska, J., Szcepanska, J., Reiter, R.J., (2011). Perspectives on the use of melatonin to reduce cytotoxic and genotoxic effects of methacrylate-based dental materials. *J. Pineal. Res.* 51, 157-162.

[46] Karaaslan, C., Suzen, S., (2015). Antioxidant properties of melatonin and its potential actionin diseases. *Curr. Top. Med. Chem.* 15, 894-903.

[47] Capote-Moreno, A.L., Gamallo, C., Muñoz-Guerra, M.F., Mera-Menéndez, F.D., Hinojar-Gutierrez, A., (2012). Single nucleotide polymorphism of hypoxia inducible factor 1-alpha in early-stage oral squamous cell carcinoma: influence in second primary tumors. *Trends. Cancer. Res.* 8, 10-22.

[48] Chaiyarit, P., Man, N., Hiraku, Y., Pinlaor, S., Yongvanit, P., Jintakanon, D., Murata, M., Oikawa, S., Kawanishi, S., (2005). Nitrative and oxidative DNA damage in oral lichen planus in relation to human oral carcinogenesis. *Cancer. Sci.* 96, 553-559.

[49] Andersen, L. P., Gogenur, I., Rosenberg, J., Reiter, R. J., (2016). The Safety of Melatonin in Humans. *Clin. Drug. Investig.* 36, 169-175.

INDEX

#

22Rv1 cells, 109
4P-PDOT, 134, 138, 216, 218

A

acetic acid, 14, 15, 30, 32, 34, 36, 38, 39, 41, 43, 44, 47, 48, 172
acetone, 9, 17, 18, 39
acetonitrile, 8, 13, 14, 26, 27, 29, 30, 31, 32, 34, 36, 37, 38, 39, 41, 43, 44, 47, 48, 49, 50
acid, x, 9, 10, 12, 13, 14, 15, 16, 17, 19, 20, 31, 32, 33, 36, 37, 38, 39, 40, 41, 44, 47, 48, 49, 50, 70, 73, 88, 96, 104, 121, 129, 136, 137, 140, 151, 161, 163, 165, 167, 174, 179, 189, 202, 203, 213, 214, 221, 225, 226, 235, 236, 243, 247, 251, 260, 299, 301, 304, 308, 309
acidic, 41, 42, 248
adaptive immune responses, 162
adaptive immunity, 167
adenocarcinoma, 107, 108, 111, 114, 183
adhesion, 105, 107, 164, 176, 177, 190, 194, 213, 225, 238, 300
adipose tissue, 110, 123, 136, 258
adiposity, 258, 261, 262, 269, 270, 273
adolescents, xii, 157, 250, 255, 271
adults, xii, 4, 82, 145, 146, 231, 232, 239, 255, 271
adverse effects, 106, 128, 241, 270, 299, 304
affective disorder, 146, 151, 228, 231
AFMK, 115, 122, 163, 167, 174
age-related diseases, 98, 174

aging, x, xiii, 98, 104, 106, 111, 112, 114, 117, 118, 119, 122, 125, 127, 128, 129, 130, 156, 174, 199, 200, 204, 230, 248, 277, 278, 285, 289, 291, 292, 293, 307
aging process, xiii, 277, 285
agomelatine, 95, 98, 138, 139, 145, 152, 153, 217, 219, 221, 222, 229, 231, 243, 244, 247, 252, 253, 274, 275
agonist, 65, 81, 98, 100, 137, 138, 141, 144, 145, 146, 147, 154, 157, 158, 217, 219, 221, 222, 223, 229, 231, 232, 243, 244, 245, 253, 263, 266, 268, 270, 290
AKT, 115, 116, 118, 122, 124, 130, 189, 203
amino acids, 136, 137, 216, 217
ammonium, 29, 30, 31, 47, 48, 49
AMP (cAMP), xi, 93, 108, 118, 133, 136, 139, 163, 211, 214, 216, 217
amygdala, 141, 143, 146, 192, 219, 221, 242, 243, 246, 252
androgen, ix, 103, 105, 106, 107, 109, 110, 113, 114, 118, 119, 120, 121, 124, 126, 127, 128, 129
androgen ablation, 105
androgen receptor (AR), 105, 107, 109, 112, 113, 114, 118, 120, 121, 124, 126, 147, 177, 206, 249, 270, 292
angiogenesis, 114, 193, 299, 309
anticonvulsant, xii, 213, 221, 233, 240, 241, 242, 243, 244, 250, 251, 252, 291
anticonvulsant treatment, 241
antidepressant, 138, 145, 146, 151, 152, 153, 217, 221, 222, 229, 231, 243, 282, 290
antiepileptic drugs, xii, 220, 233, 236, 245, 248, 250, 251
antigen, 24, 84, 118, 167, 169, 179

antigen-presenting cells (APCs), 167
anti-inflammatory agents, 283
antioxidant activity, 85, 219, 262
antioxidant enzymes, ix, 73, 103, 115, 123, 188, 191, 193, 195, 200, 206, 208, 213, 225, 238, 241, 249, 292, 297, 298, 300, 303
antioxidant properties, ix, xi, 103, 159, 163, 188, 197, 213, 237, 238, 243, 301, 304, 310
antipsychotic, vi, vii, xii, 148, 255, 256, 257, 258, 259, 260, 261, 262, 263, 265, 266, 267, 268, 269, 270, 271, 272, 273, 305
antipsychotic drugs, 258, 269, 270, 305
antitumor, 115, 170, 292
anxiety, vii, x, xii, 133, 134, 138, 146, 147, 152, 157, 205, 222, 234, 241, 245, 274, 275, 288
anxiety disorder, 275
apoptosis, vii, ix, xi, 75, 81, 92, 96, 102, 104, 105, 106, 107, 108, 109, 116, 121, 122, 123, 129, 130, 159, 163, 165, 166, 168, 170, 171, 177, 178, 179, 181, 186, 188, 193, 195, 196, 197, 198, 205, 206, 207, 218, 219, 248, 292, 297, 299
arginine uptake, 76, 77, 88, 99
ascorbic acid, 13, 14, 17, 19, 20, 108, 213
aspartate, 70, 76, 90, 92, 214, 237, 243
asphyxia, 203, 204, 288
asthma, 171, 182
astrocytes, 75, 77, 83, 95, 100, 141, 190, 192, 196, 246, 248
ATP, 74, 115, 116, 124, 129, 130, 131, 188
atypical agents, 256, 259
atypical antipsychotic, 256, 257, 258, 259, 261, 265, 266, 269
atypical antipsychotic agents, 258, 265
autoimmune diseases, xi, 159, 175
axons, viii, 69, 71, 85, 86

B

bacterial lipopolysaccharide, 70, 83
behavioral change, 241, 251
benign hyperplasia (BPH), 106, 110, 111, 114
benign prostatic hyperplasia, x, 104, 121, 122, 123
biochemistry, 59, 66, 125, 128, 224, 249, 252
biological samples, vii, viii, 2, 3, 23, 24, 26, 28
biosynthesis, 63, 71, 76, 160
bipolar disorder, 137, 143, 257, 264, 265, 269, 271, 274, 305

blood, vii, xi, 2, 4, 5, 56, 58, 63, 64, 73, 119, 142, 156, 160, 161, 168, 177, 185, 186, 191, 201, 204, 205, 207, 208, 212, 213, 223, 257, 262, 264, 265, 267, 270, 280, 288, 292, 296, 305
blood flow, 73, 280, 288, 292
blood group, 119
blood pressure, 142, 161, 191, 264, 265, 267, 270, 280, 288, 305
blood vessels, 142, 186
blood-brain barrier, vii, xi, 185, 191, 201, 204, 205, 207, 213, 280, 288
bloodstream, 212, 304
body composition, 259, 265, 271
body fat, 260, 271, 273
body fluid, 3, 4, 57, 292
body weight, 111, 112, 126, 186, 256, 258, 259, 260, 261, 262, 266, 269, 270, 271, 272, 273
bone, xii, 3, 11, 13, 36, 37, 47, 56, 66, 134, 160, 163, 166, 169, 178, 181, 186, 212, 233, 244, 296, 297, 298, 299, 303, 304, 305, 306, 308, 309
bone marrow, 3, 11, 13, 36, 37, 47, 56, 66, 160, 163, 166, 169, 178, 181, 186, 212, 296, 299
brain, vii, xi, xii, 55, 62, 63, 70, 79, 81, 83, 84, 86, 91, 93, 96, 97, 100, 102, 122, 124, 135, 136, 138, 139, 140, 141, 142, 143, 147, 149, 151, 153, 154, 155, 185, 186, 187, 188, 189, 190, 191, 192, 193, 194, 195, 196, 197, 198, 199, 200, 201, 202, 203, 204, 206, 207, 208, 211, 212, 213, 214, 215, 216, 218, 220, 221, 223, 225, 226, 228, 229, 233, 234, 235, 236, 237, 238, 240, 247, 248, 249, 251, 252, 280, 285, 288, 289, 291, 293, 294, 296, 308
brain contusion, 206
brain damage, xi, 185, 186, 188, 192, 193, 195, 196, 200, 207, 288
brain functions, 138
brain ischemia, vii, xi, 96, 185, 186, 187, 188, 189, 190, 191, 192, 194, 195, 196, 197, 201
brain structure, 191
breast cancer, 117, 119, 130, 151, 171, 174, 183, 282, 291, 293

C

calcium, xi, 134, 136, 141, 165, 177, 186, 198, 206, 211, 213, 216, 224, 237, 243, 284, 309
calmodulin, 98, 115, 134, 136, 145, 163, 167, 168, 213, 215, 224, 226, 237, 248

cancer, viii, ix, xi, xiii, 10, 43, 59, 103, 107, 110, 111, 117, 118, 121, 123, 127, 128, 134, 159, 161, 162, 163, 169, 171, 175, 180, 181, 183, 271, 279, 282, 291, 295, 297, 301, 302, 309
cancer cells, x, 59, 104, 110, 123, 163, 171
cancer therapy, 175, 309
carcinogenesis, ix, 103, 121, 171, 287, 310
carcinoma, 109, 127, 292, 306, 309
cardiovascular disease, 257, 280, 289
cardiovascular risk, 257
cardiovascular system, 216
catalase (CAT), 70, 73, 76, 97, 100, 108, 218, 236, 300, 303
cell biology, 127, 223, 306
cell culture, 3, 61, 166, 178, 189
cell death, ix, xi, 69, 75, 77, 80, 87, 95, 97, 98, 102, 104, 105, 108, 112, 115, 119, 124, 179, 185, 186, 188, 193, 195, 200, 204, 220, 240, 292, 302
cell line, 10, 43, 110, 166, 167, 168, 178, 179, 216, 293, 302
central nervous system (CNS), xii, 187, 212, 233, 237, 243
cerebellum, xi, 135, 192, 203, 211, 216, 248
cerebral arteries, 142
cerebral blood flow, 186, 191
cerebral cortex, xi, 135, 141, 191, 211, 215, 241
cerebral edema, 190, 191, 204, 280
cerebrospinal fluid, 64, 186, 192, 199, 205, 296
chemoprevention, vii, x, 104, 127, 128
chemotherapy, 171, 182, 302, 303, 306
children, xii, 4, 82, 102, 157, 235, 238, 239, 240, 241, 250, 251, 255
chromatography, viii, 1, 3, 10, 22, 25, 26, 46, 54, 57, 58, 59, 61, 62, 63, 64, 65
chrono-disruption, 262
circadian rhythm, xi, xii, 59, 67, 108, 112, 118, 125, 139, 142, 144, 145, 146, 152, 153, 157, 160, 173, 177, 178, 186, 211, 214, 216, 217, 222, 223, 227, 228, 239, 255, 261, 266, 281, 296, 305, 307
circadian rhythmicity, xii, 178, 255, 296, 307
clinical, vii, viii, xi, xii, xiii, 4, 5, 56, 58, 63, 67, 69, 71, 72, 82, 84, 102, 106, 114, 118, 120, 123, 125, 128, 129, 133, 144, 145, 146, 148, 149, 155, 157, 159, 163, 169, 170, 171, 182, 186, 189, 190, 191, 192, 193, 195, 196, 197, 219, 220, 222, 227, 229, 230, 231, 234, 236, 238, 239, 241, 245, 247, 250, 251, 252, 255, 256, 257, 258, 260, 262, 263, 266, 267, 268, 270, 272, 274, 275, 277, 278, 279, 280, 281, 282, 286, 289, 296, 299, 305, 306, 310

clinical trials, vii, xii, 169, 186, 190, 191, 192, 196, 255, 256, 266, 268
cloning, 134, 142, 148, 150, 155
clozapine, 256, 258, 265, 266, 271
collagen, 105, 107, 112, 298
combination therapy, 89, 193, 196, 197, 207
cortex, 141, 154, 189, 215, 219, 221, 242, 247, 248, 287, 291
corticosteroids, ix, 69, 84, 87, 88, 94
cytokines, vii, xi, 111, 159, 166, 168, 169, 170, 190, 194, 236, 238, 249, 281, 285, 298, 299, 300, 302, 303, 304

D

dementia, 4, 220, 289
demyelination, ix, 69, 82, 83, 84, 85, 91, 93, 102, 279
depression, 134, 139, 143, 153, 217, 222, 228, 231, 242, 243, 269, 274, 275, 305
despair, 139, 243, 253
diabetes, ix, xii, 99, 103, 104, 106, 107, 108, 109, 110, 117, 118, 119, 120, 121, 122, 123, 125, 126, 127, 128, 130, 134, 148, 255, 257, 260, 261, 268, 273, 309
diabetes mellitus (DM), 106, 107, 108, 109, 121, 154, 157, 175
diabetic patients, 106, 301
diastolic blood pressure, 265, 267
diet, 106, 110, 112, 119, 120, 126, 127, 128, 166, 259, 261, 264, 270, 272
disability, xi, 78, 89, 185, 186, 193, 222
diseases, viii, ix, xi, xiii, 69, 82, 86, 106, 128, 144, 147, 157, 162, 170, 174, 194, 211, 218, 223, 235, 269, 278, 295, 298, 304, 306, 308, 310
disorder, 71, 83, 90, 145, 218, 222, 234, 235, 239, 257, 262, 264, 265, 266, 274, 278, 280, 282, 291, 299, 303
DNA damage, 107, 179, 188, 189, 202, 241, 251, 300, 302, 310
DNA repair, 130
dopamine, 71, 90, 98, 135, 139, 140, 143, 148, 153, 154, 155, 156, 216, 256, 269, 271, 282
drug treatment, 196, 270, 272
drugs, xii, 3, 4, 55, 81, 107, 141, 144, 147, 148, 157, 186, 190, 194, 196, 211, 219, 221, 223, 233, 236, 243, 250, 251, 256, 257, 265, 267, 281, 297, 299, 301, 302

E

edema, 82, 176, 191, 194, 195, 196, 197, 206, 280, 288
elderly, 111, 112, 144, 145, 157, 162, 170, 219, 230, 235, 275
endothelial cells, 88, 121, 164, 176, 191, 204, 213
epilepsy, vi, xi, xii, 170, 211, 218, 220, 221, 223, 224, 231, 233, 234, 235, 236, 237, 238, 239, 240, 241, 242, 243, 244, 245, 246, 247, 248, 250, 251, 252, 253, 282, 290, 291
excitotoxicity, xi, xii, 72, 77, 92, 99, 102, 185, 186, 187, 189, 193, 195, 196, 202, 203, 213, 225, 233, 238, 240, 249
experimental models, v, xii, 78, 79, 85, 86, 103, 106, 208, 229, 234, 235, 236, 237, 238, 243, 245, 305
extraction procedures, viii, 2, 9, 54

F

febrile seizure, 240, 250, 251, 291
fluid, viii, 2, 4, 10, 11, 12, 29, 48, 51, 53, 77, 104, 190, 280
food, viii, 1, 4, 5, 8, 9, 15, 16, 17, 18, 19, 20, 23, 26, 31, 32, 33, 34, 38, 39, 40, 44, 52, 60, 111, 257, 258, 259, 260, 262, 271, 273
free radicals, 74, 92, 101, 107, 112, 163, 186, 197, 200, 213, 237, 240, 297, 300, 302
frontal cortex, 192, 220
fruits, viii, 2, 6, 8, 15, 16, 17, 18, 19, 20, 24, 25, 31, 32, 33, 34, 38, 39, 40, 44, 68

G

ganglion, viii, ix, 69, 70, 71, 75, 87, 90, 91, 93, 95, 96, 97, 98, 100, 101, 102, 142, 143, 147, 224
gastrointestinal tract, 3, 60, 137, 160, 212, 216, 302
gene expression, 105, 130, 215, 225, 249, 279, 285, 287, 292, 293
genital tract, 106
gland, ix, 4, 56, 63, 103, 104, 106, 107, 108, 111, 112, 120, 134, 151, 160, 186, 212
glaucoma, viii, ix, 69, 70, 71, 72, 73, 74, 75, 77, 79, 80, 81, 86, 87, 88, 89, 90, 91, 92, 93, 94, 95, 96, 97, 98, 99, 100, 101, 102
glia, 77, 89, 93, 97, 98, 141, 191, 236
glial cells, 77, 78, 83, 85, 92, 142, 190, 237
glucagon, 272
glucose, 107, 109, 111, 112, 119, 121, 127, 129, 186, 192, 195, 204, 218, 257, 258, 260, 264, 265, 267, 270, 273, 305
glutamate, xii, 72, 76, 77, 78, 79, 80, 86, 87, 88, 90, 92, 95, 96, 97, 98, 99, 100, 101, 102, 140, 189, 193, 196, 202, 203, 213, 233, 235, 238, 240, 246, 249
glutamine, 70, 77, 78, 79, 96, 97, 99, 100
glutathione, x, 70, 73, 74, 92, 102, 104, 111, 179, 188, 195, 200, 201, 202, 213, 225, 236, 237, 244, 283, 300, 303
glutathione S-transferase (GST), vii, x, 102, 104, 108, 111, 112, 115, 128
glycogen synthase kinase 3, 116, 121
gonadotropin-inhibitory hormone (GnIH), 109, 124
G-protein-coupled receptor, 136, 243, 296
GST, vii, 108, 112, 115, 128

H

head injury, 206, 207
healing, 71, 279, 287, 299, 303, 304, 305, 308
healthy volunteers, 60, 260, 261, 268, 272, 273
high-fat diet, 110, 112, 126, 127, 128, 272
hippocampus, xi, 140, 141, 142, 143, 148, 153, 154, 192, 203, 206, 211, 215, 219, 220, 221, 226, 227, 230, 231, 242, 246, 248, 249, 251, 253
homeostasis, ix, 103, 105, 106, 110, 117, 127, 186, 195, 202, 225, 303
hormone, xi, 2, 57, 61, 107, 109, 112, 113, 114, 116, 119, 121, 125, 128, 134, 144, 149, 151, 154, 159, 160, 162, 163, 168, 170, 171, 181, 238, 240, 241, 245, 258, 261, 298
human, ix, x, xiii, 3, 4, 6, 23, 24, 46, 56, 57, 58, 59, 60, 61, 62, 66, 70, 72, 75, 81, 83, 86, 93, 95, 97, 104, 110, 112, 119, 121, 123, 124, 125, 128, 129, 130, 134, 136, 143, 145, 147, 148, 149, 150, 151, 153, 155, 156, 159, 161, 164, 165, 166, 167, 168, 173, 174, 175, 176, 177, 178, 180, 181, 182, 183, 186, 190, 199, 219, 227, 228, 229, 241, 245, 246, 247, 259, 279, 281, 282, 287, 293, 296, 299, 307, 308, 310
human brain, 143, 155, 186
hyaluronic acid, 70, 73, 88, 96
hyperglycemia, ix, 104, 106, 107, 108, 109, 118
hyperplasia, vii, x, 104, 106, 110, 112, 118, 121, 122, 123, 126, 127, 129

hypothalamus, 88, 135, 141, 150, 160, 192, 215, 216, 224

I

immune function, 98, 161, 171, 181, 283, 285, 302
immune system, x, 105, 111, 113, 120, 134, 137, 152, 159, 160, 161, 162, 163, 168, 169, 173, 174, 186, 215, 216, 226, 227, 238, 296, 298, 302, 307, 308
immunity, x, 152, 159, 162, 169, 173, 178, 179, 297, 302, 308
immunomodulatory, vii, xi, xiii, 85, 95, 148, 159, 162, 168, 171, 178, 181, 238, 283, 295, 298, 301, 304, 306
immunoreactivity, 75, 83, 85, 140, 143, 154, 156, 219, 222, 230
in vitro, 3, 10, 43, 56, 60, 61, 63, 75, 76, 79, 92, 95, 110, 114, 116, 135, 161, 164, 165, 166, 168, 169, 176, 179, 181, 192, 193, 214, 217, 243, 246, 257, 293, 304, 305, 308
in vivo, 76, 78, 79, 83, 96, 97, 110, 124, 161, 164, 165, 169, 170, 178, 181, 188, 192, 193, 194, 196, 202, 217, 241, 243, 257, 270, 287, 292, 299, 302, 304, 305, 308
inflammation, ix, xi, 69, 82, 83, 85, 86, 89, 91, 99, 102, 121, 159, 161, 162, 163, 164, 165, 167, 168, 170, 171, 173, 174, 175, 176, 182, 186, 189, 190, 193, 194, 195, 197, 198, 203, 204, 206, 225, 238, 246, 247, 249, 278, 279, 281, 282, 283, 284, 285, 286, 294, 297,298, 299, 300, 301, 303
inflammatory response, xi, xii, 83, 84, 111, 159, 161, 164, 169, 190, 194, 195, 197, 204, 233, 236, 245, 261, 283, 285, 304
insulin, 106, 107, 108, 110, 111, 112, 117, 118, 119, 121, 122, 125, 126, 127, 128, 129, 130, 131, 146, 148, 257, 258, 260, 264, 267, 270, 271, 272, 273, 302
insulin resistance, 106, 110, 111, 112, 117, 118, 126, 129, 257, 258, 260, 261, 267, 269, 270, 273
insulin sensitivity, 108, 131, 146, 260, 273
insulin signaling, 108, 111, 112
intracerebral hemorrhage, 187, 199, 203
intracranial pressure, 191, 194, 206
intraepithelial neoplasia, 107, 111, 127
intraepithelial neoplasia and adenocarcinoma, 107
intraocular, viii, ix, 69, 70, 72, 87, 88, 92, 93, 94, 95, 96, 97, 98, 99, 100, 151

intraocular pressure, viii, ix, 69, 70, 72, 87, 88, 92, 93, 94, 95, 96, 97, 98, 99, 100, 151
ischemia, vii, xi, 96, 185, 186, 187, 188, 189, 190, 191, 192, 193, 194, 195, 196, 197, 198, 200, 201, 202, 203, 204, 205, 207, 208, 213, 218, 219, 225, 229, 280, 288, 292

J

JNK, 112

K

keratinocytes, 43, 58, 302

L

LNCaP, 109, 114, 118, 125
LNCaP cells, 109, 114, 118, 125
luzindole, 135, 138, 139, 140, 146, 149, 152, 189, 193, 214, 216, 218, 220, 222, 228, 244, 253

M

major depressive disorder, 145, 221, 223, 229, 274, 275
mammalian brain, 228, 235
mammalian tissues, 199, 306
melatonin and lymphocytes, 168
melatonin and monocytes/macrophages, 166
melatonin and neutrophils, 163
melatonin receptor agonists, xi, 98, 157, 158, 211, 219, 220, 222, 223, 227, 230, 239
melatonin receptor antagonist, 140, 146, 149, 157, 214, 216, 218, 220, 228, 244
melatonin receptors, v, vii, x, xi, 59, 71, 87, 93, 97, 121, 122, 133, 134, 135, 136, 137, 138, 139, 140, 141, 142, 143, 144, 148, 149, 150, 152, 153, 154, 155, 156, 163, 166, 170, 175, 211, 212, 214, 215, 217, 218, 219, 220, 221, 222, 223, 224, 226, 227, 228, 229, 231, 241,244, 253, 280, 284, 292, 296
memory, xii, 134, 138, 139, 140, 146, 147, 153, 154, 202, 220, 221, 222, 223, 230, 234, 241, 245, 285, 290
memory formation, 134, 139, 140, 154, 220, 230
metabolic, v, ix, xii, 2, 73, 103, 106, 107, 113, 117, 119, 120, 121, 127, 129, 148, 186, 235, 237, 255,

Index

256, 257, 258, 260, 261, 262, 263, 264, 265, 266, 267, 268, 269, 270, 271, 272, 273, 286, 305
metabolic disorder, x, 104, 106, 107, 271, 305
metabolic disturbances, ix, 103, 113, 117
metabolic syndrome, ix, xii, 103, 113, 119, 121, 127, 255, 256, 257, 261, 262, 265, 266, 267, 268, 269, 270, 272
methanol, 8, 9, 10, 13, 14, 15, 16, 17, 18, 20, 21, 26, 27, 29, 30, 31, 32, 33, 36, 37, 38, 39, 40, 41, 43, 44, 45, 47, 48, 49
mice, 13, 30, 34, 57, 59, 66, 67, 84, 88, 92, 95, 102, 107, 119, 120, 121, 125, 127, 138, 139, 140, 141, 146, 147, 148, 152, 154, 155, 161, 166, 169, 176, 181, 183, 191, 196, 201, 202, 204, 205, 207, 208, 216, 218, 220, 222, 223, 226, 229, 230, 231, 237, 243, 244, 247, 249, 251, 252, 253, 258, 278, 282, 284, 285, 287, 288, 289, 291, 293
microglia, 83, 84, 89, 94, 178, 190, 192, 203, 236, 308
mitochondria, 74, 116, 117, 119, 124, 129, 168, 177, 179, 188, 195, 200, 202, 213, 242, 251, 292, 303
mitochondrial function, 94, 114, 115, 123, 195, 227
mitochondrial membrane potential, 115, 200
mitogen-activated protein kinase (MAPK), 112, 119, 196, 203
MLT, 2, 3, 4, 5, 6, 7, 8, 9, 22, 23, 24, 25, 26, 27, 28, 35, 41, 42, 46, 54, 108, 109, 111, 112, 113, 114, 115, 116, 117, 160, 161, 162, 163, 164, 165, 166, 167, 168, 169, 170, 171
MLT synthesis, 112, 114, 160, 161, 169
MMA, 304
MMP, 127, 192
MMP-9, 127, 192
MMPs, 192, 300
MnSOD, 115
mood disorder, xii, 212, 218
MT_1 receptor, xi, 97, 127, 136, 137, 138, 139, 141, 142, 143, 146, 153, 211, 216, 217, 218, 220, 222, 223, 229, 230, 237, 296
MT_2 receptor, vii, xi, 136, 138, 139, 140, 141, 142, 143, 144, 145, 146, 147, 150, 154, 155, 156, 157, 168, 170, 192, 211, 214, 216, 217, 218, 219, 220, 221, 222, 223, 226, 227, 228, 229, 242, 243, 244, 266, 270, 282, 283, 292

N

nerve, ix, 70, 72, 75, 82, 83, 84, 85, 86, 101, 112, 119, 142, 223
neural function, 205, 218, 229, 289
neurodegeneration, 79, 89, 101, 203, 218, 237, 249, 289, 290, 293
neurodegenerative diseases, 94, 128, 161, 182, 219
neurodegenerative disorders, vii, x, 133, 170, 174, 282
neurogenesis, vii, xi, 134, 148, 185, 192, 193, 197, 205, 214, 218, 226, 229, 237, 248, 280, 288, 289
neurohormone, viii, 1, 71, 134, 180, 187, 197, 225, 226, 278, 286, 296, 304, 305
neuroinflammation, 93, 207, 246, 291, 293
neurological disease, 217, 234
neurological disorders, v, vii, xi, 198, 211, 212, 218, 223, 234, 274
neuronal cells, 188, 199, 214, 219
neurons, 71, 77, 78, 81, 92, 93, 98, 140, 142, 152, 153, 186, 191, 192, 193, 200, 202, 213, 214, 219, 226, 236, 237, 243, 247, 248, 249
neuropathy, ix, 69, 75, 78, 80, 81, 82, 91, 95, 97, 106
neuroprotection, 78, 79, 80, 81, 93, 95, 96, 97, 98, 102, 134, 186, 187, 193, 194, 202, 203, 206, 207, 208, 218, 219, 220, 221, 223, 225, 229, 230, 234, 242, 243, 280, 287, 288, 290, 292
neuroprotective, vi, ix, xi, xii, 70, 79, 85, 87, 90, 95, 101, 185, 186, 187, 192, 193, 195, 196, 197, 204, 207, 208, 211, 212, 218, 219, 220, 221, 229, 233, 234, 237, 241, 242, 243, 245, 275, 280, 284, 288, 289, 292, 300
neuroprotective agents, xi, 211, 218, 275
neurotoxicity, 77, 80, 90, 98, 188, 238, 240, 251, 290
neurotransmission, 231, 241, 256
neurotransmitter, 77, 87, 189, 235, 236
neutrophils, 161, 162, 163, 164, 165, 166, 167, 172, 173, 175, 177, 190, 298, 300
nitric oxide, 88, 89, 90, 92, 94, 95, 96, 97, 98, 100, 102, 117, 121, 163, 167, 178, 179, 186, 189, 201, 204, 213, 224, 237, 240, 249, 252, 292, 298
nitric oxide synthase, 88, 94, 95, 96, 97, 98, 100, 102, 121, 167, 179, 189, 201, 204, 213, 224, 237
nitrosative stress, 75, 98, 186, 195, 200, 204
non-enzymatic glycation products (AGE), 107, 130
nuclear receptors, 134, 137, 163
nucleus, 115, 141, 142, 163, 167, 192, 193, 194, 215, 283, 284, 300

Index

O

obesity, vii, x, 104, 110, 111, 112, 120, 126, 127, 260, 262, 267, 268, 269, 270, 272

occlusion, 90, 100, 187, 201, 204, 208

olanzapine, 256, 258, 259, 260, 261, 262, 263, 264, 265, 266, 267, 268, 269, 270, 271, 272, 273

optic nerve, viii, ix, 69, 70, 71, 72, 75, 81, 82, 83, 84, 85, 86, 87, 90, 94, 95, 96, 97, 100, 102

optic neuritis, ix, 69, 70, 71, 82, 83, 84, 87, 88, 89, 90, 91, 92, 93, 94, 95, 97, 98, 100, 101, 102

oral cavity, xiii, 295, 296, 297, 298, 301, 302, 304, 306, 307, 308

osteonecrosis, viii, xiii, 295, 296, 297, 299, 305, 308

osteonecrosis of the jaw, viii, xiii, 295, 297, 308

oxidation, 8, 28, 35, 59, 66, 107, 128, 177

oxidative damage, 74, 80, 86, 102, 108, 120, 124, 171, 181, 182, 202, 206, 213, 236, 238, 249, 293, 297

oxidative stress, ix, xi, xii, 73, 74, 80, 85, 91, 92, 93, 96, 98, 99, 101, 102, 104, 107, 108, 112, 113, 115, 116, 117, 119, 122, 125, 128, 129, 147, 167, 171, 179, 185, 186, 187, 201, 202, 203, 206, 218, 233, 236, 238, 240, 245, 248, 249, 252, 270, 272, 278, 283, 292, 293, 297

oxygen and glucose deprivation followed by reoxygenation, 192

P

p38, 112, 119, 196

pancreatitis, 164, 166, 169, 176, 178

pathophysiological, 75, 175, 186, 197, 208, 225, 243, 296

pathophysiology, 85, 101, 146, 147, 188, 246, 268

PC3, 109

PCa, 106, 110, 111, 113, 114, 116

periodontal, 296, 297, 299, 300, 305, 306, 309, 310

periodontal disease, 296, 297, 299, 300, 301, 305, 306, 309

periodontitis, viii, xiii, 295, 297, 298, 300, 301, 305

peroxidase (GPx), 70, 73, 108, 112, 115, 179, 200, 201, 213, 236, 241, 303

pH, 10, 11, 12, 13, 14, 15, 16, 17, 18, 19, 20, 27, 29, 30, 31, 32, 33, 34, 36, 37, 38, 39, 40, 43, 44, 45, 47, 48, 49, 104

physiology, ix, x, 65, 67, 101, 103, 115, 116, 125, 130, 159, 175, 223, 227, 249, 281

pineal gland, viii, 2, 3, 12, 13, 14, 15, 22, 30, 34, 36, 37, 38, 51, 53, 57, 59, 61, 63, 64, 66, 71, 108, 112, 114, 122, 128, 134, 143, 151, 152, 160, 161, 172, 173, 178, 186, 187, 199, 200, 212, 219, 224, 227, 230, 242, 244, 296

pineal hormone melatonin, 58, 174, 180

placebo, 83, 148, 157, 207, 241, 250, 251, 252, 259, 260, 262, 263, 264, 265, 266, 271, 272, 273, 289

PNT1A, 110, 115, 116, 117, 129

prevention, xiii, 77, 85, 92, 117, 125, 127, 168, 247, 259, 262, 263, 268, 271, 277, 301, 305, 309

pro-inflammatory, 107, 111, 117, 123, 166, 169, 170, 190, 194, 236, 238, 301, 303

prostate, v, vii, ix, 10, 43, 59, 103, 104, 105, 106, 107, 108, 109, 110, 111, 112, 113, 114, 115, 116, 117, 118, 119, 120, 121, 122, 123, 124, 125, 126, 127, 128, 129, 130, 137, 294, 302

prostate cancer, 109, 110, 117, 118, 120, 122, 123, 124, 125, 126, 127, 128, 129, 130, 294

prostate carcinoma, 106, 118

prostate gland, x, 104, 108, 111, 113, 115, 137

prostatic intraepithelial neoplasia, 114, 119, 127

proteins, 75, 76, 104, 107, 112, 122, 134, 136, 137, 163, 188, 202, 213, 214, 216, 217, 227, 236, 244, 288, 298, 302

psychiatric disorder, 143, 250, 260, 305

psychopharmacology, 147, 157, 229, 252, 274

Q

quality of life, 72, 89, 157, 241, 245, 251, 267, 303

quetiapine, 256, 265, 266, 272

quinone, 136, 144, 145, 150, 151, 163, 266, 296

R

ramelteon, 81, 144, 157, 217, 219, 221, 222, 223, 229, 231, 232, 244, 253, 263, 266, 269, 270, 271, 272, 290

reactive nitrogen species, 188

reactive oxygen species (ROS), vii, x, xi, 70, 74, 75, 104, 107, 114, 115, 116, 127, 159, 163, 165, 167, 170, 177, 178, 187, 188, 189, 192, 193, 200, 213, 214, 218, 236, 283

receptor, vii, xi, 71, 81, 89, 95, 96, 97, 98, 100, 101, 102, 105, 107, 108, 109, 110, 114, 118, 120, 121, 124, 126, 127, 129, 134, 135, 136, 137, 138, 140, 141, 142, 143, 144, 145, 146, 147, 148, 149, 150,

151, 152, 153, 154, 155, 156, 157, 158, 163, 164, 166, 167, 168, 170, 173, 175, 177, 178, 179, 180, 181, 188, 189, 192, 193, 196, 202, 205, 211, 214, 215, 216, 217, 218, 219, 220, 221, 222, 223, 224, 227, 228, 229, 230, 231, 232, 236, 237, 239, 243, 244, 246, 247, 248, 252, 253, 270, 284, 289, 290, 296, 307

receptor agonist, 81, 100, 138, 145, 146, 152, 217, 219, 220, 221, 223, 229, 231, 232, 243, 244, 253, 290

receptor type 1B (MTR1B), 109

receptors for melatonin, 163

recovery, ix, 5, 7, 8, 9, 54, 69, 83, 92, 113, 197, 202, 205, 206, 240, 288

retina, 3, 56, 61, 63, 66, 71, 72, 73, 74, 75, 76, 77, 79, 81, 84, 86, 89, 90, 91, 93, 95, 96, 99, 100, 101, 102, 134, 135, 136, 141, 143, 148, 149, 156, 160, 212, 228, 280

retinal ganglion cells, viii, ix, 69, 70, 71, 91, 97, 100, 101, 142

rodents, ix, x, xi, 70, 84, 86, 104, 110, 111, 112, 133, 134, 137, 141, 145, 146, 153, 169, 211, 218, 256, 258, 259, 261, 262, 267, 268

S

saliva, 4, 5, 22, 24, 25, 46, 57, 60, 66, 296

scavenger, xiii, 9, 74, 98, 170, 175, 187, 188, 191, 197, 213, 225, 248, 251, 295, 297, 298, 301, 302, 305, 306

schizophrenia, 134, 147, 148, 256, 257, 260, 263, 265, 266, 269, 271, 273, 305

science, 57, 61, 62, 65, 66, 227, 247

second generation, xii, 255, 256

seizure, xii, 221, 226, 233, 234, 235, 236, 237, 239, 240, 241, 242, 243, 244, 245, 247, 250, 251, 252, 282, 291

signal transduction, 112, 122, 124, 144, 148, 150, 155, 163, 175, 208, 224, 227, 228

signaling pathway, xi, xiii, 102, 114, 119, 122, 130, 148, 167, 178, 185, 192, 193, 197, 198, 206, 215, 223, 249, 292, 295, 297

sleep apnea, 239

sleep deprivation, 239, 262, 272

sleep disorders, 143, 144, 145, 170, 194, 217, 229, 239, 250, 283, 289

sleep disturbance, 157, 174, 194, 220, 222, 223, 239, 266, 305

smooth muscle, 105, 109, 110, 156

smooth muscle cells (smc), 105

spinal cord injury, 193, 205, 206, 288

stress, x, xii, 55, 73, 80, 85, 87, 91, 98, 101, 104, 107, 108, 112, 115, 116, 117, 122, 125, 129, 130, 140, 147, 154, 161, 170, 186, 188, 195, 196, 200, 201, 204, 233, 238, 283, 294

stroke, xi, 157, 185, 186, 187, 188, 189, 190, 191, 192, 194, 195, 197, 198, 199, 200, 201, 203, 204, 205, 207, 208, 209, 218, 219, 223, 229, 230, 235, 289

stroke models, 187, 190, 191, 195

suprachiasmatic nucleus, xi, 136, 153, 156, 160, 211, 214, 216, 227, 228, 230, 231, 237, 296

symptoms, xii, 72, 84, 85, 146, 148, 158, 171, 190, 222, 230, 233, 234, 237, 244, 256, 281, 303, 310

syndrome, xii, 119, 127, 170, 176, 207, 234, 236, 247, 255, 256, 257, 261, 262, 265, 266, 267, 268, 269, 270, 272, 283

synthesis, 2, 56, 60, 66, 71, 74, 76, 79, 91, 92, 93, 100, 107, 108, 109, 112, 113, 114, 115, 119, 139, 142, 143, 152, 155, 160, 161, 166, 168, 169, 170, 171, 173, 186, 187, 188, 189, 212, 213, 217, 223, 224, 230, 236, 247, 297, 306

systemic lupus erythematosus, 82, 91, 170, 181

T

T cell, 88, 163, 166, 167, 169, 170, 179, 180, 181, 190

TBI, 192, 193, 194, 196, 206, 220

temporal lobe epilepsy, 221, 231, 235, 240, 246, 247, 251, 252

testosterone, ix, 103, 105, 106, 107, 109, 111, 124, 125, 127, 130, 263

therapy, ix, xi, 69, 72, 82, 83, 86, 106, 111, 117, 118, 121, 129, 143, 144, 156, 157, 159, 170, 171, 180, 183, 189, 191, 196, 200, 201, 203, 205, 207, 208, 209, 220, 238, 251, 272, 274, 279, 282, 283, 286, 290, 299, 302, 306

tissue, vii, ix, xi, 18, 19, 26, 61, 62, 77, 92, 103, 105, 109, 111, 115, 117, 121, 129, 130, 131, 136, 143, 170, 177, 185, 186, 187, 188, 189, 190, 191, 193, 197, 200, 201, 207, 217, 219, 236, 237, 238, 268, 279, 283, 285, 288, 301, 302, 303, 305

titanium, 303

toxicity, xi, xii, xiii, 8, 78, 87, 100, 171, 185, 197, 220, 223, 230, 233, 244, 277, 295 traumatic brain injury, xi, 192, 205, 206, 207, 208, 211, 220, 238, 246, 249, 280, 288, 292

treatment, viii, ix, xi, xii, xiii, 2, 22, 59, 69, 71, 72, 75, 78, 79, 80, 82, 83, 85, 86, 87, 88, 89, 92, 94, 96, 97, 98, 103, 107, 108, 109, 110, 117, 118, 121, 122, 127, 130, 139, 140, 144, 145, 146, 151, 153, 154, 157, 159, 164, 165, 179, 182, 183, 185, 186, 189, 190, 193, 194, 195, 197, 207, 217, 218, 219, 220, 221, 222, 223, 231, 233, 235, 236, 237, 238, 239, 242, 243, 244, 245, 246, 249, 251, 252, 255, 256, 257, 259, 260, 262, 263, 264, 265, 266, 267, 268, 269, 270, 271, 274, 277, 278, 280, 281, 283, 286, 289, 291, 293, 295, 298, 299, 300, 302, 303, 304, 305, 309

trial, 87, 88, 93, 94, 97, 145, 148, 157, 182, 207, 208, 239, 250, 251, 252, 260, 262, 264, 265, 268, 270, 271, 272, 289

tryptophan, x, 56, 60, 61, 62, 67, 133, 160, 181, 225, 287

tumor, xiii, 70, 85, 111, 114, 116, 168, 170, 178, 238, 277, 285, 286, 287, 298, 301

type 2 diabetes, 99, 106, 118, 130, 148, 257

U

urine, viii, 2, 4, 5, 10, 62, 118, 187, 199, 293, 296

V

variations, 5, 67, 73, 91, 134, 137, 155, 178, 256, 267, 278, 279

vascular endothelial growth factor (VEGF), 299, 301

vertebrates, viii, 1, 59, 71, 212

visceral adiposity, xii, 255, 261, 272

vision, 72, 81, 82, 84

W

weight gain, vi, xii, 126, 222, 255, 256, 257, 258, 259, 260, 261, 262, 263, 264, 265, 266, 267, 268, 269, 270, 271, 272, 273, 305

weight loss, 131, 256, 258

worldwide, viii, xi, 69, 72, 98, 185, 186, 193, 234, 235

wound healing, 193, 287